UNIVERSAL DESIGN THEORY

PROCEEDINGS OF THE WORKSHOP
UNIVERSAL DESIGN THEORY
KARLSRUHE, GERMANY
MAY 1998

EDITED BY

HANS GRABOWSKI
UNIVERSITY OF KARLSRUHE
GERMANY

AND

STEFAN RUDE
UNIVERSITY OF KARLSRUHE
GERMANY

AND

GUNTHER GREIN
UNIVERSITY OF KARLSRUHE
GERMANY

Berichte aus dem Maschinenbau

Universal Design Theory

Edited by

H. Grabowski, S. Rude, G. Grein

Shaker Verlag
Aachen 1998

Die Deutsche Bibliothek - CIP-Einheitsaufnahme

Universal Design Theory / edited by H. Grabowski, S. Rude, G. Grein.
- Als Ms. gedr. -
Aachen: Shaker, 1998
    (Berichte aus dem Maschinenbau)
ISBN 3-8265-4265-7

Copyright Shaker Verlag 1998
Alle Rechte, auch das des auszugsweisen Nachdruckes, der auszugsweisen oder vollständigen Wiedergabe, der Speicherung in Datenverarbeitungsanlagen und der Übersetzung, vorbehalten.

Als Manuskript gedruckt. Printed in Germany.

ISBN 3-8265-4265-7
ISSN 0945-0874

Shaker Verlag GmbH • Postfach 1290 • 52013 Aachen
Telefon: 02407 / 95 96 - 0 • Telefax: 02407 / 95 96 - 9
Internet: www.shaker.de • eMail: info@shaker.de

CONTENTS

Preface

Acknowledgements

List of Participants

SESSION ONE
CLASSICAL APPROACHES OF A GENERAL DESIGN THEORY

Axiomatic Design as a Basis for Universal Design Theory *N. P. Suh*	3
General Design Theory and its Extensions and Applications *T. Tomiyama*	25
Towards a Model of Designing Which Includes its Situatedness *J. Gero*	47
The Role of Artefact Theories in Design *M. M. Andreasen*	57
Toward a Better Understanding of Engineering Design Models *Y. Jin, S. Lu*	73

SESSION TWO
DESIGN THEORIES IN SPECIAL AREAS OF ENGINEERING AND NATURAL SCIENCES

Systematic Software Construction *G. Goos, U. Aßmann*	91
Efficient Design Methods in Microelectronic/Mechatronic Systems *M. Glesner, J. Becker*	105
Microsystems Technology - A New Challenge for the Mechanical Engineer *W. Menz*	135
Metallic Structural Materials: Design of Microstructure *D. Löhe, Vöhringer, O.*	147
Ceramics from Elementorganic Compounds *F. Aldinger*	169
From Simple Building Blocks to Complex Target Molecules and Multifarious Reactions *H. Hopf*	185
Designing Molecules *R. Herges*	197

Session Three
Computer Aided Development and Application of a Universal Design Theory

Universal Design Theory: Elements and Applicability to Computers *H. Grabowski et al.*	209
TRIZ: A Systematic Approach to Conceptual Design *V. Souchkov*	223
Software-Repository for Universal Application Within the Development Process of Artefacts *F.-L. Krause et al.*	237
Design Methods before the Change of Paradigms? Design Research in Germany, a Short Synopsis *H.-J. Franke*	249

Session Four
On the Development of a Design Theory
From the Perspective of Special Sciences

Form Follows Flow *G. Henn*	275
Empirical Design Research and its Contribution to the Formulation of a Universal Design Theory *U. Lindemann, H. Birkhofer*	291
Technical Biology and Bionics – Design Strategies From the Nature *W. Nachtigall*	309
Psychological Contributions to and Demands on a General Theory of Design *W. Hacker*	331
Epistemological Remarks Concerning the Concepts „Theory" and „Theoretical Concepts" *H. Lenk*	341

Concluding Remarks	359
Author Index	369

Preface

The background of this workshop resides in the difficulties faced by German industry some years ago. German products are expensive and therefore not easy to sell on the world market. The goal must not be to produce products less expensively than others in order to be more competitive, but to develop products which cannot be produced by others. Thus, new products must constantly be invented. Underlying this need is the question of how the design process is to be organized, and how the computer can be used to support this design process.

During that time, the so-called "Berliner Kreis" was founded by German professors. The members of this association are professors of design methodology and in the field of CAD. They share the opinion that there is no real difference between the design process be it of engineering products, architecture or civil engineering, chemical, microelectronics and micromechanical products, etc. Every artificial object in the world, everything that is not natural, must be designed by humans - and the question is: Is there a common method of doing so? This is the concept we call design theory.

There is a fundamental difference between engineering design and research. In engineering design, one aims to use existing knowledge towards the development of new products. In research, one strives to gain new knowledge from observations about the natural world. Therefore, a design theory should allow us to apply the complete knowledge of mankind to the design of new products. A design theory should not restrict research on unknown effects. Nevertheless, a design theory could initiate some research in areas that should be investigated.

Thinking about a design theory involves the question of the delimitation of this system. Artefacts also occur in biology, genetics or even in the design of laws. We restricted the scope of the discussion to non-living artefacts and also excluded arts such as music - nevertheless, it would be also be very interesting to investigate the question of how music is designed.

As this is a new idea, we decided to organize a workshop and to invite experts from different regions of the world and from different disciplines - experts from Australia, America, Asia and Europe working in the fields of engineering design, chemistry, material science, architecture, microelectronics, computer science and even philosophy. The goal of this workshop was to find out about different approaches towards a design theory, to discuss the subject and to see whether a common understanding could be reached which would lead us towards the establishment and application of a Universal Design Theory.

The design of artefacts is nowadays supported by software systems. We are convinced that the development of such software systems could be influenced (by engineering the development of CAD systems) if a design theory existed.

The goal of this workshop was to discuss these issues and to possibly initiate a research program for the investigation of approaches to such a Universal Design Theory.

Hans Grabowski

Acknowledgements

Below is the list of the programme committee whose members also acted as referees for the workshop.

- **Professor Dr.-Ing. Herbert Birkhofer**
  Machine Elements and Engineering Design,
  Darmstadt University of Technology, Germany
- **Professor Dr.-Ing. Hans-Joachim Franke**
  Institue for Engineering Design, Machine- and High Precision Elements
  Technical University of Braunschweig, Germany
- **Professor Dr.-Ing. Jürgen Gausemeier**
  Computer Integrated Manufacturing, Heinz Nixdorf Institute
  University of Paderborn, Germany
- **Professor Dr.-Ing. Dr. h.c. Hans Grabowski**
  Institute for Applied Computer Science in Mechanical Engineering (RPK)
  University of Karlsruhe, Germany
- **Professor Dr.-Ing. Frank-Lothar Krause**
  Production Systems and Design Technology
  Fraunhofer Institute, Berlin, Germany
- **Professor Dr.-Ing. Udo Lindemann**
  Department of Design in Mechanical Engineering
  Technical University of Munich, Germany
- **Dr.-Ing. Stefan Rude**
  Institute for Applied Computer Science in Mechanical Engineering (RPK)
  University of Karlsruhe, Germany
- **Professor Dr.-Ing. Christian Weber**
  Department of Engineering Design/CAD
  Saarland University, Germany

The success of the workshop was ensured by the high quality of the papers submitted as well as the presentations and contributions to the discussions. First of all, we would therefore like to thank all the participants of this workshop and in particular all those authors who presented their papers.

We would also like to express our appreciation for the organizational work done by Eike Meis and El-Fathi El-Mejbri. Their conscientiousness and involvement contributed to the great success of the workshop.

In addition, we would like to thank Andrea Pugnetti and Eva Stabenow for doing a great job in simultaneously translating both the presentations and the discussions. The discussions during the workshop could not have been published in this issue without the skillful recording by Mr. Rietz as well as the transcription and editing of the conversations by Martin Hofmann. Many thanks to Frank Jenne and Volker Sommerfeld for designing the workshop logo and the cover of this issue.

The workshop was partially sponsored by the Volkswagen-Stiftung, Hannover (Germany).

List of Participants

Albers, A., Institute of Machine Design and Automotive Engineering, University of Karlsruhe, Germany
Aldinger, F., Max-Planck-Institute for Metals Research, Stuttgart, Germany
Anderl, R., Computer Integrated Design, Darmstadt University of Technology, Germany
Andreasen, M. M., Department of Control and Engineering Design, Technical University of Denmark
Aßmann, U., Institute for Program Structures and Data Organization, University of Karlsruhe, Germany
Becker, J., Institute of Microelectronic Systems, Department of Electrical Engineering and Information Technology, Darmstadt University of Technology, Germany
Bercsey, T., Institute of Machine Design, Technical University of Budapest, Hungary
Birkhofer, H., Machine Elements and Engineering Design, Darmstadt University of Technology, Germany
Burchardt, C., Department of Mechanical Engineering - Engineering Design, Otto-von-Guericke-University of Magdeburg, Germany
Ehrlenspiel, K., Department of Design in Mechanical Engineering, Technical University of Munich, Germany
El-Fathi El-Mejbri, Institute for Applied Computer Science in Mechanical Engineering (RPK), University of Karlsruhe, Germany
Franke, H.-J., Institute for Engineering Design, Machine- and High Precision Elements, Technical University of Braunschweig, Germany
Gausemeier, J., Computer Integrated Manufacturing, Heinz Nixdorf Institute, University of Paderborn, Germany
Gero, J. S., Department of Architectural and Design Science, University of Sydney, Australia
Gerst, M., Department of Design in Mechanical Engineering, Technical University of Munich, Germany
Glesner, M., Institute of Microelectronic Systems, Department of Electrical Engineering and Information Technology, Darmstadt University of Technology, Germany
Grabowski, H., Institute for Applied Computer Science in Mechanical Engineering (RPK), University of Karlsruhe, Germany
Grein, G., Institute for Applied Computer Science in Mechanical Engineering (RPK), University of Karlsruhe, Germany
Hacker, W., Chair of Cognitive and Motivational Psychology, Technical University of Dresden, Germany
Herges, R., Institute of Organic Chemistry, Technical University of Braunschweig, Germany
Heynen, C., Chair for Engineering Design, Friedrich-Alexander-University of Erlangen-Nürnberg, Germany
Hopf, H., Institute of Organic Chemistry, Technical University of Braunschweig, Germany
Jin, Y., The IMPACT Laboratory, University of Southern California, Los Angeles, USA
Kind, C., Production Systems and Design Technology, Fraunhofer Institute, Berlin, Germany
Kocherscheidt, H., BMW AG, Munich, Germany

Krause, F. H., Production Systems and Design Technology, Fraunhofer Institute, Berlin, Germany
Lenk, H., Faculty for Humanities and Social Sciences, Institute of Philosophy, University of Karlsruhe, Germany
Lindemann, U., Department of Design in Mechanical Engineering, Technical University of Munich, Germany
Löhe, D., Institute of Materials Science and Engineering I, University of Karlsruhe, Germany
Lossack, R., Department of Precision Machinery Engineering, University of Tokyo, Japan
Ludwig, A., Institute for Program Structures and Data Organization, University of Karlsruhe, Germany
Mackay, R., Directorate-General III - Industry, European Commission, Brussels, Belgium
Matthiesen, S., Institute of Machine Design and Automotive Engineering, University of Karlsruhe, Germany
Meis, E., Institute for Applied Computer Science in Mechanical Engineering (RPK), University of Karlsruhe, Germany
Menz, W., Institute for Microsystems Technology, Albert-Ludwigs-University Freiburg, Germany
Michelis, A., Institute for Applied Computer Science in Mechanical Engineering (RPK), University of Karlsruhe, Germany
Nachtigall, W., Technical Biology and Bionics, Saarland University, Germany
Nitsch, C., Dept. I: Natural and Engineering Science, Medicine, Volkswagen-Stiftung, Hannover, Germany
Penschuck, H., Dept. I: Natural and Engineering Science, Medicine, Volkswagen-Stiftung, Hannover, Germany
Philipp, M., Computer Integrated Design, Darmstadt University of Technology, Germany
Prüfer, H. P., Information Technology in Mechanical Design, Rhur-University Bochum, Germany
Riepe, B., Computer Integrated Manufacturing, Heinz Nixdorf Institute, University of Paderborn, Germany
Rude, S., Institute for Applied Computer Science in Mechanical Engineering (RPK), University of Karlsruhe, Germany
Schweinberger, D., Institute of Machine Design and Automotive Engineering, University of Karlsruhe, Germany
Tawil, M., Fritz-Süchting-Institute for Mechanical Engineering, Technical University of Clausthal, Germany
Tomiyama, T., Department of Precision Machinery Engineering, University of Tokyo, Japan
Varady, K., Institute of Machine Design, Technical University of Budapest, Hungary
Weber, C., Department of Engineering Design/CAD, Saarland University, Germany

SESSION ONE

CLASSICAL APPROACHES OF A GENERAL DESIGN THEORY

Axiomatic Design as a Basis for Universal Design Theory

Nam P. Suh

Department of Mechanical Engineering

Massachusetts Institute of Technology

Keywords: axiomatic design, axioms, design theory, systems, system design

ABSTRACT

In engineering, design is as fundamental as mechanics and thermodynamics. The fundamental principles of these disciplines must be taught to all engineers, and engineers in industry should know these fundamental principles to be proficient in her/his profession. Since the introduction of axiomatic design, much progress has been made both in pedagogy and in research. It has been taught in many schools and in many industrial firms. Engineers and designers, once they understand axiomatic design, become much more efficient, effective and in some cases, creative in generating rational designs. It has been used to create original designs, improve existing designs, and to find the cause for failures of existing products and systems. It is also a powerful tool in dealing with complex systems. The axiomatic design theory has been used to design many different kinds of products and systems, including machines, manufacturing processes and systems, software systems, organizations, and systems that consist of a combination of hardware and software. At MIT, we teach the second year undergraduate students of mechanical engineering the axiomatic design principles as the basis for a required introductory course in design. It is also taught in upper class subjects in manufacturing as well as in graduate design subjects.

In this lecture, the basic concept of axiomatic design will be briefly introduced. Then, the implications of the design axioms and the design of complex systems based on axiomatic design will be discussed. Complex systems are designed to fulfill different functions and nearly all of them perform more than one function. Each system consists of sub-systems and components. Some, like automobiles and machine tools, perform a large number of different but dedicated functions. Others, such as a factory, perform a variety of different functions during its lifetime. Engineers and designers must design, operate and maintain such systems based on a logical and consistent framework. Systems can be designed based on axiomatic design. A "flow chart" for system architecture -- a concise means of representing and documenting a system -- will be presented, which provides a complete description of the architecture of a system. The flow chart can be used to create a central control module for execution and operation of systems.

the abscissa is the time from the first initiation to the subject. It shows that at the beginning, the concept of axiomatic design sounds easy, but as one learns more and tries to solve problems, it becomes more difficult and can be discouraging. However, once one masters the subject, it becomes very easy to practice the axiomatic design.

GENERAL INTRODUCTION TO DESIGN

The field of design is undergoing rapid changes from a field dominated by empiricism to a field of science. It is driven by the need to educate future engineers more effectively, to improve industrial competitiveness of industrial firms, to utilize the computational power of computers in all aspects of design, and by the advances in the science of design. One of the most significant changes in design practice will occur when the field of design is fully endowed with a firm science base. Axiomatic design provides the basic principles that can be applied in dealing with any kinds of design problems -- design of machines, systems, software, organizations, and materials.

Much progress has been made in the use of computers in design. Computer-aided-design packages are used routinely in industry, which has improved the productivity of designers, reduced the possibility of making mistakes, and the use of common database possible. Furthermore, numerical techniques have improved the reliability of engineering products. However, in many firms, design decisions are still made based on empirical information and educated guesses. In that sense, design practice in designing equipment, products, software, processes, organizations and systems has not changed very much. As a result, many design mistakes are being made routinely today. This problem is not confined to any one country or any one company. One of the biggest challenges of the design field is to overcome the perception that the design is a subject in arts rather than a subject in science as well as in arts.

The ultimate goal of Axiomatic Design is to establish a science base for design and to improve design activities by providing the designer with a theoretical foundation based on logical and rational thought processes and tools. The goal of axiomatic design is manifold: to make human designers more creative, reduce the random search process, minimize the iterative trial-and-error process, determine the best designs among those proposed, and endow the computer with creative power through the creation of the science base for the design field. Just as many fields of technology have gone through similar stages of development, the field of design will evolve into a true discipline with scientific bases.

AXIOMATIC DESIGN FRAMEWORK[1]

THE CONCEPT OF DOMAINS

Design involves an interplay between "what we want to achieve" and "how we choose to satisfy the need (i.e., what)". To systematize the thought process involved in this interplay, the concept of domains that create demarcation lines between various design activities is a foundation of axiomatic design. The world of design is made up of four *domains*: the *customer domain, the functional domain*, the *physical domain*, and the *process domain*. The domain structure is illustrated schematically in Figure 2. The domain on the left relative to the domain

---

[1] Some of the basic materials in this section have been presented elsewhere [*Suh-90a, Suh-98*]. The major reference is [*Suh-90a*].

on the right represents "what we want to achieve," whereas the domain on the right represents the design solution of "how we propose to satisfy the requirements specified in the left domain."

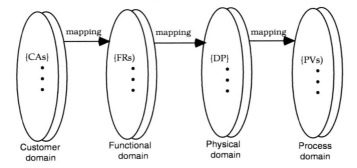

Figure 2   Four Domains of the Design World. {x} are characteristic vectors of each domain

The customer domain is characterized by customer needs or the attributes the customer is looking for in a product or process or systems or materials. In the functional domain, the customer needs are specified in terms of functional requirements (FRs) and constraints (Cs). In order to satisfy the specified FRs, we conceive design parameters, DPs, in the physical domain. Finally, to produce the product specified in terms of DPs, in the process domain we develop a process that is characterized by process variables, PVs. For example, a customer in semiconductor industry needs to coat the surface of a silicon wafer with photoresist. This is done in the customer domain. Based on this need, the engineer in an equipment company establishes the functional requirements (FRs) in terms of thickness and uniformity and also the constraints (Cs) in terms of tolerable level of contaminant particles, production rate, and cost. This is done in the functional domain. Then, the designer of equipment, based on experimental data and past experience, must conceive a design solution and identify the important design parameters (DPs) in the physical domain. The designer might choose to spray the photoresist and control thickness by spinning the disk at high speed to make use of centrifugal force. Then, the manufacturing engineer in the process domain must conceive the means of manufacturing the equipment, specifying the process variables that can provide the DPs.

Different design tasks in many different fields can be described in terms of the four design domains. In the case of the product design, the customer domain consists of the customer requirements or attributes the customer is looking for in a product; the functional domain consists of functional requirements, often defined as engineering specifications and constraints; the physical domain is the domain in which the key design parameters {DPs} are chosen to satisfy the {FRs}; and the process domain specifies the manufacturing methods that can produce the {DPs}.

All designs fit into these four domains. Therefore, all design activities, be it product design or software design, can be generalized in terms of the same principles. Because of this logical structure of the design world, the generalized design principles can be applied to all design applications and we can consider all the design issues that arise in four domains systematically and if necessary, concurrently.

Definitions

Before proceeding any further discussion of axiomatic design, it is important for us to summarize the definition of a few key words discussed in the preceding section, since axioms are valid only within the bounds established by the definitions of these key terms. Just as the words like heat and work have unique meaning in thermodynamics, which are different from their daily usage, so is the case with key words used in axiomatic design. The definitions are as follows:

- Axiom: Self-evident truth or fundamental truth for which there are no counter examples or exceptions. It cannot be derived from other laws of nature or principles.
- Corollary: Inference derived from axioms or propositions that follow from axioms or other propositions that have been proven.
- Theorems: Theorems are statement of facts that are derived from the axioms.
- **Functional Requirement (FR):** A minimum set of independent requirements that completely characterize the functional needs of the product (or software, organizations, systems, etc.) in the functional domain. By definition, each FR is independent from each other at the time the FRs are established.
- **Constraint (C):** Constraints are bounds on acceptable solutions. There are two kinds of constraints: input constraints and system constraints. Input constraints are imposed as part of the design specifications. System constraints are constraints imposed by the system in which the design solution must function. Constraints do not have to be independent from FRs and other constraints.
- **Design parameter (DP):** Design parameters are the key physical (or other equivalent terms in the case of software design, etc.) variables in the physical domain that characterize the design and satisfy the specified FRs.
- **Process variable (PV):** Process variables are the key process (or other equivalent term in the case of software design, etc.) variables in the process domain that characterize the process that can generate the specified DPs.

Most of the key words listed are associated with the Independence Axiom. Additional definitions of key words associated with the Information Axiom will be given in a later section.

Mapping from Customer Needs to Determination of Functional Requirements

The attributes desired in a product by customers (CAs) or customer needs are sometimes difficult to define or vaguely defined. Nevertheless we have to do the best we can to understand the customer needs by working with customers to define their needs. Then these needs (or the attributes the customer is looking for in a product) must be translated to functional requirements FRs. This must be done in a "solution neutral environment". That means FRs must be defined without ever thinking about something that has been already designed or what the design solution should be. If FRs are defined based on an existing design, then we will simply be specifying the FRs of that product and the result of the design endeavor will be likely to be similar to the existing product, forestalling creative thinking.

Industrial firms often use "marketing requirement specification" (MRS) as the product specification document. They are often very thick, for which the primary inputs are provided by the marketing people. Often the document is a random mixture of CAs, FRs, Cs, DPs, and

PVs. When the marketing group specifies DPs, and PVs, the design process becomes complicated since they lose freedom to come up with the best design solutions. They should limit their inputs to CAs, FRs, and Cs.

THE FIRST AXIOM: THE INDEPENDENCE AXIOM

[handwritten margin note: not an axiom, but a guideline]

During the mapping process (for example, going from the functional domain to the physical domain), we must make correct design decisions using the Independence Axiom. When several designs that satisfy the Independence Axiom are available, the Information Axiom can be used to select the best design. When only one FR is to be satisfied, the Independence Axiom is always satisfied and therefore, the Information Axiom is the only axiom the one FR design must satisfy.

The first axiom is called the *Independence Axiom*. It states that the independence of *Functional Requirements* (FRs) must be always maintained, where FRs are *defined as the minimum number of independent functional requirements* that characterize the design goals. The second axiom is called the *Information Axiom*, and it states that among those designs that satisfy the Independence Axiom, the design that has the highest probability of success is the best design. Based on these design axioms, we can derive theorems and corollaries. The axioms are formally states as

- Axiom 1: The Independence Axiom
    Maintain the independence of the functional requirements (FRs).
- Axiom 2: The Information Axiom
    Minimize the information content of the design.

As stated earlier, the functional requirements, FRs, are defined as the minimum set of independent requirements that the design must satisfy. A set of functional requirements {FRs} are the description of design goals. The Independence Axiom states that when there are two or more functional requirements, the design solution must be such that each one of the functional requirements can be satisfied without affecting the other functional requirement. That means we have to choose a correct set of DPs to be able to satisfy the functional requirements and maintain their independence.

The Independence Axiom is often misunderstood. Many people confuse between the functional independence with the physical integration. The Independence Axiom requires that the functions of the design be independent from each other, not the physical parts. This is illustrated using the beverage can as an example. Let us consider an aluminum beverage can that contains carbonated drinks. How many functional requirements must the can satisfy? How many physical parts does it have? What are the design parameters (DPs)? How many DPs are there?

According to an expert working at one of the aluminum can manufacturer, it appears that there are 12 FRs for the can. It has to contain the pressure, withstand a moderate impact when the can is dropped from a certain height, stack on top of each other, provide easy access to the liquid in the can, minimize the use of aluminum, printable on the surface, and others. However, the aluminum can consists of only three pieces: the body, the lid, and the opener tab. What the Independence Axiom requires is that the 12 FRs be independent from each other, not that there be 12 physical pieces making up the can!

Where are the DPs? According to Theorem 4, there must be at least 12 DPs. Most of the DPs are associated with the geometry of the can. Thickness of the can body, the curvatures at the bottom of the can, the reduced diameter of the can at the top to reduce the material used to make the top lid, the corrugated geometry of the opening tab to increase the stiffness, the small extrusion on the lid to attach the tab, etc. There are 12 DPs in the can design and the

Independence Axiom is satisfied by the can, according to the engineer who improved the can design after taking the axiomatic design course at MIT.

After the FRs are established, the next step in the design process is the conceptualization process, which occurs during the mapping process going from the functional domain to the physical domain. To go from "what" to "how" (e.g., from the functional domain to the physical domain) requires mapping which involves creative conceptual work. Once the overall design concept is generated by mapping, we must identify the design parameters (DPs) and complete the mapping process. During this process, we must think of all different ways of fulfilling each of the FRs by identifying plausible DPs. Sometimes it is convenient to think about a specific DP to satisfy a specific FR, repeating the process until the design is completed. One can use a database of all kinds (generated through brainstorming, morphological techniques, etc.), analogy from other examples (apparently Thomas Edison's favorite means of invention), extrapolation and interpolation, laws of nature, order of magnitude analysis, reverse engineering (copying somebody else's good idea by examining an existing product), and others. It is relatively easy to identify a DP for a given FR, but when there are many FRs we must satisfy, the design task becomes difficult and many designers make mistakes by violating the Independence Axiom.

The mapping process between the domains can be mathematically expressed in terms of the characteristic vectors that define the design goals and design solutions. At a given level of design hierarchy, the set of functional requirements that define the specific design goals constitutes a vector {FRs} in the functional domain. Similarly, the set of design parameters in the physical domain that are the "How's" for the FRs also constitutes a vector {Dps}.

The relationship between these two vectors can be written as

$$\{FRs\} = [A]\{DPs\} \quad (1)$$

where [A] is a matrix defined as the Design Matrix that characterizes the product design. Equation (1) may be written in terms of its elements as $FR_i = A_{ij} DP_j$. Equation (1) is a <u>*design equation*</u> for design of a product.

The design matrix is of the following form for a symmetrical matrix (i.e., i = j):

$$[A] = \begin{bmatrix} A11 & A12 & A13 \\ A21 & A22 & A23 \\ A31 & A32 & A33 \end{bmatrix} \quad (2)$$

where

$$A_{ij} = \partial FR_i / \partial DP_j$$

Equation (1) may be written as

$$\begin{aligned} FR1 &= A11\ DP1 + A12\ DP2 + A13\ DP3 \\ FR2 &= A21\ DP1 + A22\ DP2 + A23\ DP3 \\ FR3 &= A31\ DP1 + A32\ DP2 + A33\ DP3 \end{aligned} \quad (3)$$

For a linear design, Aij are constants, whereas for nonlinear design Aij are functions of DPs. There are two special cases of the design matrix: diagonal matrix where all Aij's except those i = j are equal to zero, and triangular matrix where either upper or lower triangular elements are equal to zero as shown below.

$$[A] = \begin{bmatrix} A11 & 0 & 0 \\ 0 & A22 & 0 \\ 0 & 0 & A33 \end{bmatrix} \quad (4)$$

$$[A] = \begin{bmatrix} A11 & 0 & 0 \\ A21 & A22 & 0 \\ A31 & A32 & A33 \end{bmatrix} \quad (5)$$

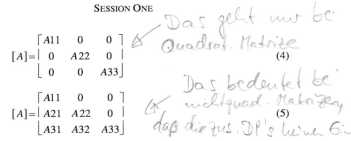

For the design of processes involving mapping from the physical domain to the process domain, the design equation may be written as:

$$\{DPs\} = [B] \{PVs\} \quad (6)$$

[B] is the design matrix that defines the characteristics of the process design.
To satisfy the Independence Axiom, the design matrix must be either diagonal or triangular. When the design matrix [A] is diagonal, each of the FRs can be satisfied independently by means of one DP. Such a design is called an *uncoupled* design. When the matrix is triangular, the independence of FRs can be guaranteed if and only if the DPs are changed in a proper sequence. Such a design is called a *decoupled or quasi-coupled* design. Therefore, when several functional requirements must be satisfied, we must develop designs that will enable us to create a diagonal or triangular design matrix.

The design matrix [A] or [B] can be made of matrix elements which are constants or functions of DPs or PVs. If the matrix is made of constants, it represents a linear design. If it is a relational matrix where its elements are functions of DPs, the design matrix may represent a nonlinear design.

The design goals are often subject to constraints, Cs. Constraints provide the bounds on the acceptable design solutions and differ from the FRs in that they do not have be independent. Some constraints are specified by the designer. Many constraints are also imposed by the environment within which the design must function. These are system constraints. Often it is best to treat cost as a constraint, but price can be treated as a functional requirement. Cost is affected by all design changes and therefore, cost cannot be made independent of other FRs in an uncoupled design. If it is decided that cost must be a functional requirement, then the best we can do is to develop a decoupled design, which also satisfies the Independence Axiom. With cost as a constraint, the design is acceptable as long as the cost does not exceed a set limit.

We can derive many corollaries and theorems based on these two axioms. For example, Theorem 1 states that *to satisfy the independence of a given set of FRs, the number of DPs cannot be less than the number of FRs.* Theorem 4 states that *in an ideal design, the number of DPs is equal to the number of FRs.* When the number of DPs is less than that of FRs, the design is always coupled. Many theorems and corollaries can be used as design rules for specific cases. In Appendix A, the theorems and corollaries are given.

FRs and DPs (as well as PVs, the characteristic vector for the process domain) can be decomposed into a hierarchy. However, contrary to the conventional wisdom on decomposition, they cannot be decomposed by remaining in one domain. One must zigzag between the domains to be able to decompose the FRs, DPs, and PVs. Through this zigzagging we create hierarchies for FRs, DPs, and PVs in each design domain. For example, if one of the FRs for a vehicle is "move forward," we cannot decompose it without deciding first in the physical domain "how we propose to go forward." If we choose a horse and buggy as a means of moving forward, the next layer of FRs will be different from when an automobile is chosen as the DP to satisfy the FR. In other words, to create a FR, DP, and PV hierarchies, we must map into the domain on the right ("how domain") first from the domain on the left ("what domain"), and then come back to the domain on the left ("what domain") to generate the next level FRs, etc.

DECOMPOSITION, ZIGZAGGING AND HIERARCHY

The design at a given level is completed when we map from the Functional Domain to the Physical Domain. However, the design may not be implementable because the details are missing. In this case, we need to decompose the high level FRs into lower level FRs, and similarly, the highest DPs to lower level DPs. This decomposition process must proceed layer by layer until the design can be implemented.

We must zigzag between the domains in order to decomposed these characteristic vectors. That is, we start out in the "what" domain and go to "how" domain. From FR in the functional domain, we go to DP in the physical domain. Then, we come back to the functional domain to create FR1 and FR2 that collectively satisfy the highest level FR and the corresponding DP. Then we go to the physical domain to find DP1 and DP2, which satisfy FR1 and FR2, respectively. This process continues until the FR can be satisfied without further decomposition. This process is pursued until all the branches reach the final state.

MODELING TO DETERMINE THE EXACT EXPRESSION FOR X OF THE DESIGN MATRIX

So far, the design matrix was formulated in terms of X and 0. In some cases, it may be sufficient to complete the design with simply X's and 0's. In many cases, we may take further steps to optimize the design. After the conceptual design is done in terms of X and 0, we need to model the design more precisely to optimize the design based on the laws of nature. Through modeling we can replace X's with exact equations which may be constants in the case of a linear design or functions that involve DPs.

REQUIREMENTS FOR CONCURRENT ENGINEERING

So far we have not discussed the mapping from the physical domain to the process domain, i.e., manufacturing process design in the case of product development. After certain DPs are chosen, we have to map from the physical domain to the process domain (i.e., process design) by choosing the process variables, PVs. During mapping, the process design must also satisfy the Independence Axiom. Sometimes we may simply use existing processes or invent new processes. When the existing processes must be used to minimize capital investment in new equipment, the existing process variables must be used and thus act as constraints in choosing DPs. In developing a product both the product design and the process design (or selection) must be done at the same time. This is sometimes called "concurrent engineering" or "simultaneous design".

For concurrent engineering to be possible, both the product design represented by Eq. (1) and the process design represented by Eq. (6) must satisfy the Independence Axiom. That means, the product design matrix [A] and the process design matrix [B] must be diagonal or triangular so that the product of these matrices [C]=[A][B] must be diagonal or triangular. (Note: each element is $C_{ik} = S_j C_{ij} C_{jk}$ summed over j.) Table 1 shows the characteristic of the matrix [C] depending on the kinds of the matrices [A] and [B] are. For example, to get an uncoupled concurrent design, both matrices must be diagonal. If one is diagonal and the other is triangular the resulting product of matrices is triangular. If both [A] and [B] are triangular, they must be the same kind, either both upper triangular denoted by [UT] or low triangular [LT]. If one is [LT] and the other is [UT], the product is a full matrix [X]. Therefore, when [A] and [B] are triangular matrices, both of them must be either upper triangular or lower triangular for the manufacturing process to satisfy the independence for functional requirements. This is stated as Theorem 9 (Design for Manufacturability).

	[A]	[B]	[C] = [A]{B}
1. Both diagonal	[\]	[\]	[\]
2. Diag x Full	[\]	[X]	[X]
3. Diag x triang.	[\]	[LT]	[LT]
4. Tria. x Triang	[LT]	[LT]	[LT]
5. Tria. x Triang	[LT]	[UT]	[X]
6. Full x Full	[X]	[X]	[X]

Table 1   The characteristic of concurrent engineering matrix [C]. Note that only (1), (3), and (4) are acceptable designs from the concurrent engineering point of view.

THE SECOND AXIOM: THE INFORMATION AXIOM

In the preceding sections, the Independence Axiom was discussed and its implications were presented. In this section, we shall now discuss how we can choose the best design. Even for the same task defined by a given set of FRs, it is most likely that every designer will come up with different designs, all of which may be acceptable in terms of the Independence Axiom. Indeed there can be a large number of designs that can satisfy a given set of FRs. However, one of these designs is likely to be superior to others. The Information Axiom provides a quantitative means of measuring the merits of a given design, which can be used to select the best among those acceptable. In addition, the Information Axiom provides the theoretical basis for design optimization and also robust design.

The Information Axiom state that the one with the highest probability of success is the best design. Specifically, the Information Axiom may be stated as:

- Axiom 2: The Information Axiom
    Minimize the information content

Information content I is defined in terms of the probability of satisfying a given FR. If the probability of success of satisfying a given FR is p, the information I associated with the probability is defined as

$$I = -\log_2 p \qquad (7)$$

The information is given in units of bits. The logarithmic function is chosen so that the information content will be additive when there are many functional requirements that must be satisfied at the same time.

In the general case of n FRs for an uncoupled design, I may be expressed as

$$I = \sum_{i=1}^{n} \left[\log 1/p_i\right] \qquad (8)$$

where p_i is the probability of DPi satisfying FRi and log is either the logarithm based on 2 (with the unit of bits) or the natural logarithm (with the unit of nats). Since there are n FRs, the total information content is the sum of all these probabilities. The Information Axiom states that the design that has the smallest I is the best design, since it requires the least amount of information to achieve the design goals. When all probabilities are equal to one, the information content is zero, and conversely, the information required is infinite when one or more probabilities are equal to zero. That is, if probability is low, we must supply more information to satisfy the functional requirements.

The definition given in Eq. (7) is the same as that used in information theory, which is also related to the negative entropy. However, there are important differences between the information used in information theory and axiomatic design. The major difference is that in information theory and thermodynamics, the total probability of an ensemble of events is always equal to zero, because there are a finite number of events that can be anticipated in information theory and natural sciences. In the case of axiomatic design, since there is an infinite number of different designs, the sum of probabilities (i.e., total probability) is not equal to zero.

A design is called *complex* when its probability of success is low, that is, when the information content required to achieve the FRs is high. This occurs when the tolerances for FRs of a product (or DPs in the process design) are small, requiring high accuracy. This situation also arises when there are many parts since as the number of parts increases, it also increases the possibility that some of the components do not meet the specified requirements. In this sense, the quantitative measure for complexity is the information content. According to Eq. (8), complex systems may require more information to make the systems function. A physically large system is not necessarily complex if the information content is low. Even a small system can be complex if the probability of its success is low. Therefore, the notion of complexity is tied to the tolerance for the FRs: the tighter the tolerance, the more difficult it becomes to satisfy the FRs.

In the real world, the probability of success is governed by the intersection of the tolerance defined by the designer to satisfy the FRs and the tolerance (or the ability) of the system to produce the part within the specified tolerance. For example, if the design specification for cutting a rod is 1 meter plus or minus one micron and the available tool (i.e., system) for cutting the rod consists of only a hacksaw, the probability of success will be extremely low. In fact, the information required to achieve the goal would approach infinity as long as the only system available to cut the rod is the hacksaw. Therefore, this may be called a complex design. On the other hand, if the rod needs to be cut within an accuracy of 10 cm, the hacksaw may be more than adequate and therefore, the information required is zero. In this case, the design is simple.

The probability of success can be computed when the designer specifies the *Design Range (dr)* for the FR and when the *System Range (sr)* that the proposed design can provide to satisfy the FR is determined. Figure 3 illustrates these two ranges graphically. The vertical axis (the ordinate) is for the probability density and the horizontal axis (the abscissa) is for either FR or DP, depending on the mapping domains involved. When the mapping is between the functional domain and the physical domain as in product design, the abscissa is for FR, whereas for the mapping between the physical domain and the process domain as in process design, the abscissa is for DP. In Figure 3, Design Range and the System Range are plotted as a probability density versus the specified FR. The overlap between the design range and system range is called the *common range (cr)*, and this is the only region where the functional requirements are satisfied. Consequently, the area under the Common Range divided by the area under the System Range is equal to the design's probability of success of achieving the specified goal. Then, the information content may be expressed as [Suh-90a]:

$$I = \log \left(A_{sr} / A_{cr} \right) \quad (9)$$

where A_{sr} denotes the area under the System Range and A_{cr} is the area of the Common Range. Furthermore, since $A_{sr} = 1.0$ in most cases and there are n FRs to satisfy, the information content may be expressed as

$$I = \sum_{I}^{n} \log (1/A_{cr})_i \quad (10)$$

Often design decisions must be made when there are many FRs that must be satisfied at the same time. The Information Axiom provides a powerful criterion for making such decisions without the use of arbitrary weighting factors used in other decision making theories. In Eq. (8), each information content term corresponding to each FR is simply summed up with all other terms without multiplying it with a weighting factor for two reasons. First, if we sum up the information terms, each of which has been modified by multiplying with a weighting factor, the total information content does no longer represent the total probability. Second, the intention of the designer and the importance assigned to each FR by the designer are represented by the design range. If it is a critical FR that must be satisfied within a tight tolerance, the designer would give a narrow design range.

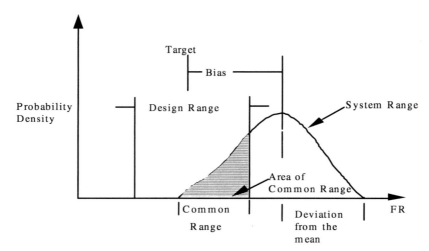

Figure 3   *Design Range, System Range, and Common Range in a plot of the probability density function (pdf) of a functional requirement. The deviation from the mean is equal to the square root of the variance.*

When there is only one FR, the independence axiom is always satisfied. In the one-FR case, the only task left is the optimization of a given design based on the Information Axiom. Various optimization techniques have been advanced to deal with optimization problems involving one objective function. However, when there are more than two FRs, some of these optimization techniques do not work. In order to satisfy a design with more than one FR, we must first develop a design that is either uncoupled or decoupled. If the design is uncoupled, it can be seen that each FR can be satisfied and the optimum points can be found, since there is one DP that controls the FR. If the design is decoupled, the optimization technique must follow a set sequence. The second axiom on information provides a metric that enables us to measure the information content and thus be able to judge a superior design.

REDUCTION OF THE INFORMATION CONTENT -- ROBUST DESIGN

The ultimate goal of design is to reduce the additional information required to make the system function as designed to zero, i.e., minimize the information content as per the Information Axiom. To achieve this goal, the design must satisfy the Independence Axiom. Then, the variance of the system range can be made small and the bias can be eliminated so that the system range lies inside the design range, reducing the information content to zero (see Figure 3). A design that can accommodate large variations in design parameters and process variables and yet satisfy the functional requirements is called a robust design.

There are four different ways of achieving the goal of reducing the bias and the variance of a design and develop a robust design, *provided that the design satisfies the Independence Axiom.*

In one-FR design, the bias can be changed by changing the appropriate DP, since FR is a function of DPs and since we do not have to worry about its effect on other FRs. Therefore, it is easy to eliminate the bias when there is only one FR.

When there are more than one FRs to be satisfied, we may not be able to eliminate the bias unless the design satisfy the Independence Axiom. If the design is coupled, each time a DP is changed to eliminate the bias for a given FR, the bias for other FRs changes also, making the design uncontrollable. If the design is uncoupled design, the design matrix is diagonal and the bias associated with each FR can be changed independently as if the design is an one-FR design. When the design is decoupled design, the bias for all FRs can be eliminated by following the sequence dictated by the triangular matrix.

Variance is the distribution of the difference between the target value and the actual outcome. The variance is caused by a number of factors such as noise, coupling, environment, and random variations in design parameters. Therefore, in most situations, the variance must be minimized. The variation can be reduced in a few specific situations [*Suh-98*]. In a multi-FR design, the pre-requisite for variance reduction is the satisfaction of the First Axiom -- the Independence Axiom.

DESIGNING WITH INCOMPLETE INFORMATION

During the design of products, processes, software, systems and organizations, we encounter situations where the necessary knowledge about the proposed design is insufficient and thus design must be executed in the absence of complete information. The basic questions are: "Under what circumstances can design decisions be made in the absence of sufficient information?" and "what kinds of information are the most essential information in making design decisions?"

The information we need is indicated by the design equations. The information on the characteristic vectors, i.e., what they are, etc., are needed. Given an FR, the most appropriate DP must be chosen, the possibility of which increases with the size of the library of DPs that satisfy the FR. Similarly, given a DP, the more PVs we have, the larger will be the options we have. Once DPs and PVs are chosen, information on all the elements of the design matrix, which define the relationship between "what we want to achieve" and "how we want to achieve", must be available.

One of the central issues in the design process is: "What are the minimum information that is necessary and sufficient in making design decisions given a set of {DPs} for a given set of {FRs}. The necessary information depends on whether or not the proposed design satisfies the Independence Axiom. In the case of a coupled design, which violates the Independence Axiom, all the information associated with all the elements of the design matrix is required. That is, design cannot be done rationally without complete information in the case of coupled designs. Similarly, even in the case of uncoupled design that satisfies the Independence Axiom, the information is required for all the diagonal elements of the design matrix. The information required for the uncoupled case is less than the coupled design case, since there are no off diagonal element.

In the case of decoupled design, information on the off-diagonal elements may not be required to satisfy the given set of {FRs} with a given set of {DPs} if the {FRs} can adjusted through a "feedback" or "closed" loop mode. However, we need to know the diagonal elements Aii in all cases to be able to choose proper values of DPs to satisfy the functional requirements. It will be also necessary to know the off-diagonal elements Aij when the system must operate in an "open"loop mode. Even in the absence of complete information on off-diagonal elements, we can proceed with the design even if the diagonal elements are known and if the magnitudes of the off-diagonal elements are smaller than those of the diagonal elements, i.e., Aii>Aij, provided that the FR can be measured and the information can be used to change the appropriate DP. This can be done since the value of FR1 can be set first and then, the value of FR2 can be set by varying the value of DP2, regardless of the value of A21. When DP2 is chosen, we must be certain that it does not affect FR1, but it is not necessary that any information on A21 is available, if DP2 has the dominant effect on FR2, i.e., A22>A21. Similarly, as long as DP3 does not affect FR1 and FR2, the design can be completed, even if we do not have any information on A31 and A32 if FR3 can be measured in a "closed" loop mode. This is the only case when design can proceed in the absence of complete information. This is stated as Theorem 17.

SYSTEM DESIGN AND SYSTEM ARCHITECTURE[2]

All major engineering tasks exist in a systems context. Machines, airplanes, software systems, and automobile assembly plants are such systems — albeit systems of different kinds — and each has sub-systems and components. Such human-made systems must be designed, fabricated, and operated to achieve their intended functions and nearly all of them perform more than one functions. Some, like the automobile, perform a large number of different but dedicated functions. Others, such as a job-shop factory, perform a variety of different functions during their lifetimes.

From the functional point of view, systems can be distinguished in a variety of different ways: large systems from small systems, static systems from dynamic systems, fixed systems from flexible systems, passive systems from active systems, and automated systems from manual systems. Some systems are open systems — systems whose constituents change throughout their lifetimes — in contrast to closed systems that are made up of the same components at all

---

[2] Some of the materials presented in this section was also presented elsewhere [Suh-95b, Suh-97].

times. Examples of open systems are factories, universities, and many machine tools; closed systems include inertial guidance systems and television sets.

From the axiomatic design point of view, the design of systems is not fundamentally different from the design of simple mechanical products and software. All of them — systems, machines, and software — must satisfy functional requirements, constraints, the Independence Axiom, and the Information Axiom. However, the specific database (i.e., the relationship among functional requirements, design parameters, and process variables) will be different depending on what the system must do. Also, the governing physical laws may also be different depending on the functional requirements (FRs) the system must satisfy, the nature of the design parameters (DPs) chosen, and the process variables (PVs) employed to achieve the design objectives.

Since systems can be complex with many functional requirements, physical components and many lines of computer codes, the design of the system must be documented to capture all the relevant information about the system. The document must show the complete structure and the relationships among the decomposed FRs, DPs, and PVs. The documentation should provide all relevant information: specific functional requirements for which the system is designed, the rationale for choosing specific design parameters and process variables, and the causality relationship between the FRs, DPs, and PVs. Such documentation is also needed in tracking the effect of engineering change orders, since a change of particular FRs or DPs will have effects on multiple FRs.

Axiomatic design enables us to represent systems using a comprehensive and yet simple system architecture [Suh-97].

When a system design is represented by the FR, DP, and PV hierarchies with the corresponding design equations and matrices, elements of the design matrix can be used to form modules that can yield the desired FR given the input DP. Thus, the FR- and DP-hierarchies can be converted into a diagram called the module-junction diagram [Kim-91, Suh-97]. From the junction-module diagram, the flow chart, which is the system architecture, can be constructed. The system architecture is a concise and powerful tool that provide a roadmap for implementation of the system design.

Consider for example a system design represented by the following design equations:

$$\begin{Bmatrix} FR1 \\ FR2 \end{Bmatrix} = \begin{bmatrix} X & 0 \\ 0 & X \end{bmatrix} \begin{Bmatrix} DP1 \\ DP2 \end{Bmatrix}$$

$$\begin{Bmatrix} FR11 \\ FR12 \end{Bmatrix} = \begin{bmatrix} X & 0 \\ X & X \end{bmatrix} \begin{Bmatrix} DP11 \\ DP12 \end{Bmatrix}$$

$$\begin{Bmatrix} FR21 \\ FR22 \\ FR23 \end{Bmatrix} = \begin{bmatrix} X & 0 & 0 \\ X & X & 0 \\ 0 & 0 & X \end{bmatrix} \begin{Bmatrix} DP21 \\ DP22 \\ DP23 \end{Bmatrix}$$

$$\begin{Bmatrix} FR111 \\ FR112 \end{Bmatrix} = \begin{bmatrix} X & 0 \\ 0 & X \end{bmatrix} \begin{Bmatrix} DP111 \\ DP112 \end{Bmatrix}$$

$$\begin{Bmatrix} FR121 \\ FR122 \\ FR123 \end{Bmatrix} = \begin{bmatrix} X & 0 & 0 \\ X & X & 0 \\ X & 0 & X \end{bmatrix} \begin{Bmatrix} DP121 \\ DP122 \\ DP123 \end{Bmatrix} \qquad (11) \dots$$

$$\begin{Bmatrix} FR211 \\ FR212 \\ FR213 \\ FR214 \end{Bmatrix} = \begin{bmatrix} X & X & 0 & 0 \\ 0 & X & X & 0 \\ 0 & 0 & X & 0 \\ 0 & 0 & 0 & X \end{bmatrix} \begin{Bmatrix} DP211 \\ DP212 \\ DP213 \\ DP214 \end{Bmatrix} \quad \ldots (11)$$

$$\begin{Bmatrix} FR1231 \\ FR1232 \end{Bmatrix} = \begin{bmatrix} A & 0 \\ B & C \end{bmatrix} \begin{Bmatrix} DP1231 \\ DP1232 \end{Bmatrix}$$

$$\begin{Bmatrix} FR12321 \\ FR12322 \\ FR12323 \end{Bmatrix} = \begin{bmatrix} X & 0 & 0 \\ 0 & X & 0 \\ 0 & 0 & X \end{bmatrix} \begin{Bmatrix} DP12321 \\ DP12322 \\ DP12323 \end{Bmatrix}$$

DPs of the above design may consists of either hardware or software.

The system architecture (sometimes called the flow chart of the system architecture) for this system defined by Eqs. (11) is shown in Figure 4, which represents the system architecture of the design represented by the design equations given in Eq. (11). Figure 4 shows how the system represented by Eq. (11) should be structured in terms of its operational sequence. This is the system structure or the flow diagram.

The system architecture is made up of modules M. Modules represent the rows of the design matrix. For example, the modules that relate FR1231 and DP1231, and FR1232 and DP1232 of Eq. (11), respectively, are given by:

M1231 = A DP1231
M1232 = B DP1231 +C DP1232   (12)

where A, B and C are the elements of the design matrix shown in Eq. (11). The flow diagram of the system architecture is a complete representation of the system, showing how the modules must connected in the system. It also states how the system must be operated to obtain the desired performance from the system in terms of the stated FRs.

In Figure 4, the following symbols in circles are used to denote the control of the software flow chart: **S** in the circle is for simple summation of FRs for an uncoupled design, **C** in the circle is for sequential control of DPs as suggested by the design matrix for a decoupled design, and **F** in the circle is for coupled designs indicating that it requires feedback. Figure 5 shows these control symbols and associated modules.

When the child FRs are uncoupled, their parent FR is satisfied by combining all the outputs of its child modules in any random sequence. This is the summation junction, S. When they are decoupled, their parent FR is determined when the output of the left-hand side module is controlled first and then the right-hand side module is executed next. This is the control junction, C. The coupled junction is a feedback junction F that requires that the output of the right-hand side module be fed back to the left-hand module, requiring a number of iterations until the solution converges. Many feedback junctions may never converge. It should be noted that the coupled junction must be avoided as per the Independence Axiom. When there are feedback junctions, the program will quickly become unmanageable.

The flow chart shown in Figure 4 shows that this design is an acceptable system design since the design satisfies the Independence Axiom at all levels of the design hierarchy. The lowest DP's are the input variables to the system that ultimately control the final output of the system.

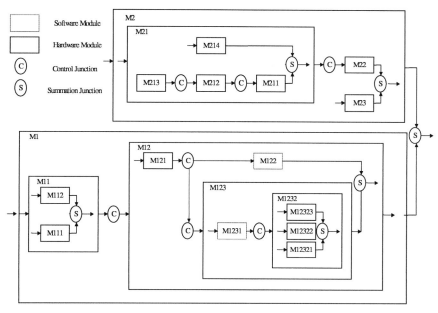

Figure 4 *The flow chart of the design of the system given by Eqs. (5). This flow chart represents the system architecture of the system. This system is made up of only summation and control junctions and thus, satisfies the Independence Axiom.*

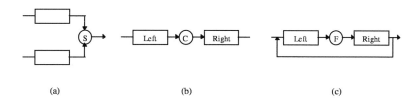

Figure 5 *Junction Properties of the Module-Junction Structure Diagram (a) Summing Junction (Uncoupled Case), (b) Control Junction (Decoupled Case), (c) Feedback Junction (Coupled Case). [Kim-91, Suh-97]*

COMPARISON OF AXIOMATIC DESIGN WITH VARIOUS METHODOLOGIES

Often questions are asked as to how axiomatic design differs from other design methodologies. When they ask these questions, they have many different methodologies in mind, including statistical process control (SPC) techniques, the Taguchi methodology [*Tagu-87*], and the Altshuller inventive problem solving methodology [*Alts-96*]. The following comments are offered as general comments:

- Axiomatic design deals with principles and methodologies rather than simply algorithms or methodologies. Based on the two axioms, it derives theorems and corollaries, and also develops methodologies based on functional analysis and information minimization which lead to robust design.
- Axiomatic design is applicable to *all* designs: products, processes, systems, software, organizations, materials, and business plan, whereas many of other methodologies are confined to product design.
- All methodologies, including the Taguchi method, must satisfy the design axioms for them to be valid. For example, the Taguchi method is valid only on designs that satisfies the Independence Axiom. So far, there seems to be no contradiction between Altshuller's methodologies and the design axioms since Altshuller's methodologies provide a means of identifying appropriate DPs.
- The Taguchi method does instruct how to make design decisions. It is a method of checking and improving a finished design.
- Both axiomatic design and the Taguchi method lead to robust design for designs that satisfy the Independence Axiom.
- Robust design cannot be done by applying the Taguchi method if it violates the Independence Axiom. (See the example involving design of an automatic transmission given in Chapter **SYSTEM DESIGN AND SYSTEM ARCHITECTURE**.)
- Although many efforts are being made in industry to improve a bad design using optimization techniques, the design that violates the Independence Axiom cannot be improved. Optimization of bad designs lead to optimized bad designs.

CONCLUDING REMARKS

The field of design is an academic and intellectual discipline that has the abstracted generalized principles that can form the science base. The process of design also requires the empirical know-how and database that have values in specific design situations. The basic principles of design as embodied in the axiomatic design principles and the design axioms can be applied to any specific fields where the synthetic solutions are needed to satisfy a set of functional requirements. In this lecture, the basic concepts and methodologies of axiomatic design, including the concepts of domains, mapping, the two design axioms (the Independence Axiom and the Information Axiom), decomposition, hierarchy, and zigzagging, are presented. Finally the design of systems is discussed, including the concept of system architecture, which should be useful in all systems design.

REFERENCES

Alts-96   Altshuller, G.: And Suddenly the Inventor Appeared. Technical Innovation Center, Worcester, MA, 1996.
Kim-91   Kim, S. J.; Suh, N. P.; Kim, S.-G.: "Design of software systems based on axiomatic design", CIRP Annals, Vol. 40, No. 1, 1991 [also: Robotics & Computer-Integrated Manufacturing, Vol. 3, 1992].
Sloc-98   Slocum, A.: Private email, MIT, March 2, 1998.
Suh-90a   Suh, N. P.: The Principles of Design, Oxford University Press, 1990.
Suh-90b   Suh, N. P.; Sekimoto, S.: "Design of Thinking Design Machine", Annals of CIRP, Vol. 1, 1990.

Suh-95a	Suh, N. P.: "Axiomatic Design of Mechanical Systems", Special 50th Anniversary Combined Issue of the Journal of Mechanical Design and the Journal of Vibration and Acoustics, Transactions of the ASME, Volume 117, pp 1-10, June 1995.
Suh-95b	Suh, N. P.: "Design and Operation of Large Systems", Journal of Manufacturing Systems, Vol. 14, No.3, pp 203-213, 1995.
Suh-97	Suh, N. P.: "Design of Systems," Annals of CIRP, Vol. 46, pp 75-80, 1997.
Suh-98	Suh, N. P.: Axiomatic Design: Advances and Applications, Oxford University Press, To be published in 1998.
Swen-96	Swenson,A.; Nordland, M.: Unpublished Report of Saab, 1996.
Tagu-87	Taguchi, G.: Systems of Engineering Design: Engineering Methods to Optimize Quality and Minimize Cost, American Supply Institute, 1987.
Thom-95	Thomas J.: „The Archstand Theory of Design for Information", Ph.D. Thesis, Massachusetts Institute of Technical, Department of Civil Engineering, February 1995.
Wats-69	Watson, J. D.: The Double Helix, Athenaeum, New York, 1969.

APPENDICES

Some of these theorems are derived in this book as well as in the references given. For those theorems not derived in this book, the readers may consult the original references.

COROLLARIES [*SUH-90A*]

- Corollary 1   (Decoupling of Coupled Designs)
  Decouple or separate parts or aspects of a solution if FRs are coupled or become interdependent in the designs proposed.
- Corollary 2   (Minimization of Frs)
  Minimize the number of FRs and constraints.
- Corollary 3   (Integration of Physical Parts)
  Integrate design features in a single physical part if FRs can be independently satisfied in the proposed solution.
- Corollary 4   (Use of Standardization)
  Use standardized or interchangeable parts if the use of these parts is consistent with FRs and constraints.
- Corollary 5   (Use of Symmetry)
  Use symmetrical shapes and/or components if they are consistent with the FRs and constraints.
- Corollary 6   (Largest Tolerance)
  Specify the largest allowable tolerance in stating FRs.
- Corollary 7   (Uncoupled Design with Less Information)
  Seek an uncoupled design that requires less information than coupled designs in satisfying a set of FRs.
- Corollary 8   (Effective Reangularity of a Scalar)
  The effective reangularity R for a scalar coupling „matrix" or element is unity.

THEOREMS OF GENERAL DESIGN (MOST OF THESE THEOREMS ARE FROM [*SUH-90A*]

- Theorem 1   (Coupling Due to Insufficient Number of Dps)
  When the number of DPs is less than the number of FRs, either a coupled design results, or the FRs cannot be satisfied.

- **Theorem 2** (Decoupling of Coupled Design)
  When a design is coupled due to the greater number of FRs than DPs (i.e., m. > n), it may be decoupled by the addition of new DPs so as to make the number of FRs and DPs equal to each other, if a subset of the design matrix containing n x n elements constitutes a triangular matrix.
- **Theorem 3** (Redundant Design)
  When there are more DPs than FRs, the design is either a redundant design or a coupled design.
- **Theorem 4** (Ideal Design)
  In an ideal design, the number of DPs is equal to the number of FRs.
- **Theorem 5** (Need for New Design)
  When a given set of FRs is changed by the addition of a new FR, or substitution of one of the FRs with a new one, or by selection of a completely different set of FRs, the design solution given by the original DPs cannot satisfy the new set of FRs. Consequently, a new design solution must be sought.
- **Theorem 6** (Path Independence of Uncoupled Design)
  The information content of an uncoupled design is independent of the sequence by which the DPs are changed to satisfy the given set of FRS.
- **Theorem 7** (Path Dependency of Coupled and Decoupled Design)
  The information contents of coupled and decoupled designs depend on the sequence by which the DPs are changed to satisfy the given set of FRs.
- **Theorem 8** (Independence and Tolerance)
  A design is an uncoupled design when the designer-specified tolerance is greater than

$$\left(\sum_{\substack{j \neq i \\ i=1}}^{n} (\partial FR_i / \partial DP_j) \Delta DP_j \right)$$

  in which case the nondiagonal elements of the design matrix can be neglected from design consideration.
- **Theorem 9** (Design for Manufacturability)
  For a product to be manufacturable, the design matrix for the product, [**A**] (which relates the **FR** vector for the product to the **DP** vector of the product) times the design matrix for the manufacturing process, [**B**] (which relates the **DP** vector to the **PV** vector of the manufacturing process) must yield either a diagonal or triangular matrix. Consequently, when any one of these design matrices, that is, either [**A**] or [**B**], represents a coupled design, it is difficult to manufacture the product. When they are triangular matrices, both of them must be either upper triangular or lower triangular for the manufacturing process to satisfy the independence for functional requirements.
- **Theorem 10** (Modularity of Independence Measures)
  Suppose that a design matrix [**DM**] can be partitioned into square submatrices that are nonzero only along the main diagonal. Then the reangularity and semangularity for [**DM**] are equal to the product of their corresponding measures for each of the non-zero submatrices.
- **Theorem 10a** (R and S for Decoupled Design)
  When the semangularity and reangularity are the same, the design is a coupled design.
- **Theorem 11** (Invariance)
  Reangularity and semangularity for a design matrix [**DM**] are invariant under alternative orderings of the FR and DP variables, as long as orderings preserve the association of each FR with its corresponding DP.

- Theorem 12 (Sum of Information)
  The sum of information for a set of events is also information, provided that proper conditional probabilities are used when the events are not statistically independent.
- Theorem 13 (Information Content of the Total System)
  If each DP is probabilistically independent of other DPs, the information content of the total system is the sum of the information of all individual events associated with the set of FRs that must be satisfied.
- Theorem 14 (Information Content of Coupled versus Uncoupled Designs)
  When the state of FRs is changed from one state to another in the functional domain, the information required for the change is greater for a coupled process than for an uncoupled process.
- Theorem 15 (Design-Manufacturing Interface)
  When the manufacturing system compromises the independence of the FRs of the product, either the design of the product must be modified, or a new manufacturing process must be designed and/or used to maintain the independence of the FRs of the products.
- Theorem 16 (Equality of Information Content)
  All information contents that are relevant to the design task are equally important regardless of their physical origin, and no weighting factor should be applied to them.
- Theorem 17 (Design in the absence of complete Information)
  Design can proceed even in the absence of complete information only in the case of decoupled design if the missing information is related to the off-diagonal elements and if the output FR can be measured in a "closed" loop mode so as to use the measure output of FR to change the corresponding DP to satisfy the FR.

THEOREMS FOR DESIGN OF LARGE SYSTEMS [SUH-95B, SUH-98]

- Theorem 18 (Importance of High Level Decisions)
  The quality of design depends on the selection of FRs and the mapping from domains to domains. Wrong selection of FRs made at the highest levels of design domains cannot be rectified through the lower level design decisions.
- Theorem 19 (The Best Design for Large Systems)
  The best design among the proposed designs for a large system that satisfy n FRs and the Independence Axiom can be chosen if the complete set of the subsets of {FRs} that the large system must satisfy over its life is known *a priori*.
- Theorem 20 (The Need for Better Design for Large Systems)
  When the complete set of the subsets of {FRs} that a given large system must satisfy over its life is not known *a priori*, there is no guarantee that a specific design will always have the minimum information content for all possible subsets and thus, there is no guarantee that the same design is the best at *all times*, even if there are designs that satisfy the FRs and the Independence Axiom.
- Theorem 21 (Improving the Probability of Success)
  The probability of choosing the best design for a large system increases as the known subsets of {FRs} that the system must satisfy approach the complete set that the system is likely to encounter during its life.
- Theorem 22 (Infinite Adaptability versus Completeness)
  The large system with an infinite adaptability (or flexibility) may not represent the best design when the large system is used in a situation where the complete set of the subsets of {FRs} that the system must satisfy is known *a priori*.

- Theorem 23 (Complexity of Large Systems)
  A large system is not necessarily complex if it has a high probability of satisfying the {FRs} specified for the system.
- Theorem 24 (Quality of Design)
  The quality of design of a large system is determined by the quality of the database, the proper selection of FRs, and the mapping process.

THEOREMS FOR DESIGN AND OPERATION OF LARGE ORGANIZATIONS
(MOSTLY FROM [SUH-95B, SUH-98])

- Theorem 25 (Efficient Business Organization)
  In designing large organizations with a finite resource, the most efficient organizational design is the one that specifically allows reconfiguration by changing the organizational structure and by having a flexible personnel policy when a new set of FRs must be satisfied.
- Theorem 26 (Large System with Several Sub-Units)
  When a large system (e.g., organization) consists of several sub-units, each unit must satisfy independent subsets of {FRs} so as to eliminate the possibility of creating a resource intensive system or coupled design for the entire system.
- Theorem 27 (Homogeneity of organizational structure)
  The organizational structure at a given level of the hierarchy must be either all functional or product-oriented to prevent the duplication of the effort and coupling.

General Design Theory and its Extensions and Applications

Tetsuo Tomiyama

RACE (Research into Artifacts, Center for Engineering)

The University of Tokyo

Komaba 4-6-1, Meguro-ku, Tokyo 153-8904, Japan

Keywords: General Design Theory, Function-Behavior-State Modeling, Knowledge Intensive Engineering, Modeling of Synthesis

INTRODUCTION

This paper reviews General Design Theory (GDT) and its extensions. It also discusses how GDT evolved and arrived at the concept of knowledge intensive engineering as applications. Since GDT was first proposed in 1977 in Japan and in 1980 in English, GDT has never received an appropriate evaluation with respect to its original two goals (described below) [*Tomi-90*]. GDT was considered something too abstract and practically meaningless primarily due to its formalism based on axiomatic set theory by the majority of the community except for a few researchers [*e.g. Reic-91*]. However, due to the AI boom during the 1980s and the movement toward scientific understanding of design and manufacture initiated by the awareness about manufacturing competitiveness in the late 1980s, the general acceptance of such abstract design theories as GDT has improved.

While GDT scientifically pursues better understanding of design, GDT further aims at two engineering goals [*Yosh-81*]. One is that GDT serves as a guiding principle for building future (intelligent) CAD systems. The other goal is that GDT provides a designer with a practical, efficient design methodology for better design. GDT is not purely a design theory, however, because it mathematically defines design knowledge that can be operated as mathematical sets. In this sense, GDT is a theory about design knowledge or even knowledge in general.

This paper demonstrates how GDT achieved its original two goals by illustrating GDT's contributions to the development of advanced CAD technology and to innovative design from research results of our group at the University of Tokyo. In particular, we will derive a design process model, called refinement design process model, from comparison of theoretical results of GDT with empirical findings obtained from what we call design experiments. The refinement design process model can be further elaborated as a guiding principle to develop an advanced CAD system. Design on this CAD system can be knowledge intensive activities, hence, called knowledge intensive design.

Engineering design consists of a variety of thought processes, but most of them can be categorized into either analysis or synthesis. Existing research attempts to model design processes have such assumptions as „design as problem solving," „design as decision making," and „design by analysis," and did not explicitly address „design as synthesis." Compared with analysis, synthesis is less understood and codified as a model. For instance, formal design methods proposed by German researchers frame engineering design as a process that begins with decomposition of given specifications into functional structure followed by embodiment using physical effects. This view of engineering design lacks notion of designer's thought process. This lack of understanding about synthesis leads to inefficiency in design, bad design solutions, or even unsolvable design problems. It is, therefore, crucial to have a model of synthesis. The paper also describes our fundamental view on synthesis and how to tackle the problem of modeling synthesis.

The rest of the paper is organized as follows. Chapter **GENERAL DESIGN THEORY** briefly reviews GDT which deals with design knowledge and formalizes it based on axiomatic set theory. GDT identifies two situations of knowledge; namely, ideal and real. The ideal knowledge is a situation in which design is considered as a mapping from the function space to the attribute space, while in the real knowledge design is a stepwise refinement process from function to attribute through physical behavior. The physical behavior level descriptions are described in the *metamodel space*.

Chapter **DESIGN EXPERIMENT AND DESIGN PROCESS MODEL MODELS** outlines our experimental work on design processes. It begins with design experiments by which design protocols are extracted from real design sessions. From the protocol data, we obtain a cognitive design process model that will be further elaborated and formalized as a computable design process model. Design knowledge obtained from design experiments is analyzed as well.

In Chapter **A NEW DESIGN PROCESS MODEL**, we attempt to test the experimental results against the theoretical results of GDT. Although in principle we can conclude that theoretical results agree with experimental findings, there are some problems. For instance, in GDT design is supposed to begin with specifications. However, this is not the case in real design practices. Design instead begins with requirements and continues to complete functional, behavioral, and attributive information about design objects. Therefore, we propose a new design process model, refinement design process model, that has better agreement with experimental findings.

Chapter **MODELING OF SYNTHESIS** outlines our more recent effort to model synthesis in an on-going project called „The Modeling of Synthesis" supported by the Japan Society for Promotion of Science (JSPS). This project, JSPS-RFTF 96P00701, aims at establishing a scientific model of synthesis taking about all findings obtained from our previous work.

Chapter **IMPLICATIONS AND APPLICATIONS OF GENERAL DESIGN THEORY** discusses implications of GDT as a guiding principle to develop a future CAD system. There are three issues. First, we need a mechanism for physics centered modeling and multiple model management. For this purpose, we developed a straightforward implementation of the metamodel concept. Second, we need a mechanism for function modeling. The FBS (Function-Behavior-State) modeling is proposed for modeling function. Without such a methodology for function modeling, design methodology cannot be used in real design activities. These advanced modeling techniques help a designer to focus on functional information that plays a dominant role in conceptual design stages and to utilize design knowledge intensively. An advanced CAD system equipped with intensive design knowledge, called a *knowledge intensive engineering framework* (KIEF), is currently developed to enable innovative design [Tomi-94].

Chapter **SUMMARY** concludes the paper.

GENERAL DESIGN THEORY

In this chapter, we review results and implications of General Design Theory (GDT). GDT's major achievement is a mathematical formulation of the design process [*Yosh-81, Tomi-87*]. GDT deals with concepts that only exist in our mental recognition. GDT tries to explain how design is conceptually performed with knowledge manipulation. In this sense, GDT is not merely a design theory but an abstract theory about (design) knowledge as well.

DESIGN IN THE IDEAL KNOWLEDGE

GDT begins with a manifesto that our knowledge can be mathematically formalized and operated. This is represented by three axioms that define knowledge as topology and operations as set operations. GDT regards a design process as a mapping from the function space to the attribute space, both of which are defined over the entity concept set. Based on axiomatic set theory, we can mathematically derive interesting theorems that can well explain a design process.

GDT makes a distinction between an entity and an entity concept. An entity is a concrete existing object, and an entity concept is its abstract, mental impression conceived by a human being. An entity concept might be associated with its properties, such as color, size, function, and place. These properties are called abstract concepts and include attributes.

- **Axiom 1 (Axiom of recognition):** Any entity can be recognized or described by attributes and/or other abstract concepts.
- **Axiom 2 (Axiom of correspondence):** The entity set S' and the set of entity concept S have one-to-one correspondence.
- **Axiom 3 (Axiom of operation):** The set of abstract concept is a topology of the set of entity concept.

Axiom 2 guarantees the existence of a superhuman who knows everything. Axiom 3 signifies that it is possible to logically operate abstract concepts as if they were just ordinary mathematical sets. Accordingly, we get set operations, such as intersection, union, and negation. We can then introduce ideal knowledge that knows all the elements of the entity set and that can describe each element with abstract concepts without ambiguity. Theorem 1 mathematically describes this situation.

- **Theorem 1:** The ideal knowledge is a Hausdorff space.

In GDT, a design specification is given as an abstract concept that the design solution must belong to. Thus, the specifications can be given by describing an entity with only abstract concepts (for example, functionally). The function space is the entity concept set with a topology of functions, and the attribute space is the one with a topology of attributes. Therefore, a design specification is a point in the function space, and a design solution is a point in the attribute space. The most significant result of having the ideal knowledge that can be further proven from the three axioms is that design as a mapping from the function space to the attribute space successfully terminates when the specifications are described.

- **Theorem 2:** In the ideal knowledge, the design solution is immediately obtained after the specifications are described.

Because we know everything perfectly in the ideal knowledge, when we finish describing the specifications, they converge to a point in the function space. Because the function space and the attribute space are built on the same entity concept in the attribute space, this point (that is, an entity concept) can also be considered in the attribute space. Thus, the design solution will be fully described by attributes; that is, the design in the ideal knowledge is a mapping process from the function space to the attribute space (Figure 1).

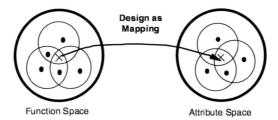

Figure 1   Design Process in the Ideal Knowledge

DESIGN IN THE REAL KNOWLEDGE

The situation in the ideal knowledge does not apply to real design in many points. First, design is not a simple mapping process but rather a stepwise refinement process where the designer seeks the solution that satisfies the constraints. This can be observed in design experiments described in the next chapter. Second, the ideal knowledge does not take physical constraints into considerations, and it can produce design solutions such as perpetual machines.

These restrictions are considered in the real knowledge, where design is regarded as process in which the designer builds the goal and tries to satisfy the specifications without violating physical constraints. To formalize the real knowledge, we first define a physical law as a description about the relationship among physical quantities of entities and the field. The concept of physical laws is one of the abstract concepts formed when one looks at a physical phenomenon as manifestation of physical laws. Physical laws constrain entities in the real world; in other words, physical laws must explain how a feasible entity behaves. This fact can be proven as a theorem.

- **Theorem 3:** The set of physical law concepts is a base of the attribute concept topology of the set of (feasible) entity concepts.

This theorem states that attributes can be measured by using physical laws. An interesting fact about the real knowledge is that we can prove finiteness or boundedness of our knowledge with the following hypothesis.

- **Hypothesis:** Finite subcoverings exist for any coverings of the set of feasible entity concepts made of sets chosen from the set of physical law concepts.

Basically, this hypothesis says that a feasible entity is explicable not by an infinite number but a finite number (as small as possible) of physical laws. From this hypothesis, we can prove an interesting theorem that explains that an attribute has a value if it is possible at all to measure the distance between entities. We can also mathematically prove Theorems 4 and 5.

- **Theorem 4:** The real knowledge is a compact Hausdorff space.
- **Theorem 5:** In the real knowledge, if we can produce a directed subsequence from the given design specifications, this subsequence converges to a single point.

Theorem 5 indicates that in the real knowledge, a design process can be regarded as a convergence process if we can pick up some (but not necessarily all) specifications that seem to yield meaningful solutions.

The next step is to formalize design processes in the real knowledge. For this purpose, the concept of metamodels is formally introduced, where a metamodel is a finite set of attributes, and the metamodel set is the set of all metamodels. Such a metamodel can evolve during a design process by increasing the number of attribute concepts.

- **Theorem 6:** If we evolve a metamodel, we get an entity concept as the limit of evolution.

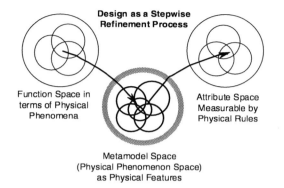

Figure 2   *Design Process in the Real Knowledge*

Theorem 6 is a corollary to Theorem 5, and it does not guarantee that we obtain a design solution as the result of this evolution; however, we might arrive at an entity that is only an approximate solution. This approximate solution is described with attributes that can be measured by physical laws and can be physically realized. In other words, we are not allowed to consider objects that contradict physical laws.

Theorem 6 also indicates that a design process is a stepwise transformation process, and solutions are obtained in a gradual refinement manner because the metamodel can be evolved only by increasing the number of attributes. Figure 2 depicts a design process in the real knowledge in which we design a design object from functional specifications through physical behaviors of the design object arriving at attributive descriptions. In other words, in the real design, design is a stepwise transformation process from the function space to the attribute space via the metamodel space. Figure 3 illustrates the evolution of a metamodel.

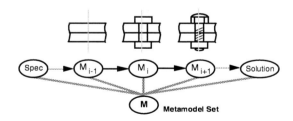

Figure 3   Metamodel Evolution

DESIGN EXPERIMENT AND DESIGN PROCESS MODEL MODELS

DESIGN EXPERIMENTS

To test the theoretical results of GDT, we conducted an experimental investigation of design processes [*Take-90a*] and proposed a cognitive design process model [*Take-90b*]. A design experiment is a kind of psychological experiment in which a pair of designers is asked to design an artifact that is not so familiar to them. The whole designing session is video-taped and analyzed with protocol analysis methods. In some experiments, the subjects were allowed to use a CAD tool that could record drawing operations for more accurate protocol analysis. In total, twelve design experiments were conducted.

THE COGNITIVE DESIGN PROCESS MODEL

One of the major results obtained from the experiments is a cognitive model of design processes when examining a design process from a problem-solving point of view. This model is constructed from unit design cycles. (For this analysis, we analyzed 494 protocols that were obtained three design experiments to design a box handling mechanism for automatic cigarette vending machines. Each design experiment lasted in average three hours. Two of them were conducted by two pairs of professional designers and one by a pair of a professional designer and an engineering student.)

A unit design cycle consists of five subprocesses:
- *awareness of the problem*: to pick up a problem by comparing the object under consideration with the specifications;
- *suggestion*: to suggest key concepts needed to solve the problem;
- *development*: to construct candidates for the problem from the key concepts using various types of design knowledge (when developing a candidate, if something unsolved is found, it becomes a new problem that should be solved in another design cycle);
- *evaluation*: to evaluate candidates in various ways, such as structural computation, simulation of behavior, and cost evaluation (if a problem is found as a result of the evaluation, it becomes a new problem to be solved in another design cycle); and
- *conclusion*: to decide which candidate to adopt, modifying the descriptions of the object.

Utterances in the protocol data are categorized into these subprocesses, and then a design cycle is composed of these subprocesses. Basically, a single design cycle solves a single problem, and sometimes new problems that must be solved in other design cycles arise during the suggestion and evaluation subprocesses.

We can distinguish two levels in the design process when we consider the designer's mental activity. One is the object level, where the designer thinks about design objects themselves, for example, what properties the design object has and how it behaves in a certain condition. The other is the action level, where the designer thinks about how to proceed with his/her design, that is, what s/he should do next. The designer seems to perform his/her design mutually using these two types of thinking. When looking at the design cycles from this aspect, they also contain these two levels. Figure 4 depicts the cognitive design process model.

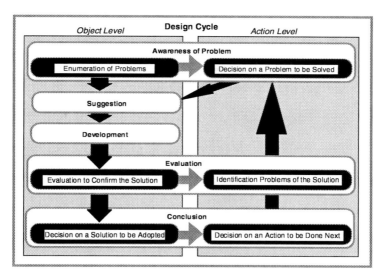

Figure 4   Cognitive Design Process Model

THE COMPUTABLE DESIGN PROCESS MODEL

The cognitive design process model was further logically formalized [*Take-90b*] as a computable design process model based on which a design simulator was developed [*Take-92*]. Figure 5 illustrates a computable design process model that has two levels of inferences. The object level inference deals with facts about the design object, whereas the action level inference guides the object level inferences. At the object level, the model formalizes design processes as combinations of three types of reasoning, i.e., deduction, abduction, and circumscription. Deduction is a process to derive facts from known facts about a design solution. Abduction is a process to find possible ways to obtain descriptions about a candidate solution from the desired properties. Circumscription is a process to reform knowledge to resolve contradiction within the knowledge base. The action level inference has only deductive reasoning to determine object level operations from the conditions of the object level inference and the action level knowledge. The action level knowledge is general and applicable to any design.

The design simulator (Figure 6) further has a multiple world management system based on ATMS (Assumption-based Truth Maintenance System) for design process management. The design simulator traces back design processes using design knowledge extracted from protocol analyses. The computable design process model allows to set a sound, firm foundation for developing design process management facilities for future CAD systems.

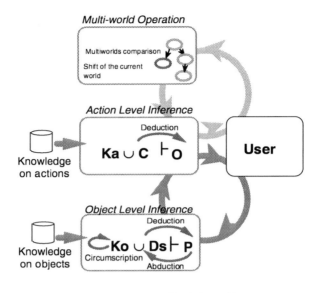

K_a: Knowledge on actions K_o: Knowledge on objects
O: Operations D_s: Design Solution
C: Conditions P: Properties of design solution

Figure 5 Computable Design Process Model

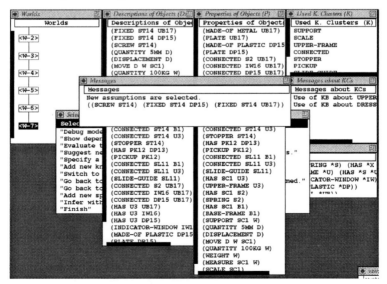

Figure 6 Screen Hardcopy of the Design Simulator

ANALYSIS OF DESIGN KNOWLEDGE

Further analysis of the design protocols resulted in categorization of design knowledge. (We studied another set of three design experiments to design a scale. Each design experiment lasted in average five hours. Again, two of them were conducted by two pairs of professional designers and one by a pair of a professional designer and an engineering student.)

We first identified six categories of design knowledge; i.e., knowledge about entities, functions, attributes, topological relationships, connection methods, and manufacturing methods. Between these categories, there are eight types of primitive transitions, which comprise design process knowledge. There were 134 of these transitions extracted from design protocols. The transitions are:

- from functions to entities
- from entities to functions
- from attributes to entities
- from entities to attributes
- from attributes to attributes (via entity)
- from topologies to relationships
- from entities to manufacturing methods, and
- from manufacturing methods to attributes.

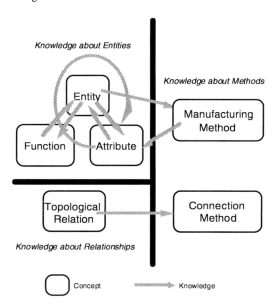

Figure 7   Design Process Knowledge as Relationships among Concepts

In addition, there is an independent type of knowledge concerning design strategy at the action level. Figure 7 depicts the six categories of design knowledge and the eight transitions among them. Table 1 shows how these transition types of design knowledge are differently used in the cognitive design process model.

A NEW DESIGN PROCESS MODEL

TESTING THEORY AGAINST EXPERIMENTS

Chapter **GENERAL DESIGN THEORY** briefly reviewed GDT and discussed the ideal knowledge and the real knowledge. Chapter **DESIGN EXPERIMENT AND DESIGN PROCESS MODEL MODELS** outlined the results of our experimental study on design processes. The cognitive design process model was proposed based on observations obtained from design experiments. A design cycle of the cognitive model is a unit step of the metamodel evolution: In this regard, the cognitive design process model does agree with the theoretical result that design is a stepwise, evolutionary transformation process which characterizes the real knowledge.

Knowledge Type	Awareness of problem	Suggestion	Development	Evaluation	Total
Function to Entity		30			30
Entity to Function	5	1	1	1	8
Attribute to Entity		1			1
Entity to Attribute	7	15		16	38
Attribute to Attribute		1	3		4
Topological Relation to Connection Method		24			24
Entity to Manufacturing Method		1	2	6	9
Manufacturing Method to Attribute		5	2	10	17
Design Strategy		3			3
Total	12	81	8	33	134
(%)	9.0	60.4	6.0	24.6	

(Numbers of Knowledge Chunks Observed in Three Design Experiments)

Table 1   Knowledge Utilization in Design Processes

However, does this conclude that a design process is a mapping from the function space via the metamodel space to the attribute space (function to attribute through physical behavior)? If so, Table 1 should suggest that design knowledge almost all fall into the category of „Function to Attribute." Unfortunately, such a category does not even exist. This means that design is not a simple mapping process from functions to attributes. The transitions in Table 1 imply more complicated thought process.

Both the ideal knowledge and the real knowledge assume that design begins with specifications and ends up with solutions. This might be qualitatively true, but specifications are not always described as functions. In the next section, we argue for an alternative model of a design process.

DESIGN NEVER BEGINS WITH SPECIFICATIONS

If design is a mapping from the function space onto the attribute space as was the case in the ideal knowledge, functional specifications are mapped or transformed to attributive descriptions that suffice for physically manufacturing artifacts. However, usually design specifications include not only functions but also attributes (such as weight and dimensions) and physical behaviors (such as specifications regarding the type of energy source).

In addition, the design protocol data indicate that the specifications are incomplete, inconsistent, or sometimes even infeasible. Table 1 shows that one type of design knowledge can be used in the opposite direction (e.g., from function to attribute and vice versa). A candidate solution is gradually refined through the five subprocesses of design cycles each of which is dominated by different kinds of design knowledge. This suggests that design specifications are incomplete, inconsistent, or sometimes infeasible. This is why during the design experiments we could observe that design processes had different kinds of exception handling operations such as backtracking, addition or relaxation of specifications, and conflict resolution. On one hand, in the early stage of design, designers may spend most of their effort in identifying hidden or unclear specifications. In later stages, on the other hand, specification relaxation can be carried out to balance between, e.g., cost and performance.

In other words, design in the sense of the ideal knowledge does not exist. (In fact, if the designer is a superhuman who knows about all the entities both functionally and physically, the design immediately terminates at the moment the specifications are described. Such a superhuman designer does not really design; what he does is merely data retrieval.) Specifications could be incomplete (or partial), inconsistent, and as a result infeasible, before the design terminates. In other words, a real design process is a process in which the designer has to augment, correct, relax, refine, and compromise the given specifications. Since we are designing an artifact that will eventually exist in a physical world, is it ever possible to describe every specification before we experiment and test the idea by some physical method (including physical experiments, analysis by hand or by computer, and so forth)?

We have to accept the reality that specifications are never complete, consistent, or feasible before the design is completed. Thus, design never begins with specifications.

DESIGN PROCESS TO COMPLETE SPECIFICATIONS

If design does not begin with specifications, then what is an alternative approach to modeling design processes? Certainly, all the mapping models (Figures 1 and 2), the metamodel evolution model, and the cognitive design process model capture some aspects of design but not all. Figure 8 illustrates one better model, called *refinement model,* in which design is modeled as a process to complete design specifications. This model has both theoretical backgrounds of GDT's real knowledge (i.e., consideration of function, behavior, and attribute and set operations) and agreements with experimental findings such as the cognitive design process model. The refinement model is different from the mapping models in that design does not begin with functional specifications but with *requirements* described in terms of functions, behaviors, and attributes. The initial requirements can be incomplete, inconsistent, or infeasible.

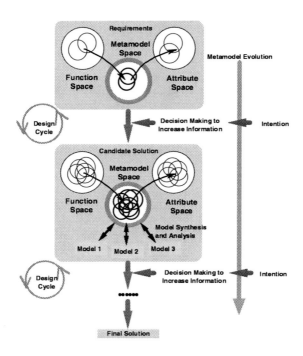

Figure 8   A New Design Process Model: Refinement Design Process Model

During the design process, these three types of descriptions about a design object are gradually increased, refined, corrected, and tested for consistency, completeness, and feasibility, as the cognitive design process model through design cycles. There are five subprocesses of a design cycle to perform these operations; i.e., awareness of problem, suggestion, development, evaluation, and conclusion. In the subprocess of awareness of problem, the designer tries to find anything that is not complete, consistent, or feasible. For example, a function concept that has no realization method can be a problem to be solved. In the suggestion subprocess, candidate solutions to this problem are suggested. For instance, the designer may add behavior and attribute information corresponding to the function concept. Each candidate will be developed in the development subprocess by augmenting, e.g., attributive information from the added behavior information. The next evaluation subprocess tests the added information for completeness, consistency, and feasibility. If the result is acceptable, the added information becomes a permanent decision; if not, other alternative solutions will be considered in the next design cycle. The conclusion subprocess is, therefore, a decision-making phase.

The suggestion subprocess is a synthesis phase in design, while the development and evaluation subprocesses are analysis phases. It is expected that models of the design object are heavily used during these subprocesses. This implies that the classic dichotomy „Design by Synthesis vs. Design by Analysis" is not valid; rather, a design process is a combination of these two.

In the refinement model, not only the design object but also the specifications are gradually (step by step) refined, improved, relaxed, expanded, or modified and become complete. Specifications become complete only when the design terminates arriving at the final design solution. This suggests that future CAD should accept such a nature of evolutionary specifications.

MODELING OF SYNTHESIS

The work described in the previous chapters was an attempt to cognitively and computationally model design processes. The computable design process model resulted in the design simulator which succeeded in playing back design protocols obtained by design experiments. Although this simulator employs not only deductive reasoning but also non-deductive reasoning such as abductive reasoning, circumscription, and meta-level reasoning, it could only „simulate" design processes rather than automatically design by itself. The result was far away from a model of synthesis.

---

(1) **Observation of phenomena**
A phenomenon is observed as objectively as possible. Observations are recorded to form findings.

(2) **Extraction of facts**
Facts are extracted from the observations.

(3) **Formation of hypotheses or selection of axioms**
Facts can be used to reason out rules or hypotheses. For instance, experiments can lead to formation of experimental equations. In obvious cases, a set of known axioms is selected instead of hypotheses. The smaller number of axioms is preferred.

(4) **Derivation of theorems from the axioms**
Theorems are derived from the hypotheses or axioms deductively. This process may break down the original problem (i.e., derivation of theorems) into smaller subproblems (the „divide-and-conquer strategy").

(5) **Verification of theorems against facts**
The derived theorems are tested against the facts to check the explanability of the theorems.

(6) **Verification of theorems against other known axioms**
The derived theorems are again tested against other known sets of axioms. This test verifies if the theorems are compatible with the known axioms or at least if they do not violate the known axioms. If the hypotheses obtained in step (3) pass tests (5) and (6), they become axioms.

Figure 9   Classic Analytical Thought Process

What was wrong then? A quick answer is that the abductive reasoning used for the simulator was performed by simple backward reasoning. Obviously this resulted in very simple, known design solutions. Also, the model and assumptions of design knowledge, including the ontology and reasoning method, could have been wrong. The cognitive design process model, such as unit design cycles, could have been wrong and insufficient, too. Among others, however, the most suspicious is the knowledge-centered view of design.

In order to better understand design, we started the Modeling of Synthesis project to establish a scientific model of synthesis. The followings are the preliminary results of the project [Tomi-97].

---

(1) **Describing requirements**
Requirements for the synthesis are described.

(2) **Selection of axioms**
Synthesis requires, by nature, various viewpoints to be considered.

(3) **Derivation of solutions from requirements and axioms**
The basic reasoning mode could be abduction, rather than deduction, if there exists an algorithm to arrive at solutions. Just like analysis, the divide-and-conquer strategy might be used, but since the number of sets of axioms could be larger than analysis, trade-off and negotiation among different solutions are important.

(4) **Derivation of theorems from the axioms**
Theorems are derived from the axioms deductively. This process corresponds to the development subprocess in the cognitive design process model. Deduction and the divide-and-conquer strategy are central.

(5) **Verification of theorems against facts**
The derived theorems in the previous step are tested against the requirements.

(6) **Verification of theorems against other known axioms**
The derived theorems are again tested against other known sets of axioms. This test verifies if the theorems are compatible with the known axioms (for instance, not violating any constraints not considered at (2)).

Figure 10   Synthesis Oriented Thought Process

Figures 9 and 10 compare the analytical oriented thought process and synthesis oriented thought process. Analysis is a process to extract facts from observations and to try to give best explanations (theorems) of these facts from a set of axioms or hypotheses generated from the facts. According to Pierce, step (3) in Figure 9 is an abductive process [Hart-31, Burk-58], while step (4) is often conducted deductively.

In contrast, synthesis is considered an opposite thought process to analysis. In the context of design, analysis is a process to derive attributive descriptions about the design object's behavior and function, whereas synthesis is to derive attributive descriptions about the design object from functional requirements to the design object. This justifies the claim that synthesis is an opposition of analysis. The problem here is that such a thought process as synthesis cannot be framed in the thought process model described in Figure 9. Also, it is known extremely difficult to explicitly describe knowledge and its operations for synthesis in a deductive manner needed for step (3) in Figure 9. Note that deduction significantly dominates this analysis oriented thought process. In order to be efficient, deduction should be done with as few axioms as possible. This inevitably compartmentize engineering design knowledge and such compartmentized domains become independent from each other.

Figure 10 shows a synthesis oriented thought process model. First, synthesis begins with describing requirements, rather than observation of phenomena to be explained. Then, axioms on which synthesis is based are selected. Synthesis involves multiple viewpoints as axioms, so the number of axioms is inevitably large. It is interesting to notice in this model that abduction plays a critical role in step (3) as well as deduction in step (4). In this sense, this model considers that synthesis is not just an opposition of analysis, but rather analysis and synthesis are complementary to each other. Just like analysis, synthesis also requires verification of solutions against axioms.

This model is similar to the computable design process model (Figure 5) but it captures the nature of synthesis that synthesis inevitably involves multiple viewpoints as axioms. Also, the amount of axioms used in these two processes are significantly different. Since analysis is a deduction oriented process and deduction should be done with as few axioms as possible for the sake of theoretical simplicity and efficiency, the smaller amount of axioms is the better for analysis. On the contrary, since synthesis is an abduction oriented process and abduction should be done with as many axioms as possible for generating various types of candidate solutions, the larger amount of axioms is better for synthesis. Another difference is subjectiveness of these processes. Analysis is, by nature, an objective process. However, in the context of design, the designer is requested to set the viewpoint for analyzing the design objects. It means synthesis is considerably subjective.

Figure 11 compares analysis and synthesis in a logical framework in which A, F, Th denote axioms, facts, and theorems, respectively. σ signifies inference rules (such as *modus ponens*).

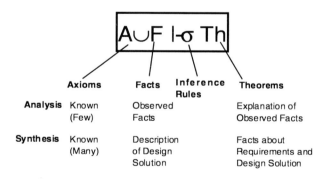

Figure 11 Comparison of Analysis and Synthsis

Synthesis or, in particular, engineering design is a model-based reasoning process. While analysis is based on a single set of axioms, synthesis requires a number of sets of axioms, which the comparison between analysis and synthesis tells us. Figure 12 depicts a framework for model-based reasoning. First, the physical world is observed. By doing so, a designer creates a mental model through which models in „media" are generated. These models include a drawing and a design object itself. Second, the designer abstracts these models and codifies them in a logical world. The designer first generates object dependent models based on the set of axioms of his choice. These models serve as reference models for logical reasoning in the logical workspace. As Figure 12 suggests, the designer's thought process (such as of Figures 9 and 10) controls logical reasoning in the workspace and the object dependent models serve as „models" in the logic sense. In other words, the truth value in the logical workspace is determined by the object dependent models.

Figure 12 suggests that a model of synthesis must include features of multiple model-based reasoning (as opposed to single model-based reasoning). Each model is formulated based on a set of axioms (or models) of the identical design object. Since these models represent the same design object, they are not independent with each other and we need a mechanism to maintain consistency among them. This is the issue the metamodel system exactly addresses (see Section PHYSICS CENTERED MODELING FRAMEWORK).

To maintain consistency, we have to represent the relationship among models and it means we need basic ontology for this purpose. In this research, we define ontology as a basis for describing the recognition of the world for one purpose. This definition implies all of the descriptions of the world are subjective in certain sense and there doesn't exist a neutral ontology. However, if the purpose is similar, we believe we can make a more general and shareable ontology to guarantee interoperability among various kinds of models in the logical workspace. (Ontology is enthusiastically looked at by the AI community.)

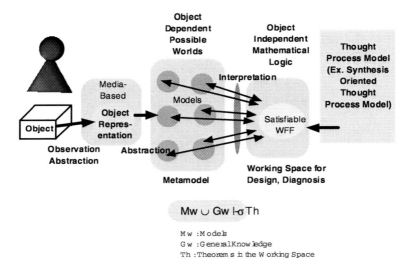

Figure 12   A Hypothetical Reasoning Model of Synthesis

The framework depicted in Figure 12 also captures the fact that some design knowledge is hard to formalize. In other words, abstraction from media based representation to codified models is a research issue. In other words, synthesis is very much subjective. This is most significant when dealing with the designer's intention. Function modeling is one approach to model the designer's intention. There are a number of reports on function modeling [*Umed-97*]. Most of these methods treat function design as a process that begins with function analysis followed by functional decomposition which always refer to entity descriptions. While these methodologies advocates embodiment only after functional decomposition, observations of function design reveal that designers always need to have references to physical entities. Functional decomposition based purely on functional representation is almost impossible. We need a similar approach that subsumes functional modeling in a multiple model-based reasoning context.

IMPLICATIONS AND APPLICATIONS OF GENERAL DESIGN THEORY

One of the major conclusions of GDT is that during design, models of a design object must be intensively used to evaluate functionality and performance of candidate solutions. For designing artifacts that shall physically exist, evaluation of models is mostly based on „physics" in a broad sense. This implies that a CAD system should incorporate a modeling framework that is centered around physics.

Chapter **A NEW DESIGN PROCESS MODEL** proposed a new design process model, the refinement design process model. This has two implications for future CAD studies. One is that we need a framework to describe functional and behavioral information, besides attributive information based on which traditional CAD systems are built. The other addresses how to represent design process information for design process management.

PHYSICS CENTERED MODELING FRAMEWORK

In engineering design, models based on „physics" are used to evaluate functionality and performance of a design object. This includes a geometric model, an FEM model, a function model, a control model, a dynamics model, and any other models used by modeling „agents." As a result, model management becomes a crucial task for a future CAD [*Tomi-89*]. We have developed a framework, called metamodel, which is based on common ontology and terminology for model management to allow agents to communicate with each other. It is called metamodel, not only because it models various kinds of models, but also it follows the metamodel concept introduced in the real knowledge of GDT.

The metamodel mechanism is a straight-forward implementation of the refinement design process model. It allows the designer to use functional models (see the next section), behavioral models (i.e., models that represent physical behaviors), and attributive models (such as a geometric model) in an integrated manner. A design object is gradually refined as is the case with the refinement model until its descriptions become complete, consistent, and feasible. This can be checked by various kinds of modelers integrated in the system. The qualitative physics based reasoning system supports the designer in conceptual design. The designer combines elementary building blocks, called *physical features*, and the reasoning system can derive and check feasible behaviors of the object.

FUNCTION MODELING

For modeling function, we employ the FBS (Function-Behavior-State) modeling [*Umed-90*]. One of the advantages of the FBS modeling is that it allows explicit representation of functions with respect to physical behaviors and states that can be simulated on the qualitative reasoning system. The FBS modeling can be, thus, applied to conceptual design in which information about function together with physical behavior and attribute is crucial. This is a fairly straightforward interpretation of the refinement model.

We define a function as a subjective description of a behavior, while a behavior is defined as a set of state transitions. A function is represented in the form of „to do something." Figure 13 shows our representational scheme named FBS diagram. In an FBS diagram, functions of a design object are represented by a functional hierarchy constructed subjectively and each subjective function is represented with respect to objective behavior constrained by physical laws (F-B relationships). The relationships between behaviors and states of objects (B-S relationships) are represented and formalized as Qualitative Process Theory's views that can be reasoned out by the qualitative reasoning system mentioned in the previous section.

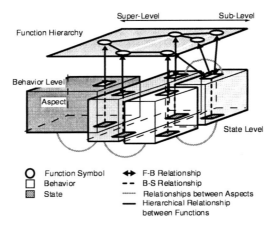

Figure 13   FBS Diagram

KNOWLEDGE INTENSIVE ENGINEERING

Engineering design requires a wide variety of knowledge about product life cycle aspects including manufacturing, operations, maintenance, disposal, and recycling. This idea resulted in the concept of *knowledge intensive engineering*. Knowledge intensive engineering is a new way of engineering activities in various product life cycle stages conducted with more knowledge in a flexible manner to create more added value. Applied to design, we arrived at the idea of „knowledge intensive" design of „knowledge intensive" machines on a „knowledge intensive" CAD [*Tomi-93*].

We are currently developing a computational framework for knowledge engineering called KIEF (Knowledge Intensive Engineering Framework). Figure 14 shows its architeture in which the model management is performed by the metamodel mechanism. This is a straightforward implementation of the refinement design process model [*Tomi-96*].

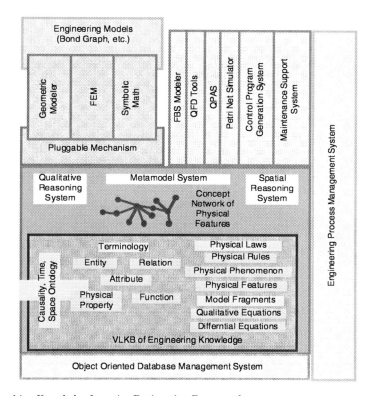

Figure 14   Knowledge Intensive Engineering Framework

SUMMARY

This paper reviewed GDT and proposed a new design process model called refinement design process model that explains real design processes better than GDT's old design process models. In this model, design is a process to complete design specifications. At the beginning, specifications can be given in terms of functions, behaviors, and attributes. During the design process, these descriptions about a design object are gradually refined for consistency, completeness, and feasibility. When the design finishes, we obtain a complete set of specifications that includes sufficient information for manufacturing.

This model also suggests that we need modeling techniques for physics centered design and for modeling functions and intentions. We reviewed the metamodel mechanism that allows physics centered design in a multiple model environment. These advanced modeling techniques helped us to arrive at the idea of knowledge intensive engineering. We are currently developing a framework for knowledge intensive engineering, KIEF, to give a better support based on GDT. However, GDT and its theoretical results do not give a full account of synthesis. A new model of synthesis should be developed taking into consideration a variety of issues including the multiple viewpoints and the role of abduction for synthesis.

ACKNOWLEDGEMENT

Many thanks go to my current and former collegues who helped to develop the ideas described in this paper. In particular, I would like thank Prof. Hiroyuki Yoshikawa, Prof. Takashi Kiriyama, Prof. Hideaki Takeda, Dr. Yasushi Umeda, Masaki Ishii, Dr. Masaharu Yoshioka, and Dr. Yoshiki Shimomura.

REFERENCES

Burk-58 Burks, A. (ed.): *The Collected Papers of Charles Sanders Peirce*, Vol. VII-VIII, Harvard University Press 1958.

Hart-31 Hartshorne; Weiss, P. (eds.): *The Collected Papers of Charles Sanders Peirce*, Vol. I-VI, Harvard University Press, 1931-1935.

Reic-91 Reich, Y.: Design theory and practice II: A comparison between a theory of design and an experimental design system, Technical Report EDRC 12-46-91, Engineering Design Research Center, Carnegie Mellon University, Pittsburgh, PA 1991.

Take-90a Takeda, H.; Hamada, S.; Tomiyama, T.; Yoshikawa, H.: A cognitive approach to the analysis of design processes, *Design Theory and Methodology —DTM '90 —*, DE-Vol. 27, in Rinderle, J.R. (ed.) ASME, pp. 153-160, 1990.

Take-90b Takeda, H.; Tomiyama, T.; Yoshikawa, H.: Logical formalization of design processes for intelligent CAD systems, *Intelligent CAD, II*, in Yoshikawa, H. and Holden, T. (eds.) Amsterdam: North-Holland, pp. 325-336, 1990.

Take-92 Takeda, H.; Tomiyama, T.; Yoshikawa, H.: A logical and computable framework for reasoning in design, *Design Theory and Methodology — DTM '92 —*, DE-Vol. 42, in Taylor, D.L. and Stauffer, L.A. (eds.) ASME, pp. 167-174, 1992.

Tomi-87 Tomiyama, T.; Yoshikawa, H.: Extended general design theory, *Design Theory for CAD*, in Yoshikawa, H. and Warman, E.A. (eds.) Amsterdam: North-Holland, pp. 95-130, 1987.

Tomi-89 Tomiyama, T.; Kiriyama, T.; Takeda, H.; Xue, D.; Yoshikawa, H.:Metamodel: A key to intelligent CAD systems, *Research in Engineering Design* 1(1), 19-34, 1989.

Tomi-90 Tomiyama, T.: Engineering design research in Japan, *Design Theory and Methodology – DTM '90 –*, DE-Vol. 27, in Rinderle, J.R. (ed.), ASME, pp. 219-224, 1990.

Tomi-93 Tomiyama, T.; Umeda, Y.: A CAD for functional design, *Annals of CIRP* **43**(1), 143-146, 1993.

Tomi-94 Tomiyama, T., Umeda, Y.; Kiriyama, T.: A framework for knowledge intensive engineering, *Lecture Notes of the Fourth International Workshop on Computer Aided Systems Technology (CAST '94)*, University of Ottawa, Ont., Canada 1994.

Tomi-96 Tomiyama, T.; Umeda, Y.; Ishii, M.; Yoshioka, M.; Kiriyama, T. : Knowledge Systematization for a Knowledge Intensive Engineering Framework, in Tomiyama, T., Mäntylä, M., and Finger, S. (eds.): *Knowledge Intensive CAD*, Volume 1, London: Chapman & Hall, pp. 33-52, 1996.

Tomi-97 Tomiyama, T.; Murakami, T.; Washio, T.; Kubota, A.; Takeda, H.; Kiriyama, T.; Umeda, Y.; Yoshioka, M. : The Modeling of Synthesis – From the Viewpoint of Design Knowledge, in Riitahuhta A. (ed.), *WDK 25, Proceedings of the 11th International Conference on Engineering Design in Tampere 1997*, Vol. 3, Laboratory of Machine Design, Tampere University of Technology, Tampere, Finland, pp. 97-100, 1997.

Umed-90 Umeda, Y.; Takeda, H.; Tomiyama, T.; Yoshikawa, H.: Function, behaviour, and structure, *Applications of Artificial Intelligence in Engineering V*, Vol. 1, in Gero, J. (ed.) Berlin: Springer-Verlag, pp. 177-193, 1990.

Umed-97 Umeda, Y.; Tomiyama, T.: Functional Reasoning in Design, *IEEE Expert*, Vol. 12, No. 2, pp. 42-48, 1997.

Yosh-81 Yoshikawa, H.:General design theory and a CAD system, in Sata, T. and Warman E.A (eds.), *Man-Machine Communication in CAD/CAM*, Amsterdam: North-Holland, pp. 35-58, 1981.

Discussion

Question - H.-J. Franke
There seems to be a sharp change compared to earlier descriptions of the „general knowledge theory" in what you told us, especially your proposal to a more cognitive-based approach to the theory.

Answer - T. Tomiyama
Well, the conclusion from the cognitive experiment was that the theory in general may explain cognitive findings but in some places we found some incompatibilities like e.g. we found that design does not begin with specifications. Specifications are something we arrive at the final state of design, in fact. In the first place, requirements will be given. So we had to modify our design-process-model to include these experimental findings. I don't think it is a big change, but it is an improvement of our theory.

Question - C. Weber
You have very important terms in your contribution, these are „function", „entity" and „attribute". Could you tell us, what is meant by these terms, because small differences are very important in our discussion.

Answer - T. Tomiyama
Attributes are any properties or measurements like length, weight, colour, whatever can be physically measured. The entity concept is a kind of identifyer referring to an entity, a physical entity. In our modelling function is a subjective discription of objective behaviour. So function could be different, if you are in a different situation.

Question - F.-L. Krause
General Design Theory is a theory about design knowledge rather than design. What is then the theory about design?

Answer - T. Tomiyama
What I meant is theory of design processes. General Design Theory as is does not explain processes.

Towards a Model of Designing Which Includes its Situatedness

John S. Gero

Department of Architectural and Design Science

University of Sydney

Keywords: design model, function-behaviour-structure, situatedness

INTRODUCTION

Our knowledge of designing comes from many sources. [We will use the word 'designing' as the verb and the word 'design' as the noun in order to distinguish between these two, rather than use the word 'design' for both and then utilise the context to disambiguate the meanings.] Until relatively recently knowledge of designing has come from either well-informed conjectures about how humans design or introspection by designers [*Asim-62, Jone-62, Broa-73*]. The introduction of formal methods from logic, mathematics and operations research into models of designing opened up numerous alternate approaches to the treatment of design processes [*Alex-64, Mitc-77, Radf-88*]. More recently, concepts from artificial intelligence have extended the range of approaches available to describe and model designing – both human designing and certain design processes carried out inside computers [*Coyn-90*]. However, approaches based on artificial intelligence concepts are still largely based on conjectures about putative human designing behaviour. Most recently, tools from cognitive science have started to provide some more insight in human designing [*Akin-86, Laws-90, Cros-96*]. As we find out more about how humans design we are able to construct models of increasing explanatory power; models which form the basis of computational systems which either mimic designing or provide aids to designing. As discussed below, models with explanatory capabilities can become theories. In all of this endeavour, however, a number of assumptions have remained constant. Primary amongst these is the notion that designing is an act comprised of different processes, implying that time is involved. Thus, designing is more than conceptual leaps of the „aha" kind; this is not to imply that such acts do not play a role in designing but rather that designing is much more than conceptual leaps.

The *Shorter Oxford English Dictionary* defines theory in a number of ways:
- a scheme or system of ideas or statements held as an explanation or account of a group of facts or phenomena;
- a hypothesis that has been confirmed or established by observation or experiment and is propounded or accepted as accounting for the known facts;
- a statement of what are held to be the general laws, principles or causes of something known or observed;
- systematic statement of the principles of something; and

- a hypothesis proposed as an explanation, hence a mere hypothesis, speculation or conjecture („theory" used loosely).

It is hard to claim that a theory of designing could satisfy any of the first three of these definitions since insufficient is known and agreed upon about the acts of designing to provide details of the phenomena to be accounted for. Thus, a theory of designing is likely to belong to either the fourth or fifth definitions. Yoshikawa [*Yosh-81*] in describing the development of his General Design Theory [*Yosh-79*] stated that one problem was „... due to the fact that the process of design has been less understood. It is difficult to abstract the process itself from the practical designing activity even for designers." However, Yoshikawa's General Design Theory clearly fits within definition 4, whereas the vast majority of other theories would best fit into definition 5, ie speculation or conjecture.

A model is a representation of some thing and as such is descriptive. As a consequence the model itself makes no claim for explanatory power. However, the boundary between a theory (of the definition 4 or 5 kind above) and a model is not as clear as the above descriptions would suggest. Often, models of designing have extensional explanatory capacities or may simply be used as an explanation. In which case the model becomes the theory. We shall adopt this last view: a model with explanatory capacities becomes a theory of the fifth kind. Hence, we shall use the term model in this paper to refer specifically to a theory of this kind except that we shall claim that such a theory, when based on the evidence increasingly available to us moves towards a theory of the first kind, ie, a scheme or system of ideas or statements held as an explanation or account of a group of facts or phenomena.

In the remainder of this paper we commence by outlining some important models of design before introducing some recent insights into human designing which have the potential when incorporated to explain some of the missing characteristics of earlier theories and models of designing. We then proceed to show how such theories and models can be augmented to include some of these characteristics. In particular we introduce the notions of „situatedness", „emergence" and „constructive memory" as important concepts for any theory of designing.

MODELS/THEORIES OF DESIGNING

We make the same assumption as before: designing is a sequence of acts which may be described through processes. The act of designing has attempted to be modeled at various levels of abstraction. Perhaps the earliest of the widely accepted models of designing is by Asimov [*Asim-62*] who divided all the designing processes into three classes:
- analysis
- synthesis
- evaluation.

He and others ordered these as processes as shown in Figure 1.

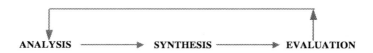

Figure 1   *The analysis–synthesis–evaluation model.*

This model has been used to explain what it is that designers do when they are designing. The processes involved in this view of designing use a terminology which is no longer widely accepted. The term „analysis" has been replaced by „formulation" or similar terms and „analysis" is now used to refer to a precursor of evaluation.

Whilst there have been numerous related models developed and a number of formal theories as well as formal methods, the next one we shall describe is the function–behaviour–structure (F–B–S) model which abstracts the processes of designing even further [*Gero-87, Gero-90, Umed-90*] The F–B–S model provides a framework into which design processes can fit, Figure 2.

Figure 2 The F–B–S model, where the behaviour is bifurcated into expected behaviour, B_e, and behaviour derived from structure or actual behaviour, B_s; D represents the documentation, ie the formal output of designing; $\rightarrow$ = transformation and $\leftrightarrow$ = comparison.

The F–B–S model provides the framework for the following eight design processes:
1. formulation: F $\rightarrow$ B_e
2. synthesis: B_e $\rightarrow$ S via B_s
3. analysis: S $\rightarrow$ B_s
4. evaluation: B_s $\leftrightarrow$ B_e
5. documentation: S $\rightarrow$ D
6. reformulation - 1:   S $\rightarrow$ Sí
7. reformulation - 2:   S $\rightarrow$ B_e
8. reformulation - 3:   S $\rightarrow$ F via B_e

Processes 1 through 5 match well those which have appeared in earlier models. The class of processes represented by processes 6 through 8, although recognised, have not been well articulated in most models, partly because they have not be well understood. We will come back to these later in this paper. It is in these last three processes that the role of situatedness is dominant, although it is not the only place where it can occur.

DESIGNING AS SEARCH

Search as a computational process underlies much of the use of artificial intelligence techniques when applied to designing [*Coyn-90, Russ-95*]. The basic and often implicit assumption in designing as search is that the state space of possible designs is defined a priori and is bounded. The state space to be searched maps onto structure space in the F–B–S model and the criteria used to evaluate states map onto behaviours. The designing processes focus on means of traversing this state space to locate either an appropriate or the most appropriate solution (depending on how the problem is formulated). The advantages of modeling designing as search include the ability to search spaces described symbolically rather than only numerically. However, the assumption that the space is defined prior to searching relegates this model to detail or routine designing.

DESIGNING AS PLANNING

Planning here is taken from its conception in artificial intelligence as the determination of the sequence of actions required to achieve a goal state from starting state. It is a natural consequence of the existence of a well-structured search space. Planning has been used to model design (Gero and Coyne, 1987). It also takes the same assumptions that designing as search does and therefore can only be considered as a model to detail or routine designing.

DESIGNING AS EXPLORATION

Designing as exploration takes the view that the state space of possible designs to be searched is not necessarily available at the outset of the design process. Here designing involves finding the behaviours, the possible structures and /or the means of achieving them, ie. these are only poorly known at the outset of designing (Logan and Smithers, 1993), [*Gero-94*]. Designing as exploration provides another dimension to the F-B-S framework which connects with the ideas of conceptual or non-routine designing: not specifying or even being able to specify at the outset all that needs to be known to finish designing to produce a design. Designing has long been recognized as belonging to the class of problems called "wicked" problems [*Ritt-73*], exploration is an attempt to deal with this issue.

OTHER MODELS OF DESIGNING

Other models of designing based on artificial intelligence or cognitive science concepts are generally either a specialization or a generalization of the models described above. Often they focus on some aspect of the model, commonly it is a procedural aspect. Of particular interest here are two concepts: „reflection in action" and „emergence". The first of these refers to the notion that a designer does not simply design and move on but rather reflects on what he is doing and as a consequence has the capacity to reinterpret it. Schon [*Scho-83*] has called this a designer „carrying out a conversation with the materials". Implicit in these important ideas are the seeds for what will be described later. Emergence, which is a related concept to reflection, is „seeing" what was not intentionally put there [*Gero-96, Holl-98*]. Reflection and emergence have increasing evidentiary support from recent studies of designers.

RECENT INSIGHTS INTO DESIGNING

Protocol studies of human designers are beginning to provide descriptions of some of the phenomena which have long been recognised but not adequately described. Further, some results from cognitive science on human intellectual behaviour potentially have a direct bearing on designing and provide the basis for the examination of such behaviour in designing. We will primarily look at two such results and attempt to incorporate them in a model of designing. The first relates to Asimov's basic model while the second is somewhat more general and relates to the F–B–S model.

EXTENDING ASIMOV'S ANALYSIS–SYNTHESIS–EVALUATION MODEL

Protocol studies of designers carrying out designing which have produced results of a task analysis of the processes have indicated that Asimov's model does not adequately capture some of the base activity. Let us examine Figure 3 where the transitions between the three phases of Asimov's model are plotted across each tenth of the design session as a percentage of the total activity.

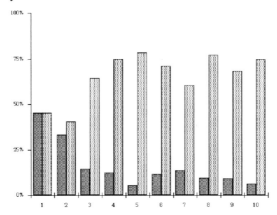

Figure 3   *Transitions between analysis, synthesis and evaluation phases, plotted for each tenth of the design session as a percentage of total activity. Dark shading: evaluation –> analysis; light shading: evaluation –> synthesis [McNe-98].*

We can see that in the beginning of the design session that the designer not only follows evaluation by analysis but for an equal amount of time follows evaluation by synthesis. Already this behaviour is different to that „predicted" by Asimov's model. As the design session proceeds so this behaviour increasingly diverges from Asimov's model. Thus, in the last 75% of the time of the design session the predominant behaviour is not that predicted by Asimov at all since it is: evaluation followed by synthesis. The revised Asimov model now looks like that in Figure 4.

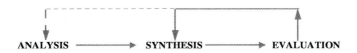

Figure 4   *Revised Asimov model taking account of recent studies of human designers.*

SITUATEDNESS IN DESIGNING

Situatedness [*Clan-97*] holds that „where you are when you do what you do matters" This is in contradistinction to many views of knowledge as being unrelated to either its locus or application. Much of artificial intelligence had been based on a static world whereas design has as its major concern the changing of the world within which it operates. Thus, situatedness is concerned with locating everything in a context so that the decisions that are taken are a function of both the situation and the way the situation in constructed or interpreted. The concept of situatedness can be traced back to the work of Bartlett [*Bart-32*] and Dewey [*Dewe-1896*] who laid the foundations but whose ideas were eclipsed for a time. Situatedness allows for such concepts as emergence to fit within a well-founded and explanatory framework. Figure 5 demonstrates situated emergence – the notion of how a situation affects what can be „seen". The emergent white vase does not appear when the situation changes. Further, situatedness can be used to provide the basis of conceptual designing when we introduce another idea from cognitive science, namely that of „constructive memory".

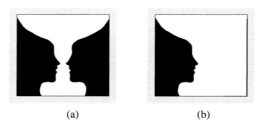

(a)   (b)

Figure 5   (a) *Two black human-like heads in profile, reflections of each other create the situation where a white vase can be seen to emerge; (b) a single black human-like head on the same background does not create the same situation and therefore no emergent vase can be found.*

Constructive memory holds that memory is not a static imprint of a sensory experience that is available for later recall through appropriate indexing [*Rose-88*]. Rather the sensory experience is stored and the memory of it is constructed in response to any demand on that experience. In this manner it becomes possible to answer queries about an experience which could not have been conceived of when that experience occurred. „Sequences of acts are composed such that subsequent experiences categorize and hence give meaning to what was experienced before" John Dewey [*Dewe-1896*]. This view of memory fits well with the concept of situatedness. Thus, the memory of an experience may be a function of the situation in which the question, which provokes the construction of that memory, is asked.

One area of design research based on cognitive studies of designers designing that is beginning to be examined is the use of sketches in designing. Protocol analysis is the primary tool to examine such cognitive processes in designing [*Ecke-88, Gold-91, Scho-92, Suwa-96, Suwa-98a*]. Schon and Wiggins [*Scho-92*] found that designers use their sketches as more than just external memory, they used them as a basis for reinterpretation of what had been drawn: this maps on to emergence and theirs and other studies provide strong evidence for this form of situated designing. Suwa, Purcell and Gero [*Suwa-98b*] have found that designers when sketching revisit their sketches after a while they sometimes make unexpected discoveries, Figure 6. They concluded that „sketches serve as a physical setting in which design thoughts are constructed on the fly in a situated way".

These two short introductions to situatedness and constructive memory allow us to now utilise these ideas in the development of a model of designing that includes its situatedness.

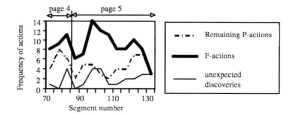

Figure 6   *Correlation between unexpected discoveries and functional cognitive actions (F-actions) as opposed to purely perceptual actions (P-actions) in a design session. Segment number refers to the segments in the protocol and the page number refers to the pages of sketching [Suwa-98b]*

A MODEL OF DESIGNING WHICH INCLUDES ITS SITUATEDNESS

The F–B–S model provides a framework that remains unchanged by the introduction of situatedness into designing. The most obvious and most interesting place to locate situatedness is in the reformulation phase. Earlier we had listed eight design processes that we claimed covered designing. Of these eight processes three were concerned with reformulation (labelled processes 6, 7 and 8). The first of these, reformulation – 1, occurs when the structure state space is modified. Here the role of the situation is to provide opportunities to source new structure variables, Figure 7. Typical design processes here include analogy and induction. However, they are both dependent on the perception of what can be the source in both analogy and induction.

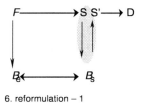

Figure 7   *Transforming S to S' is based on the situation.*

If we indicate such a situated transformation by $^1\tau_S$ we can model this as:

$$S' = {}^1\tau_S (S)$$

Design process 7, reformulation – 2, involves redefining what the expected behaviours are going to be, Figure 8. This is often very much a function of what sorts of structures have been proposed since they bring with them their own ancillary behaviours. Alternately, new behaviours may be derived by analogy. Existing behaviours may be dropped if they are shown to play no discriminatory role.

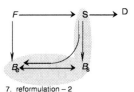

Figure 8   *Transforming the expected behaviours is a function of the situation which exists in terms of the structure synthesised up to this point and the discriminatory capacities of the existing behaviours.*

If we indicate such a situated transformation by $^2\tau_S$ we can model this as:
$$B_e' = {}^2\tau_S (B_e)$$
Design process 8, reformulation – 3, involves redefining what the functions are to be, Figure 9. Redefining functions for an artifact has the potential to change both the expected behaviours as well as the resulting structure. New functions are derived from the situation.

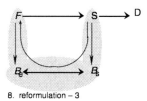

Figure 9   *Redefining the functions or purposes of the artifact is dependent on the situation.*

ACKNOWLEDGEMENTS

This work has benefited from discussion with many members and visitors of the Key Centre of Design Computing and is supported in part by grants from the Australian Research Council.

REFERENCES

Akin-86   Akin, O.: Psychology of Architectural Design, Pion, London 1986.
Alex-64   Alexander, C.: Notes on the Synthesis of Form, McGraw Hill, New York 1964.
Asim-62   Asimov, M.: Introduction to Design, Prentice-Hall, Englewood Cliffs, New Jersey 1962.
Bart-32   Bartlett, F. C.: Remembering: A Study in Experimental and Social Psychology, Cambridge University Press, Cambridge, 1932 reprinted in 1977.
Broa-73   Broadbent, G.: Design in Architecture, John Wiley, New York 1973.
Clan-97   Clancey, W. J.: Situated Cognition, Cambridge University Press, Cambridge 1997.
Coyn-90   Coyne, R. D.; Rosenman, M. A.; Radford, A. D.; Balachandran, M. B.; Gero, J. S.: Knowledge-Based Design Systems, Addison-Wesley, Reading 1990.
Cros-96   Cross, N.; Christiaans, H.; Dorst, K. (eds): Analysing Design Activity, Wiley, Chichester 1996.
Dewe-1896   Dewey, J.: The reflex arc concept in psychology, Psychological Review **3**: 357ñ370, 1896 reprinted in 1981.
Ecke-88   Eckersley, M.: The form of design processes: a protocol analysis study, Design Studies **9**(2): 86–94, 1988.

Gero-87   Gero, J. S.: Prototypes: A new schema for knowledge-based design, Working Paper, Design Computing Unit, Department of Architectural Science, University of Sydney, Sydney, Australia 1987.
Gero-90   Gero, J. S.: Design prototypes: a knowledge representation schema for design, AI Magazine, **11**(4): 26–36, 1990.
Gero-94   Gero, J. S.: Towards a model of exploration in computer-aided design, in J. S. Gero and E. Tyugu (eds), Formal Design Methods for CAD, North-Holland, Amsterdam, pp. 315ñ336, 1994.
Gero-96   Gero, J. S.: Creativity, emergence and evolution in design: concepts and framework, Knowledge-Based System **9**(7): 435ñ448, 1996.
Gold-91   Goldschmidt, G.: The dialectics of sketching, Creativity Research Journal 4(2): 123–143, 1991.
Holl-98   Holland, J.: Emergence, Addison-Wesley, Reading Massachusetts 1998.
Jone-62   Jones, J. and Thornley, D. (eds): Conference on Design Methods, Pergamon, Oxford 1962.
Laws-90   Lawson, B.: How Designers Think, Butterworths, London 1990.
McNe-98   McNeill, T.; Gero, J. S.; Warren, J.: Understanding conceptual electronic design using protocols analysis (submitted), 1998.
Mitc-77   Mitchell, W. J.: Computer-Aided Architectural Design, Van Nostrand Reinhold, New York 1977.
Radf-88   Radford, A. D.; Gero, J. S.: Design by Optimization, Van Nostrand Reinhold, New York 1988.
Ritt-73   Rittel, H.; Webber, M.: Dilemma in a general theory of planning, Policy Sciences **4**: 155ñ160, 1973.
Rose-88   Rosenfield, I.: The Invention of Memory, Basic Books, New York 1988.
Russ-95   Russell, S.; Norvig, P.: Artificial Intelligence, Prentice-Hall, Upper Saddle River, New Jersey 1995.
Scho-83   Schon, D.: The Reflective Practitioner, Harper Collins, New York 1983.
Scho-92   Schon, D.; Wiggins, G.: Kinds of seeing and their functions in designing, Design Studies **13**(2): 135–156, 1992.
Suwa-96   Suwa, M.; Tversky, B.: What architects see in their design sketches: implications for design tools, Human Factors in Computing Systems: CHI'96, ACM, New York, pp. 191–192, 1996.
Suwa-98a   Suwa, M.; Gero, J. S.; Purcell, T.: Analysis of cognitive processes of a designer as the foundation for support tools, in J. S. Gero and F. Sudweeks (eds), Artificial Intelligence in Design'98, Kluwer, Dordrecht (to appear) .
Suwa-98b   Suwa, M.; Purcell, T.; Gero, J. S.: Macroscopic analysis of design processes based on a scheme for coding designers' cognitive actions, Design Studies (to appear) 1998.
Umed-90   Umeda, Y.; Takeda, H.; Tomiyama, T.; Yoshikawa, H.: Function, behavior and structure, in J. S. Gero (ed.), Applications of Artificial Intelligence in Engineering V: Design, Springer-Verlag, Berlin, pp. 177–193, 1990.
Yosh-79   Yoshikawa, H.: Introduction to general design theory, Seimitsukikai **45**: 906–926, 1979.
Yosh-81   Yoshikawa, H.: General design theory and a CAD system, in T. Sata and E. Warman (eds), Man-Machine Communication in CAD/CAM, North-Holland, Amsterdam, pp. 35–53, 1981.

Discussion

Question - S. Rude

You distinguished between functions, behavior and structure, but only on the level of behavior did you further distinguish between expected and real behavior. Why don't you also make a distinction between expected and real functions in the function space and also between expected and real structures?

Answer - J. Gero

I do so for a very simple reason, that I know how to do it at the behaviours. Structures and function, the expected behaviour and the real behaviour are not that simple to this ambiguate. When you change the function of something without changing the object, you change the function without changing the structure, is marketing. Think about it. That is what marketing does. The difference between the predicted structure and the actual structure, I think, is a very, very important area, which is not well developed at all. So, it is because I don't know how to do it, not because it shouldn't be done.

Question - H.-J. Franke

I believe your proposal to the importance of reformulation of problems. But I had a catastrophic vision. Perhaps the only universal method is trial-and-error.

Answer - J. Gero

I see no connection between your statements, because trial-and-error is a process.

Question - H.-J. Franke

Reformulation and learning in process, it leads to making a new attempt, and you mean this is a totaly other thing than trial-and-error?

Answer - J. Gero

Yes, because, let me restate my first statement, trial-and-error is one of a number of processes. I can conceive of and what we have implemented many other processes than trial-and-error. So for example, the motion of situated learning. You learn as you do and then you change what you have learned based on what you are doing. It is not trial-and-error, it is a formal process.

Question - H.-J. Franke

Perhaps an ordered trial-and-error?

Answer - J. Gero

There is a very simple idea from the social scientists called Maslow, who said that if the only thing you have is a hammer, it is amazing how everything looks like nails. If you only believe in trial-and-error, everything looks like trial-and-error.

The Role of Artefact Theories in Design

Mogens Myrup Andreasen

Professor, Ph.D.

Department of Control and Engineering Design

Technical University of Denmark

Keywords: Theory of technical systems, design theory, domain theory, design co-ordination

ABSTRACT
This positioning paper reflects personal experiences and considerations related to design research and theories. The multiple object of design theory is pointed out and the possible research questions related to these multiple aspects. The complexity leads for need for simplification, linking together and unification, but we also need a metric: What is a good theory?
An important theory dimension in synthesis is the artefact theory, leading to the question of a design language. Such a language might be the core of a universal design theory.

INTRODUCTION
In a paper on design theory there will be many "believes", therefore the contributors background and experiences are important.

Since I was confronted with German design methodology and Hubka's theories in the late 60'ies, our department has developed a comprehensive theory foundation and teaching system, which has rapidly expanded the last decenium. Because of our close co-operation with Institute of Product Development, a consulting foundation at the Technical University of Denmark, we cover "the whole range" from theory formulation, design research, application research, tool formulation and installation and follow up in companies. A very important basis for inspiration and insight into design theory is my annual Ph.D.-courses on Design Methodology and research, for Nordic Ph.D.-students, and my supervision and examination of Ph.D.-students.

Based upon our deep concern and focus on design science's theory basis and research methodology, latest expressed in the ICED97 pamphlet on design research, the questions opened by this workshop are challenging and gives an excellent chance to reflect and balance.

The challenges of finding a new, universal design theory are:
- that we need a substantial empowerment of our design ability, seen at individual level, among teams, in companies and in the educational system.
- creation of a strong, unified theory basis, empowering research and teaching.
- finding better and applicable understandings of the way, humans are designing, and bringing this into our design theories.

In the following, I will try to formulate questions and some of the answers, related to a theory basis for *designing* and *designs*.

INTERPRETATIONS OF DESIGN THEORY UNIVERSALITY
In the following the question of universality will be decomposed into the following questions:
- What theory objects do we focus on?
- What explanations do we expect?
- How can we relate different theories?
- What is a good theory?
- What is the role of an artefact theory?
- Does a design language exist?
- Which approaches will lead to a universal theory?

WHAT THEORY OBJECTS DO WE FOCUS ON?
A theory is related to a phenomenon and gives explanations of the phenomenon. A theory of designing is a theory which facilitates the synthesis or raise the chance that we find a good solution.
There seems to be different theory sources in design:
- the way humans solve problems, decide, organise, plan, document etc.
- the artefact to be designed and its nature.
- the factors influencing design: the designer's knowledge, skills, attitude, modelling possibilities, technical means like CAD, management, available information etc.

Besides these sources and types of design theories there exist many other views, see section **WHAT EXPLANATIONS DO WE EXPECT?**. One may argue, that the core of design research is a theory explaining the synthesis, but also here we are forced to look upon two sources of explanation: the way humans solve problems and the artefact's nature.

So it seems that the core of design disappear, and we might only find explanations of single views or „shadows". One dimension in these views or shadows is the explanation of design we get, when we look upon artefact theories, i.e. how artefacts are synthesised.

WHAT EXPLANATIONS DO WE EXPECT?
The core of design is *synthesis*, i.e. the creation of artefacts, plans or programs, which serve the satisfaction of human needs. In relation to industries, which depends on product innovation and development, three patterns of synthesis may be distinguished [*Andr-91*]:
- *problem solving*, i.e. the cognitive, human activity with characteristics like information handling, creativity, learning, experience, values, motivation etc. The result is pointed out as means, plans etc.
- *engineering design*, i.e. a synthesis procedure, where the nature of the artefact is taken into account (decomposition, composition, function, structure, part, form, etc.). The result is a specification of the artefact.
- *product development*, i.e. a synthesis procedure, leading to a new business, i.e. a product, running production system, and established sales channels as basis for a cash flow.

Problem solving is a building block in both engineering design and product development, and engineering design may be seen as the neckbone of product development.

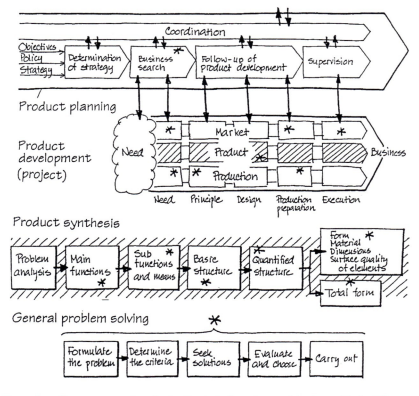

Figure 1 Four patterns of synthesis of technical artefacts and business, [Andr-91].

Traditionally all these models from which there exist hundreds in current literature are presented as an sequence of activities sometimes with defined milestones, but it seems too simplistic. Resent research on *design co-ordination* [Andr-94, Duff-93] propose to see design co-ordination as having its origin in design complexity. A coordination framework propose the following complexities or frames for product development:
- Complexity of the *activities*, i.e. the necessary parallel, sequential or interrelated pattern of design activities.
- Complexity of the *artefact*, i.e. the decomposition and composition of the artefact based upon different views (function, organ, part).
- Complexity of *product life cycle*, i.e. the focus upon and knowledge about such life phases, which have to be taken into account in designing.
- Complexity of *goals*, i.e. the overall and broken down goals.
- Complexity of *tasks*, i.e. the tasks (often allocated to persons) which shall be fulfilled during design (for instance cost reduction) and has to be linked to activities.
- Complexity of *resources*, i.e. the organisation, teams, suppliers, individuals and boards contributing to the design.

- Complexity of *disciplines*, i.e. the theories and methodologies necessary for solving sub-problems, for instance mechanical, electrical, chemical or software related problems.
- Complexity of *aspects*, for instance aspects of product life (service, disassembly), product aspects (like reliability, stability, strength...) and other (patents, liability,...), covering all DFX-types.
- Complexity of *life systems*, i.e. the complexity of each system's characteristics, for instance production type, layout, tooling, quality control etc.

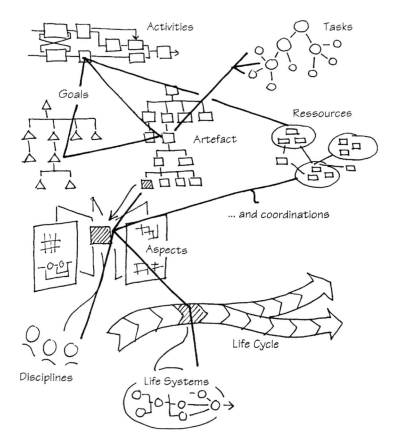

Figure 2   The Design Co-ordination Framework [Duff-93, Andr-94] shows design complexities and their different co-ordinations.

These dimensions, showing complexity, shall be co-ordinated. For instance a plan may be seen as a co-ordination between activities, resources, the artefact complexity and time.

Each dimension shown here may also be seen as a dimension or phenomenon, which is part of designing and needs an explanation. Therefor designing has many dimensions and therefor it needs *explanations in many theories*.
Among these explanations, two has our major focus:
- how humans and teams utilise their cognitive abilities in designing.
- how the artefact theory influences and may be respected by the theory of designing.

Both of these viewpoints will be treated in the following.

HOW CAN WE RELATE DIFFERENT THEORIES?

A theory is related to a specific phenomenon and often it leads to models of the phenomenon. Above we saw more than ten phenomenon related to design, each of which could call for a theory, and we may add many approaches and theories to the way we explain how the human designer is working.

From literature we know, that there are many theories. Some are embedded in normative instructions for designing, others are explicitly formulated. Some are formulated in weak, daily life terms, others are based upon proper definitions. Many theories belong to neighbour sciences, which does not have the goal of synthesis, like cybernetics, ergonomics, ecology, economy, systems theory, information theory, social sciences, philosophy, management etc. One may see design sciences as a science which is not a science in itself, but drawn upon other sciences [*Ferr-90*].

From literature we also know, that there are many "practical advises" which from the authors side are not presented as a theory, but on the other hand is believed to be a universal explanation, like the school of morphology, functional reasoning seen as general tool, right/left-brain explanations of humans abilities, pseudo psychological explanations of creativity and even religious influenced believes.

Today Ph.D.-students and other researchers are confronted with a vast, divergent jungle of theories and sometimes they are confused picking up non compatible elements from different theories as a basis for research leading to weak or non-valid results. Do we have to accept, that there exist many theories of designing? Are some of these theories later to be seen as nonsense while other will survive as fittest?

The integration of different theories is a complicated scientific question. A theory may be seen as a view or modelling, and such views may have each their validity, but may not be integrated or related. An example: Man/Machine-Interaction has surprising many theoretical views [*Mark-95*]:
- a *technical*, where focus could be system thinking, functional reasoning, domain theory oriented or action based.
- a *human embodiment* view, where ergonomics and sensomotorical foci may be used.
- a *human psychological* view, where focus could be cognition psychology, association or perception psychology, learning, linguistics, personality psychology, social psychology, etnologics, or a dramaturgical view.

Each of these views may lead to a better understanding, better synthesis or even better qualities of the interaction, but each view is a separate world, with models, terminology and theories. Some of the issues belongs to recognised "*design schools*", for instance "User Software Engineering", "Cognition Systems Engineering", "Co-operative Design", the Verplank school with focus on metaphors, scenarios and observations, "Intermance Design" and others. Some researchers has tried to let these schools 'compete' i.e. to evaluate their efficiency or productivity. Weak points in such research are the difficulty of mastering more methodologies, and the rather unknown relation between problem type and the methods yield.

The example above raise questions related to unification. The lesson seems to be, that theories belonging to different scientific approaches cannot be integrated, but may be utilised parallel on the same problematic, giving different answers and insights.

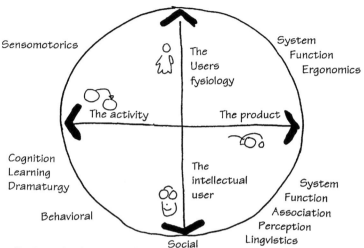

- Design schools:
  - User Software Engineering
  - Cognitive Systems Engineering
  - Co-operative Design
  - The Verplank School (metaphors)
  - Informance Design.

- Different scientific approaches cannot be integrated, but may be utilized parallel on the same problem.

Figure 3   Theories of Man/Machine-Interaction put into a framework. The theories may not be integrated, but used supplementary to each other, [Mark-95].

WHAT IS A GOOD THEORY?

Design research is a young science, which has not yet found it's scientific paradigms. The origins of design theories in different countries are very different and the importance paid to scientific reasoning and the applied scientific methods also differs very much from university to university.

It seems to be a universal agreement (and any science is such an agreement?) in the design research society, that scientific validity and quality should be measured by the applicability and effects of the research results in practice. Our situation today is, that only a minor part of methods and tools used in industry has a scientific origin and background, and only few companies pay interest in the available techniques.

One could argue that a good theory is one that sells and has only little flaws and critiques. For a long period the major concern in design methodology was the method's ability to create new solutions. Today the design scientist has taken over the industrialist's credo: cost/quality/time, but is does not open the doors to the companies R & D departments.

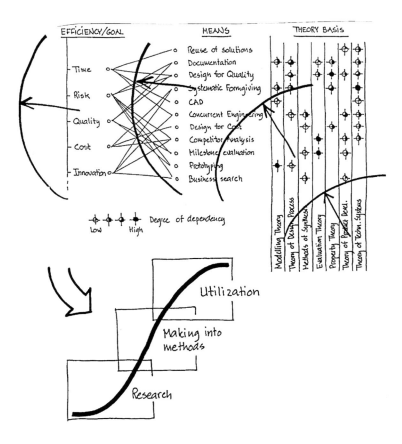

Figure 4   *The development of scientific results from a theory basis through company applied means to results [Hein-94] may be seen as a S-curve of development from research through method development into utilization, [Andr-97].*

The ideal development of a scientific result in this area is the crystallisation and structuring of a theory, its transformation into methods and techniques, fitted and mediated for the user, and implemented, utilised and followed up in the industrial enterprise [*Andr-97*]. This S-curve development shows, that a critical phase, when many scientists loose their interest (and their credibility) is in the tool making phase. If the methodology is not brought into use in industry, how can we then judge the theory's goodness?

Reaction from industry today is that there are too many tools and methods. This mental overloading could be relaxed by means like:
- creating methodologies which are integrated, based upon common concepts and carried by integrating IT-tools.
- creating designs which has the methods embedded as working mode, as mindset, with skills based upon training and high learning ability.

The first proposal has very much to do with reduction of complexity, which should be the first characteristic of any design method. The demand for universality seems to indicate, that universal methods have higher right in their use and teaching. This right should be carefully justified by the applicants.

Following the idea of Yoshikawa [*Yosh-87*], that the factory of the future shall be friendly to humans in the broadest sense, we may reason, that a good method should lead to intellectual challenging jobs and creative work. In Europe we are facing a crisis because of negative attitudes to technology and engineering. It is worthwhile to see new creative *métiers* or work patterns developed, based upon taking *design and innovation* serious, by selecting, training and utilising talents, supported by strong methods and tools.

WHAT IS THE ROLE OF AN ARTEFACT THEORY?

When we here open the question of the role of an artefact theory for a universal design theory, we face what seems to be a paradox: we have a high range of technical and technological theories, explaining the behaviour of structural beams, semiconductors, transfer of laser light in light conductors, humans response to interfaces on computers etc. etc. But we do not have a general agreement upon an artefact theory, which support the synthesis of artefacts.

The key effort of German design methodology was for a long period the identification of a set of design characteristics, which could describe design solutions and could be used for synthesis, for instance in a classification manner or for the use in variation methods. Because of several different and non-compatible proposals and because the proposed characteristics seemed to lack generality, these efforts were never matured.

An important dimension in machine design is the functional reasoning [*Andr-92*]. We utilise the causality between structural characteristics and the systems behaviour and seek up structures which are in accordance with the required behaviour.

Technical systems have more than one type of behaviour, actually several:
- When we use the system, a transformation goes on. This transformation (energy, data, material) is the purpose of the system and we specify the required output and the qualities related to the activity of using the system or product: Cost, time, quality, efficiency, flexibility, risk and environmental effects.
- The system or product in it self has functionality and subsystems, which we call organs, has functionality, which means that they are able to create effects. Also the product has behaviour in the form of properties which the user appraise as qualities. Many classes of properties has to be designed in into the product: productability, assemblability, low cost, reliability, liability, environmental friendliness and many other.

- The product consists of parts and assemblies. Each part has behaviour (strength, surface properties, dimensions etc.), which are utilised as wirk elements in the organs mentioned above, [Andr-98]. Certain part qualities (tolerances, surface properties etc.) are not ideally obtainable in production, but shows a distribution. Maintaining functionality's in spite of other distributions is the role of robust design.

So, the Domain Theory, [Andr-80] based upon Hubka [Hubk-73], operates with three issues of the constitual elements and their behaviours in machines. The three issues are causally interrelated. The three issues create a *genetic* modelling of the design, allowing us to create a product model (The Chromosome, [Ferr-90] which shows all three domains.

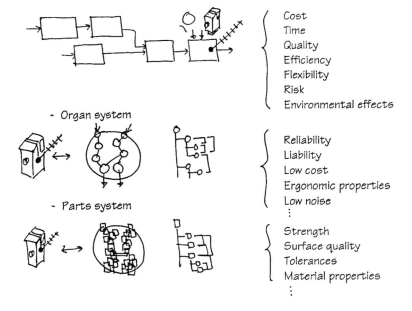

Figure 5   Artefact Theory as proposed by Hubka [Hubk-73] and later Andreasen in a revision as the Domain Theory. Three views explain three different structure behaviour relations of a technical system.

The Domain Theory is one of several published theories of technical systems. Because of the structure/behaviour-relations and the causal links between the domains, this theory heavily influence a *theory of designing*.

One may ask what is general aspects of the Domain Theory. Or more precisely: What theories do we need about what we design? Some proposals:
- Designing is performed in a context of need satisfaction. How do we registrate and transform needs into a product goal specification?
- Designing is performed in a business context. How do we "see" a business opportunity and transform it into a business goal specification?
- The artefact is a man/machine-system. We need theories for such systems and their role in transformations. And we need theories for man/machine-interaction design.
- The artefact is a machine system. We need a theory of structure/behaviour relations.
- The product will perform a life cycle from raw materials and machining of parts through assembly, test, distribution, sales, installation and use to service and disposal. We need theories and models of each of these "meetings" [*Oles-92*].
- Any machine system has to be designed from several viewpoints: Function, control, safety, reliability etc. Therefore we make view models derived from the constitual model to be able to treat each viewpoint in suitable models. We need a theory of views.
- In designing machine systems we need computer support. It means that the designer and the computer shall share models and knowledge about the artefact. In what language(s) is that feasible? (See section **DOES A DESIGN LANGUAGE EXIST?**).

So the artefact theory has influence on our ability to reason about the design, it delivers us models and possibility to simulate properties and it gives us an understanding of hopefully all influences to be respected.

DOES A DESIGN LANGUAGE EXIST?

One may distinguish languages for articulating the design activity and the artefact to be designed. The last topic is the one in focus here.

We have used models in a long period for capturing and defining the artefact, which we design, and for simulation and reasoning about the properties of the design. These models have covered our needs in a rather uneven way [*Roth-86*] and many models have been rather specialised and difficult to understand, leaving the interpretation to the human mind.

Today we face the situation that we want computers to support design. It means that the human operator shall share, at least in the interaction, models with the computer. We have suddenly recognised the need for semantics and syntax and we recognise a close link between models and language.

Humans have the ability to reason and fill gaps in sparse and incomplete modelling types. The computer need to „learn" the meaning of the model or language elements. It seems to be possible to develop a design language [*Mort-98*] which allows the designer to synthesise design models, which
- allow alternation between entirety/elements modelling and abstract/concrete modelling.
- allow alternation between different classes of models and model representations.
- allow gradual determination (German = Prägung) of the design model content.
- allow that modelling is based upon entities, which have functional and structural significance.
- allow that part of the design reasoning can be captured and documented in design models.

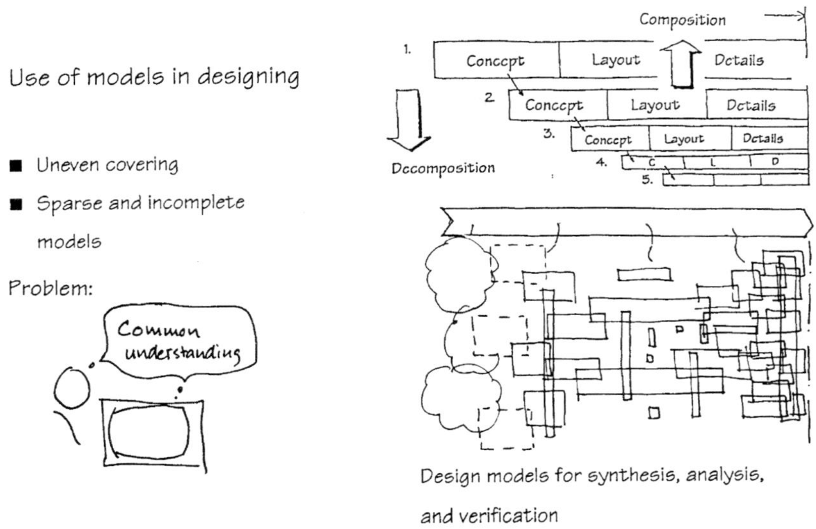

Use of models in designing

- Uneven covering
- Sparse and incomplete models

Problem:

Figure 6   Roth [Roth-86] pointed out that current design models course the determination (Prägung) of the design characteristics (Konstruktionsmerkmale) in an uneven, scattered way.

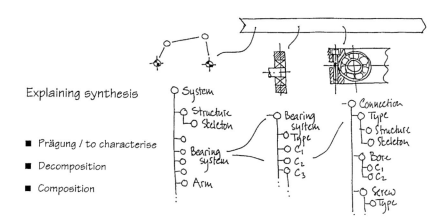

Explaining synthesis

- Prägung / to characterise
- Decomposition
- Composition

Figure 7   A design language [Mort-98] may be seen as a continued modelling where synthesis adds to the characterisation (Prägung)

Our research [Mort-97, Mort-98] indicates that there exist no general design language, not even for mechanical systems. We need distinct set of characteristics (German = Merkmale) for each class of organs, determined by their type of physical effect. Work on design features seems to lead to identification of language elements.

As pointed out by [Duff-93], design and designing may be seen in four different domains: A reality domain, a phenomenon modelling and theory domain, an informative modelling domain, and a domain where the modelling is brought into a computer. Ideal seen there is a mapping between these domains. Especially the mapping chain reality - phenomenon - information model is important here. Only a phenomenon model or theory of artefacts can be transformed into an information model. Each of the four domains have their own language, which shall not be mixed with language elements from other domains. It means that it is not to be expected that a general design theory is created in an information theory language. There has to be links through a phenomenon model or theory to reality.

A Research approach

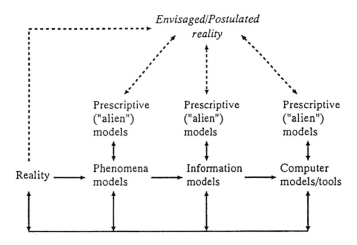

Aspects:

- Different languages.

- A chain of proofs.

- Research types.

Figure 8   *A research approach [Duff-93], in which phenomena seen as reality by theories are transformed into phenomena models, informative models and computer models and tools. Each one has ist own language and models.*

Maybe the research on design language will show such a meta-modelling pattern that a new general way of reasoning and synthesising evolve, which we will see as a universal design theory.

WHICH APPROACHES WILL LEAD TO A UNIVERSAL THEORY?

Whatever we interpretate universal as
- unified, it means bridging different theories and approaches
- integrating and integrated, it means bringing theories into a common, interrelated pattern
- all round applicability, it means a standard, all round tool or technique

we need to consider by which approach we get to better theories. The effect of a theory is much more important than its elegance.

The problems of designing in practise have been the object of rationalisation efforts for a long period, and computer support is the most distinct means for rationalisation. Design is not showing the traits of industrialisation [Andr-98] like controlled progression, predictable output and quality, reproducible steps, planned reuse etc. Therefor there is still to expect a reasonable high potential for rationalisation.

A core element in synthesis seems to be functional reasoning, which is domain independent. Therefore the so-called Function/Means pattern [Andr-80, Andr-98] could be seen as a mechanism, which is linking the design of a composed system, even if it consist of quite different elements from electronics, software, micro-mechanics or 'normal' mechanics.

CONCLUSION

This positioning paper has raised issues related to the question of existence of a universal design theory. Universality shall have a purpose, both empowering of practice and research might have effects. Especially a better understanding of the human mind might have effects.

Research on unification needs a promising approach. In this paper hopefully important aspects of theory creation and approaches have been introduced, allowing the discussion of promising approaches, which is the purpose of the workshop.

REFERENCES

Andr-80   Andreasen, M. M.: Machin Design Methods Based on a Systematic Approach. Contribution to a Design Theory. Dissertation. Department of Machine Design, Lund Institute of Technology, Sweden, 1980 (in Danish).

Andr-91   Andreasen, M. M.: Design Methodology Journal of Engineering Design, Vol. 2, No 2, 1991.

Andr-92   Andreasen, M. M.: The Theory of Domains. Proceedings of Workshop: Understanding Function and Function-to-Form evolution. D.C. Ulman, L.T.M. Blessing, K.M. Wallace. CUED/EDC/TR 12, Cambridge, 1992

Andr-94   Andreasen, M. M.; Bowen, I.; MacCallum, K. I.; Duffy, A. H. B.; Storm, T.: Design Coordination Framework. Proceeding of CIMMOD/CIMDEV Workshop, Torino, Italy, 1994.

Andr-95   Andreasen, M. M.: What is the right balance between industrialisation and mental work? Contribution to Designers Day (in Danish). Technical University of Denmark, 1995

Andr-97   Andreasen, M. M.: Design Research and Industrial Applicability. Invited paper at ENDREA kick-off meeting, Stockholm, 1997. Unpublished.

Andr-98   Andreasen, M. M.: Conceptual Design Capture. Proceedings of Engineering Design Conference'98, Brunel University, 1998.

Duff-93   Duffy, A. H. B.; Andreasen, M. M.; MacCallum, K. I.; Reijers, L. N.: Design Coordination for Concurrent Engineering. Journal of Engineering Design, Vol. 4, No 4, 1993.

Duff-95 Duffy, A. H. B.; Andreasen, M. M.: Enhancing the evolution of design science. Proceedings of ICED95, Hubka, V. (ed). Heurista, Zürich, 1995.

Ferr-90 Ferreirinha, P.; Møller, T. G.; Hansen, C. H.: TEKLA, a language for Developing Knowledge Based Design Systems. Proceedings of ICED 1990, Dubrownik, Heurista, Zürich, 1990.

Hubk-73 Hubka, V.: Theorie der Maschinensysteme, Springer-Verlag, Berlin, 1973. Later published as: Hubka, V.; Eder, E.: Theory of Technical Systems, Springer, New York, 1988.

Mark-95 Markussen, T. H.: A theory based action plan for Interaction Design. Ph.D. thesis (in Danish) Department of Control and Engineering Design, Technical University of Denmark, 1995.

Mort-97 Mortensen, N. H.: Design Characteristics as Basis for a Design Language. Proceedings of ICED'97, ed. Asko Riitahuhta, Tampere University, 1997.

Mort-98 Mortensen, N. H.: Design Modelling in a Designer's Workbench - Contribution to a Design language. Ph.D. dissertation. Department of Control and Engineering Design, Technical University of Denmark, 1998.

Oles-92 Olesen, J.: Concurrent Development in Manufacturing - Based upon dispositional mechanisms. Diss. Department of Control and Engineering Design, Technical University of Denmark, 1992.

Roth-86 Roth, K.: Modelbildung für das methodische Konstruieren ohne und mit Rechnerunterstützung., VDI-Z, Vol. 128, No 1/2, 1986.

Yosh-87 Yoshikawa, H.: Computers in Manufacturing. Visions and Realities about the Factory of the Future. Computers in Industry, No. 8, 1987.

Discussion

Question - J. Gero
You've said that you need to invent the design-language for each class of entity we are going to design. Is this putting the cart before the horse? What about the motion that design produces things that were not there before. How do you invent a language for something that you have not yet produced? Or are you limiting your view to things that you already know about? The example, I am thinking of, is the most famous one, the Sony Walkman. Before the Sony Walkman, there were no such things. Therefore you couldn't possibly have described it. According to an artefact-theory of the kind embodying that statement, you can't design it.

Answer - M. M. Andreasen
Well, I think that you agree that the Sony Walkman is a good example of what you presented, namely the question of the transformation between function, behaviour and structure. You may say that any designer familiar to a tape recorder could „spell" the basic structure of a walkman. I believe the Sony Walkman was an extrapolation of artefacts, with already existed, but fulfilling extreme demands to be small, elegant, and portable. What was done was to question the functionality and to see the need in a new light and then to make the inbustion. It could have been supported by an artefact modelling.

I have no explanations to how to cope with new things. I do not know if the spelling is feasable. But if you want to share what you are doing with a computer, then the computer should be able to spell the artefact in your way, as well as I can see. Therefore we have to create a science of artefacts, to be able to spell many different types of artefacts. For instance we have to create a structure theory of microelectronics, because it will be a new language, not comparable to theory of mechanisms or other known artefact theories.

Question - H. Grabowski
This leads us to the question: Do you need research or do you need the knowledge about research results to use for designing?

Answer - J Gero
This of course is a question for many countries, who are supporting research. The current view, of course, is that you don't need research any more in basic theory, because it is not much more to discover, which is an intriguing idea that is quiet contrary to the axioms of science. The axioms of science, which there are two I believe, the first is the things are knowable and the second is that there is an infinite amount to know. Both of them are axioms that is they are not provable. I believe, if you look at the history, in research it is not possible to know, when you do the research, how valuable it would be. And therefore also scientists and the global society in which we live must continue to do research. So I would say, you need both.

Question - F.-L. Krause

I liked very much your statement that we may have too many theories, but currently, in this group here, we are looking maybe for one. And I think you gave an idea how we could overcome this problem, because you talked about languages. And maybe we are able to have a kind of generative theory to generate the relevant theories which we need to have for the relevant artefacts we are talking about. And when we are able to express that in the languages which we have to define, maybe this could be a way.

Answer - M. M. Andreasen

Yes, I know that you know much more about design language than I do. I am surprised, how naive we have been earlier in relation to many types of models we have used for designing, how little they can explain and how much we have to add by our imagination to the models. I think the approach to look upon the language science and transfer results to the design area, and understand how to „spell" artefacts, it means to create semantics and syntax, will be a very promising one. Maybe our understanding of the spelling will lead to a better understanding of what is going on in designing.

Toward a Better Understanding of Engineering Design Models

Yan Jin and Stephen Lu

The IMPACT Laboratory, University of Southern California

DRB101, Los Angeles, CA 90089-1111, USA

Phone: +1-213-740-9574, Fax: +1-213-740-6668

yjin@usc.edu, sclu@usc.edu

Keywords: Design, design theory, modeling, analysis, process, decision-making, knowledge.

ABSTRACT

Developing models to explain and guide engineering design has been a topic for the research community for years. Various models of design have been proposed by researchers in the light of their own disciplinary perspectives and experiences. Partially due to the fact that different models are based on different views of design and employ different modeling approaches, the relationships between the models are not well understood. Our on-going research attempts to develop a common framework to correlate different design models through the analysis of seven of selected models. This framework consists of a general characterization of engineering design problems and a set of modeling dimensions against which the design models can be assessed. In this paper, we present a framework of analysis of design models as the first step toward a common framework for design modeling.

INTRODUCTION

Design is the most central activity that defines the engineering profession [Simo 69]. Within the engineering community, however, many recognize the fact that design is a poorly understood activity in education, research and practice [Dym-94]. While methods have been proposed by researchers and practitioners, engineering design is still highly experience-based. One of the difficulties that the design community is facing is the perceived lack of rigor. This perception has led to calls for more systematic, organized, formalized, and scientific approaches to study and practice design [Dym-94].

Problems arise in attempting to formalize design. Compared with other disciplines, our understanding of design is at a fledgling stage. We have not arrived at a state in which there is a coherent tradition of scientific research and practice that is matured enough to embody the law, theory, and application [Coyn-90]. In seeking this maturity, researchers from different disciplines have taken a modeling approach to studying design. Models are abstractions of reality. They are content with explanations and predictions within a subset of connected phenomena. Some examples of design models presented thus far include *systematic design* [*Pahl-96*], *axiomatic design* (Suh 1990), *quality function deployment (QFD)* [*Haus-88*], *total design* (Pugh 1990), *decision theoretic model* [*Haze-97*], *decision-support model* [*Mist-97*], and *general design theory* [*Yosh-81*].

These models all represent some abstractions of design behavior based on respective researchers' scientific and practical background. While all of them describe or prescribe certain design behaviors, our review of the models has shown that little cross reference exist between the models. It is not clear whether two models address the same design issues in different ways, or same concepts are applied to describe or prescribe different design behaviors, or if there are contradictions between the models. We argue that the lack of understanding of the relationships between different design models represents a barrier that prevents researchers from communicating with each other and consequently hinder the progress of our study of engineering design. There is a strong need for developing a better understanding of relationship among different design models.

Our on-going research attempts to develop a conceptual framework that exposes the disciplinary backgrounds, assumptions, and focuses of different models and explains what each model is about and why the models sometimes seem to be at odds with one another. We selected seven models mentioned above as the basis of our study. In the following, we first present a framework analysis that we developed for analyzing and linking the models. After that, we discuss how different models deals with two important elements of engineering design, i.e., product specification and design concept generation.

DESIGN AND DESIGN MODELING – A FRAMEWORK OF ANALYSIS

While everyone has some kind of design experience, be it design of *Lego* structures or space shuttles, not many people can well explain "what is design." This question has been explored by design researchers and practitioners for years [*Fing-89*]. Although many answers are present in the literature, it is hard to find one that can be accepted by all others. Instead of attempting to define "what is design", we look into the features of design problems and try to identify fundamental elements involved in engineering design. These elements of design provide a basis for conception of design and collectively form a framework of analysis for our study of design models.

The second question we came across in our study of design models was "what is design modeling?" We found some of the design models aimed at developing systematic and practical processes for designers to follow [*Pahl-96*, Pugh 1990, *Haus-88*]; some at proposing generally useful principles to guide design decision making (Suh 1990, *Alts-94*); and others at explicating types of knowledge involved in solving design problems [*Yosh-81*, *Gero-88*]. We view design modeling as the exploration of the space of design activities or behaviors. Different models explore different part of the space and have different niches. The challenge is, however, how can we link the diversely explored spaces into a coherent one to support the next level exploration.

INTERNAL AND EXTERNAL FEATURES OF DESIGN PROBLEMS

Engineering design problems range from relatively small component design to the design of large scale systems such as automobiles and space shuttles. Design problems can be characterized in different ways, such as creative design vs. routine design and simple design vs. complex design. In our study of engineering models, we take a rather broad scope to view design problems, it starts from understanding customer needs, through developing product definition (or specification), concepts, and details, to delivering products to customers. In an attempt to characterize design problems in a uniform way and reflect problem features in different design models, we found following generic internal and external features of design problems useful.

Internal features of a design problem are associated with the problem itself and are not influenced by external social, economical, or political situations. *Ambiguity*, *complexity* and *uncertainty* are the three internal features we identified as internal features of design problems.

Ambiguity

Ambiguity of a design problem has to do with lack of clarity or consistency in defining the design problem. For most, if not all, design problems, customer needs are often ambiguous that cannot be specified clearly. In some cases, even designers' goals or values are ambiguous in the sense that it is not clear whether profit or reputation is more important. Ambiguity is related to, but distinguishable from, uncertainty described below. Uncertainty has to do with imprecision in estimates of future consequences conditional on present design decisions. Uncertainty can be reduced by the unfolding of information over time. Ambiguity, on the other hand, refers to lack of understanding or lack of belief (or confidence) in existence of design problems or problem attributes such as customer needs and functional requirements. Ambiguity cannot be resolved by simply acquiring more information [*Marc-75*]. We argue that in the face of ambiguity, design problems, including customer needs and design specifications, are actually defined by a concept construction process through defining and inventing together with an information collection process through discovering.

Dealing with ambiguity requires construction of conceptual structures. Several of the models we studied attempt to provide constructive methods to reduce ambiguity of design problems and to lead to clearer design problem definitions. *QFD* [*Haus-88*], *total design* (Pugh 1990), and *systematic design* [*Pahl-96*] are the examples.

Complexity

Most engineering design problems involve multiple functional requirements and require multi-disciplinary knowledge. These design problems are inherently complex. The complexity of a design problem can be related to the amount of knowledge that we need to bring to bear to solve the problem. If a design problem involves more functions, components, relationships among components, it is likely to be more complex. The complexity of most system design problems is beyond the comprehension of individual designers. It has two effects on designing. First, high degree of complexity makes it hard for designers to understand the design problems, although the problems are clearly defined. As a result, designers may not be competent enough to explore and generate possible solution options. Second, given that options are generated, the large number of combinations of the options makes it very difficult to evaluate the options and make decisions.

There are two basic issues must be address to deal with complexity. The first is how can we *expand* the option "territory" in the face of complexity, and the second is how can we *converge* in the face of a huge combination of options, assuming we can expand. Generally, knowledge and constraints can be applied to deal the two issues, respectively. Different models have taken different approaches to address the two issues, as discussed in the following section.

Uncertainty

Engineering design is usually carried out under uncertainty. Uncertainty is particularly associated with making design decisions. One important and difficult task for designers is to estimate what will be the consequences, say on cost, quality, or profit of product, if a specific design choice is made. Uncertainty brings imprecision into the estimates. Uncertainty has to do with information. It can be measured by the difference between the information needed for making a pure rational decision and the information available. The level of uncertainty can be reduced by acquiring information to reduce the difference. It can be said that it is the uncertainty that made experience so important for design, since it helps designers to make more precise estimates by reducing uncertainty using the information acquired in the past.

There can be two ways to deal with uncertainty of design problems. One is based on the logic of *consequence*, and the other is based on the logic of *appropriateness* [Marc-89]. The approach of logic of consequence attempts to develop analysis methods to acquire or generate more information for making more precise estimates. The *decision theoretic design model* [Haze-97] falls into this category. On the other hand, the approach of logic of appropriateness attempts to use experience and natural principles to reduce the uncertainty by eliminating the needs of making estimates, e.g., reducing the number of options to be evaluated by including only appropriate ones and ignore the others. We will discuss how the selected models deal with uncertainty in the following section.

Engineering design is not carried out in vacuum for pure technical purposes. It is carried out in changing social, economical, and political environments. Among various possible situations, the most relevant one for characterizing a design problem is how the design problem is perceived by customers and competitors. In our study, we found that perceiving design problems at different market position in terms of competition, namely, *pre-competition*, *imperfect-competition*, and *perfect-competition*, may lead to different conceptions of design. *Market-position* thus can be considered as an important external feature of design problems.

Market-position

The market-position of a design problem is important because it influences designer's focus and consequently methods employed in design. When a design problem is in a *pre-competition* situation, fulfilling functional requirements is almost the only focus of designers. In this case, designer's value is closely linked with functional fulfillment, since as long as you can build it and it works you can sale it. Design in military areas and monopoly markets can be considered *pre-competition* problems. More function-centered methods are needed for this type of design.

As technology for solving a design problem matures and spreads, dealing with competitions becomes as important as fulfilling functional requirements. Along with the collapse of monopoly, maintaining reputation and customer relations takes place in solving design problems. In this case, we say that the design problem migrates into a market of *imperfect-competition*. Designer's value is linked to customer satisfaction as well as functional fulfillment. To solve this kind of design problems, one needs methods to communicate with customers, innovate new ideas to define variations, and design quality into the products. Current emphasis on customer needs and product quality in automobile and new consumer electronics reflects this external feature of these design problems.

When a design problem and its related technologies become so matured that no ambiguity exists in defining the design problem and both complexity and uncertainty become relatively low, a pure demand driven *perfect-competition* situation forms for the design problem. In this case, price becomes the only competition means and designers need methods to renovate their design in order to cut cost and increase the profit margin solely based on the market demand.

FUNDAMENTAL ELEMENTS OF DESIGN

Design refers to different things in different models. In some models, design is a whole process from market-based product definition to the delivery of products [*Pahl-96*, Pugh 1990). Other models treat product specification as given and define design as a process to transform or map specifications to product attributes [*Yosh-81*, *Mist-97*]. To analyze design models, we take a broader view of engineering design. We consider engineering design as a collection of *processes* of producing a *product* that are participated by three types of *participants* who possess different *goals and means*, as shown in Figure 1. We found this characterization useful because different models have different focuses on these elements. Revealing the difference in focus among different models helps us understand the models better.

Participants

We distinguish between three types of participants of engineering design, namely, Customer, Designer, and Producer. *Customer* is the source of needs, and in many cases stimulation, of design. Although sometimes designers may generate good product ideas, these product ideas are actually based on observations of customer needs in the market place. Customer can be individuals, groups of individuals, or even the whole market, depending on what products are under consideration. For example, ship design usually has individuals, i.e., ship owners, as customers, while design of coffee maker must deal with a wide range of market. As the *market-position* of a specific design problem shifts from *pre-competition* to *imperfect-competition*, The voice of customers is becoming more important in defining and solving the design problem.

Designer is in the center of design and participates in all design processes, as shown in Figure 1. In our study of the models, we observed that although designers play a center role in design, the notion of designer is often missing in the models. These models tend to specify what designers *should* do or follow and ignore what designers *want* to do. Designers' goals and values are not explicitly addressed. We argue that it is important for design models to recognize the position of designers and take their values into consideration explicitly. After all, designers do design more for their own purposes than for their customers. The value system and knowledge possessed by the designers have significant impact on their ways of doing design. Designers can be individuals, group of individuals, or a whole company. While specific design activities are carried out by individual designers, their values and goals can be influenced by their groups or their companies.

Producers develop physical or software products based on design. From a perspective of designing, producers provide needed information and knowledge for design evaluation. The evaluation can be carried out at a specific decision level, after a design concept is developed, or when the whole design has been finished. A *producer* can be a job shop of the designer's company or an independent company that manufactures products based on given designs.

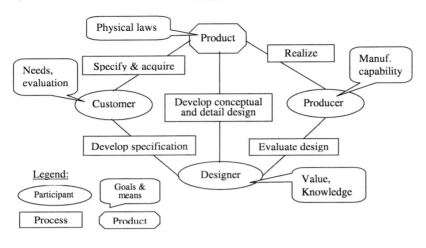

Figure 1   Fundamental elements of design

Processes

As shown in Figure 1, we distinguish between five processes in the world of engineering design, namely, *develop specification, develop concept and detail, evaluate design, realize product*, and *purchase product*. Among these processes, The first three participated by the *designer* are the essential ones for design research and are within the scope of our study. The separation of the three processes in Figure 1 does not imply that the processes are completely independent with each other, but that the interaction between the processes happens only through the *designer*.

Designing involves *defining* the problem, *generating* possible solutions, called concepts, and *selecting* solutions based on evaluation of the possible solutions. These three types of activities correspond roughly with the three processes shown in Figure 1.

Develop product specification: The ambiguity of design problems calls for constructive processes to define and clarify what need to be designed. The goal of this process is to reduce the ambiguity of design problem to the point that the concept or ideas about solutions can be considered. Uncertainty is another feature of design problem this process must address. Information must be acquired to predict what products or features of products will lead to customer's satisfaction and designer's values. Both customer and designer are the participants of this process.

Develop concept and detail: After a design problem is defined, designers must develop product concepts and details to satisfied the specification. Concept and detail development is the main part of designing. It has been the focus of design research as well. There are two fundamental activities involved in this process. One is to explore the space of concepts so that the known space of the solutions can be expanded as large as needed. The second activity is to evaluate, select, and compose solutions to achieve a desired design. The complexity and uncertainty of the design problem have made this process highly difficulty. Different design models have taken different ways to deal with the two activities as will be discussed on the following section.

Evaluate design: Design evaluation is carried out mostly by designers and producers. The goal of design evaluation is to estimate the consequences of specific design choices and choose the ones that lead to desired ones. The models and methods developed for evaluating design must address the issue of uncertainty and propose methods to increase the precision of estimates. Analysis methods and tools are often needed to make estimates more precise with limited information.

Goals and Means

All participants of design have their goals and means to carry out their activities in design. Customers as individuals have their needs and their value systems that decide what they want and what they do not. Customers as a market have their economic mechanisms that define the demand for certain products at given prices. Focusing on customer needs may lead to more customer oriented design models, such as *QFD* [*Haus-88*].

Designers have their own knowledge to explore solution spaces and values to guide evaluation of alternatives. Knowledge about products plays an important role in conceptualizing functional requirements and expanding the space of possible solutions. Knowledge about design processes helps efficient application of the knowledge about products. Designer's values, on the other hand, is the basis for designers to evaluate alternatives and converge from the expanded solution space into a desired final design. From a perspective of design modeling, focusing on knowledge means to address the issue of "how can we organize our knowledge in the way that can be best retrieved and applied to generate design solution?" This line of thinking has the potential of leading to automated design systems. General design theory [*Yosh-81*] is an example model of this type. Researchers who focus on designer's values system emphasize the role of decision-making in design. The decision theoretic model [*Haze-97*] is an example. This model believes that as long as utility function is clearly defined, choosing design alternatives to maximize the expected utility will lead to a desired design.

While it can hardly be justified that products have goals, products do have their means that governs their performance. If a product is a physical product, the means would be natural laws found in physics, material science, chemistry, etc. In case a product is a software system, then the means is defined by the computer processes including compiling and running mechanisms. Natural laws are the basis of design and must be obeyed at any stage of designing. From a design modeling point of view, the issue is how can we make use of the natural laws to guide our design. Although different types of physics or chemistry based analysis methods have been developed to support design analysis, we observed that the linkage of designing and natural laws at the design methodology level remains loose.

THREE VIEWS OF DESIGN

As described above, different design models deals with different elements of design. Some focus on *processes* and attempt to develop a set of experience based, and well-defined, *methods* for designers to follow [*Pahl-96*, Pugh 1990, *Haus-88*, *Suh-90*]. Some models emphasize importance of decision-making in design and propose to apply utility theory for design education and practice [*Haze-97*]. Researchers who aim at develop intelligent computer aided tools tend to focus on the role of designers and attempt to understand and develop a way to acquire, represent and manage the knowledge of design [*Yosh-81*]. We identified following three views of design modeling. Each view reflects researchers' disciplinary background and the purposes of their models.

Process view

Researchers taken a process view toward design treat designing as a unique or distinct process that needs to be carefully planned and systematically executed. It is implied that as long as one follow the specified process, he or she will likely to come up with a desired design result. A process based design model usually consists of a *process flow* that defines steps or phases of designing and a set of *methods* that are applied at different steps of the design process. Most process definitions and design methods are developed based on, or emerged from, engineering practices and experiential knowledge accumulated in the respective engineering domains.

The process view of design attempts to reduce design ambiguity, complexity, and uncertainty by restricting designers' activities to a set of predefined ones. This restriction can maintain the consistency of designing and keep it "on track". The premise here is that since process and methods are results of past successful practice, following these steps will lead to good design. The logic of this premise is based on appropriateness rather than consequence. The former expects good design by attempting to repeat what people have done appropriately and the latter expects a good design by try all means to estimate and evaluate possible consequences of all possible situations. On the other hand, the restriction may also limit designer's thinking space and miss possible good designs that are not "on track" of the process flow. Examples of the models that take a process view toward design include *QFD, systematic design* [Pahl-96], *total design* (Pugh 1990), and *zig-zag* part of *axiomatic design* [Suh-90]. A general feature of these design models is that they are practically easy to follow. In the next section, we will discuss some of the specific features of these different methods.

Decision view

The decision view of design emphasize the role of designers' value and believes that designers' role in design is decision-making and designing is essentially a decision-making process. The decision-oriented design models treat designing as repetitions of a simple two-step process, namely, option generation and decision-making, of which decision-making is considered the most essential part of designing. Decision-oriented models assume, or require, designers to be pure rational in the sense that designers have clear, consistent and stable preferences and utility functions; they can acquire all needed information to make estimates on all considerable choices; and they have consistent and stable rules for making choices, e.g., to maximize expected utility [*Haze-97*].

The decision view of design does not recognize the ambiguity of design problems. It attempts to manage the complexity of design by relying on designers' option generation capabilities and through rigorous option evaluation. How to deal with uncertainty is the main issue for decision-oriented models. Methods developed so far in decision theory have been proposed to guide designing [*Haze-97*]. From a pedagogy perspective, the decision view is useful to demonstrate how design decisions should be made under uncertainty. The ambiguity and complexity of the real problems, however, made it unclear how this model deals with practical design problems at specific levels.

Knowledge view
A knowledge view of design treats design as a knowledge-based problem solving process. Unlike the process view which thinks design has its unique processes and ways of doing things, the knowledge based design models focus on the uniqueness of knowledge and see little different between designing and other goal driven problem solving processes. The issues here are what knowledge is needed to carry out designing and how the knowledge should be organized to make efficient design.

Knowledge view of design modeling requires clear definition of design problems. It deals with complexity and uncertainty by acquiring and organizing knowledge about products (including fundamental natural lows) and the knowledge about design processes that controls application of the product related knowledge. Design decision-making is rather implicit in these models. It is treated as an integral part of application of knowledge. *General design theory* [*Yosh-81*] and some artificial intelligence based design models [*Gero-88*] take the knowledge view of design. One of the goals of the design models based on the knowledge view is to develop automated design systems.

APPROACHES TO DESIGN MODELING

Different views of design lead to different focus and conception of designing. To develop a design model, however, one needs to follow a specific approach to define concepts, relationships, and to develop principles and methods.

Many researchers have taken a *systematic* approach to develop design models. The systematic approach tends to think designing as a "system." Once the structure of the "system" is constructed and the processes that run the "system" are defined, the "system" will run well as long as designers "move" within the "system." The issue here is consistency and appropriateness rather than accuracy and correctness. Most of the process-oriented models characterized by process structures and experiential methods, and some of the knowledge-oriented models, characterized by knowledge structures and experiential knowledge, are taking this approach.

The *axiomatic approach* to modeling design tends to think design as a self-contained field of study, rather than an extension or an extrapolation of any other study fields. It attempts to reveal the fundamental truth of designing in designing terms. To distinguish axiomatic approach from non-axiomatic ones, we define the axiomatic approach as the one that introduces axioms at the designing level, i.e., directly related to the ways of doing design. The issues here are *truth*, i.e., argument of axioms, and *correctness*, i.e., provable theorems as design principles. The *axiomatic design model* [*Suh-90*] is an example taking this approach.

Mathematical approaches to design modeling attempt to achieve rigor and correctness in design modeling by relying on the mathematical power of formulation and derivation. A mathematical approach usually starts with a number of axioms. In contrast with the axiomatic approach, however, these axioms are rather primitive and are not at designing level. The truth of designing is treated in this approach as a result to be derived rather than an axiom. The formality and rigor of mathematical models have attracted researchers to take the mathematical approach to illustrate what is designing about and to show how the designing should be done. The mathematical models, however, tend to simplify the problem of designing by staying at relatively high level of abstraction. The complexity and uncertain in the real world made it hard to apply these models directly for design practice. The examples of pure mathematical models include *general design theory* (Yosh-81 1981), *decision theoretic model* [*Haze-97*] and *decision support model* [*Mist-97*].

DESIGN MODEL ANALYSIS – METHODS

We have discussed relationships between the selected models at the level of focus, view, and approach. To understand how different models are related to each other at the level of principles and methods of designing requires deeper analysis. In this section, we focus on two process elements shown in Figure 1, namely product specifications and development of concept and detail. We discuss how different models proposed different ways to deal with ambiguity, complexity, and uncertainty of the design problems.

DEVELOPMENT OF DESIGN SPECIFICATIONS

As described above, developing design specification is an important task of designing that reduces ambiguity by developing definitions that satisfy both designers and customers. Our study found that different models have different ways to deal with this process, partially due to their disciplinary backgrounds as well as their perception or ignorance of the market-position of design problems.

Design models that explicitly or implicitly treat design problems as in *pre-competition* situations tend to ignore the importance of designer-customer interaction for specification development. For example, *general design theory [Yosh-81]* does not include this process into their definition of design but treats it as the task for management people rather than designers. *Axiomatic design model [Suh-90]* recognizes the needs of explicit consideration of customer needs but does not focus on providing guidance clarifying the customer needs. The *decision support model [Mist-97]* focuses more on the down stream design processes and does not address customer needs explicitly.

Many sophisticated product design problems, such as improving existing products and innovating new products, are actually positioned in an *imperfect-competition* place in the market. Recognizing this competition as well as the immaturity of technology, a group of design models proposed concepts and methods for developing innovative product ideas and definitions by working with customers and analyzing markets. *QFD [Haus-88], systematic design [Pahl-96], total design* (Pugh 1990) all explicitly address the needs of considering voice of customers. The *parametric analysis* method of *total design model* and a set of methods described in *systematic design model* help us to understand who are there in the market and what kind of products is likely to be successful. *QFD's* "house of quality" systematically brings customer needs, competitors' strength and weakness, and designers' goals or values into one picture for clearly defining what is the designing problem the designer is about the address. In this sense, we view *QFD* more as a design problem definition tool than as a quality tool.

The *decision theoretic model [Haze-97]* takes an extreme position to addressing the voice of customers. Due to its pure rational decision view of design, this model does not believe the existence of customer needs. It argues that a rational preference set in terms of product attributes does not exist for collections of rational customers. Instead, the only voice of customer that can be heard is the market demand defined by past market information of the same or similar products. From this point of view, product specification can only be defined based on market demand, not the attributes of customer needs. We argue that the decision theoretic design model can be applied to product design problems that are in the *perfect-competition* market positions. The ambiguity, complexity, and uncertainty of design problems and the boundedly rational customer preferences made this model difficulty to apply for non-*perfect-competition* situations.

DEVELOPMENT OF CONCEPTS

Development of concepts and details is the center task of designing. Different models have proposed different methods and principles to deal with the complexity, uncertainty, and designer's value. To focus our discussion, we only examine how do different models develop design concepts.

As described above, we consider this process consists of two basic activities. One is to *expand* the space of possible solutions through *knowledge exploration*, and the second is to *converge* to a desired solution through *value based evaluation*. The two activities can be dealt with *sequentially*—i.e., expand first and then converge—or *iteratively*—i.e., expand and then converge repeatly step-by-step.

To establish and expand the space of possible solutions, most models treat design option space as defined by a number of primitive functions derived from functional decomposition and potential sub-solutions for each primitive functions generated by designers, as seen in morphology charts [Pahl-96]. In a *sequential* expand-converge approach, the option space is established and expanded by first decomposing the initial functions to as detail as possible and then generating as many as possible sub-solutions for each of these primitive functions. After the space is established and expanded, various design variances or concepts can be generated. The number of possible variance is definite and defined by the number of primitive functions and number of sub-solutions. *Systematic design model* [Pahl-96] and *total design model* (Pugh 1990) follow this way of expanding the space.

The goal of "converge" is to select one or several desired concepts for further designing. The converging can be achieved by excluding obviously bad sub-solutions, generating possible variances by combining the sub-solution, and then selecting one or several designed variances as concepts for further designing. This converging method is used in *systematic design model* [Pahl-96]. A modified and more dynamic "expand-converge" method, called concept generation method, was proposed in the *total design model* (Pugh 1990).

The zig-zag approach proposed in the *axiomatic design model* [Suh-90] takes another approach to generating concepts. It combines expansion and converging step-by-step. That is, at each step, when a function is decomposed, the solutions or sub-solutions are generated/expanded and evaluated/converged immediately before going to the next level of decomposition. This way, at the time when the function decomposition is finished, the convergence of ideas is also finished. *General design theory* [Yosh-81] takes a similar step-by-step expansion, called data retrieval, and converging, called evaluate, approach. The difference is that in *general design theory*, functions are assumed already decomposed. When searching for ideas to fulfill the functional requirements, each function is considered one-by-one in sequence. After the last function is considered, design converges to a desired solution.

The *decision theoretic design model* [Haze-97] does not recognize the specificity of design option space described by the dimensions of *function* and *sub-solution*. This model relies on designers to "establish and expand" their own options spaces in their minds. The "converging" is realized through a general value-based decision-making mechanism. Due to its lack of specificity as applied to the design domain, it is not clear how this approach can be implemented to guide design practice without substantial further exploration.

CONCLUSION

Each of the different engineering design models we studied captures certain specific aspect of engineering design. Clarifying the focuses, views, approaches and methods of different design models deepened our understanding of the design models and has led us to discovering the intrinsic properties of designing. Design in a broad sense can be characterized by its *participants*, *processes*, and *goals and means*, as shown in Figure 1. Dealing with ambiguity of design problems is an important task for design problem definition. Design in a narrow sense involves *expanding* option spaces by knowledge exploration and *converging* to desired solutions in the expanded space through value-based evaluation. Different design models demonstrated different methods to support *expanding* and *converging* by explicitly or implicitly dealing with complexity and uncertainty of design problems. While the discussion of this paper is still at an abstract level, our research made the first step toward developing a common framework for design modeling. As our research progresses, we expect such a common framework will contribute to the design research community by providing a basis for communication among different design models.

The authors are grateful to Prof. John Chipman, an economist at University of Minnesota, and Prof. Chunming Wang, a mathematician at University of Southern California, for the valuable discussion and critical comments they provided in the course of developing better understanding of engineering design models.

REFERENCES

Alts-94　Altshuller, H.: *The Art of Inventing (And Suddenly the Inventor Appeared)*. Translated by Shulyak. Worcester, Massachusetts 1994.

Coyn-90　Coyne, R. D.; Rosenman, M. A.; Radford, A. D.; Balachandran, M.; Gero, J. S.: *Knowledge Based Design Systems*, Addison-Wesley Publishing Company, Reading, Massachusetts 1990.

Dym-94　Dym, C.: *Engineering Design – A Synthesis of Views*, Cambridge University Press 1994.

Fing-89　Finger S.; Dixon, J. R.: "A Review of Research in Mechanical Engineering Design", in *Research in Engineering Design*, Vol.1, No.1, 1989.

Gero-88　Gero, J. S. (ed.): *Artificial Intelligence in Engineering: Design*, Elsevier/CMP, Amsterdam 1988.

Haus-88　Hauser, J. R.; Clausing, D.: "The House of Quality," *Harvard Business Review*, 63-73, May-June, 1988.

Haze-97　Hazelrigg, G. A.: *An Axiomatic Framework for Engineering Design*, Working Paper, NSF, 1997.

Kuhn-70　Kuhn, T. S.:*The Structure of Scientific Revolutions*, University of Chicago Press, Chicago 1970.

Marc-75　March, J. G.; Olsen J .P.: "The Uncertainty of the Past: Organizational Learning under Ambiguity", in *European Journal of Political Research*, Vol.2. 147-171, 1975.

Marc-89　March, J. G. "A Chronicle of Speculations: About Organizational Decision-Making", in *Decisions and Organizations*, March, J.G. (Eds). Blackwell, Oxford UK 1989.

Mist-97　Mistree, F.; Allen, J. K.: Optimization in Decision-Based Design, *Working Notes of Open Workshop on Decision-Based Design*. Orlando, Florida 1997.

Newe-72　Newell, A.; Simon, H. A.: *Human Problem Solving*, Prentice-Hall, Englewood Cliffs, New Jersey 1972.

Pahl-96　Pahl, G.; Beitz, W.: *Engineering Design - A Systematic Approach*, Second Edition, Springer, London 1996.

Simo 69　Simon, H. A.: *The Sciences of the Artificial*, MIT Press, Cambridge, Massachusetts 1969.

Suh-90　Suh, N. P.: *The Principles of Design*, Oxford University Press, Oxford 1990.

Yosh-81　Yoshikawa, H.: "General Design Theory and a CAD System", in *Man-Machine Communication in CAD/CAM*, T. Sata, E. Warman (Eds). IFIP, 1981.

Discussion

Question - F.-L. Krause

You have been the first, taking into account also a producer. And I think this seems to be very important. When we think to the Latin language, "construere" is the word which is designing and "construere" means to build something. And then we have had the talk by Professor Gero about Hamurabi. I think that the question was: Did they really make the sketch before they built it or did they build before they made a sketch? I mean that is a hen-and-egg-question, maybe, but generally speaking, there seems to be very tight influence. And I think that we are also talking about some changings and paradigms and when we look to a large German company, BMW, they recently have changed their organisation. There is one person now responsible for production and development. And I think this is, to my understanding, a very interesting change. And the question is: Are we able to take that into account when we talk about design-theory?

Answer - Y. Jin

Yes, I hope. The answer is that based on this fundamental element picture, we can see that the producer has a lot to do with the evaluation of your design. However, this doesn't really mean that customer, designer and producer have only fixed relations. Actually, there are interactions between those three blocks. The main role of the producer is: try to give the information to evaluate whether your design is adequate or not. And the governing equation for the producer is the manufacturability. You have to take those things such as manufacturability into consideration even when you do design. This picture has a good point in that those things are not taken into consideration directly by the producer by interaction between the producer and designer.

Question - A. Albers

It is a very interesting model, but in my opinion, there is a very important factor you don't have in this sheet, it is the competitor. After my industrial experience, the influence of the competitor on the individual design and, as Mr. Krause said, the influence of the producers on the individual design are extremely high, normally or very often higher than the influence of the customer. We have to accept that in practice and so I think, it is necessary to consider this point, the competitor.

Answer - Y. Jin

That is a very good point. Actually, I was thinking, we can treat market as. It is not individual, it is the market. In the market, you have competitors, you have customers. But maybe, it makes more sense to make the competitor explicit, we have not thought about that.

Comment - A. Albers

Directly to that, it is completely different, when you have an element customer or you have an element competitor, because the relationships between both are totally different and the influence on the design-process or the input in the design-process, in your processes or production-processes as well, are completely different. So, in my opinion, you have to split up.

Question - H.-J. Franke

I agree in most points with your opinions and proposals, but I missed cognitive psychological problems of designers in your picture.

Answer - Y. Jin

The cognitive thing is really happening within the box of designer's goals-and-means. So, you can model all the processes here. If you put your modeling focus on this part, you pretty much end up with a cognitive model, I mean knowledge centered model of design.

Comment - H. Grabowski

It is too early to make a conclusion, but I have learned that it is a very difficult domain we are dealing with. We have heard about four different views of the design-process which stressed different important points. The conclusion of this workshop should be: What are the basic ideas for the development of a more general design theory and how can we find a way to this kind of theory step by step. An important thing could be to find the common language in this area. Today, every researcher has his own special point of view and uses special words and expressions. Researchers probably all mean the same, but they cannot understand each other.

Summary Session 1

S. Rude

To summarize what has been presented in the first session, I would like to begin with Professor Tomiyama. He spoke about general design theory and I found it remarkable that general design theory is a theory of design knowledge and does not explain how and why we can design. I understood the key elements of the paper to be that this has to do with functions, entities and attributes. The subsequent discussion showed that functions are the object behavior, the entities are refer to the real objects and the attributes are the measurable object characteristics. So general design theory does not describe the design process itself, which nevertheless was described to some extent as a stepwise refinement from functions through physics to the attribute space. Again, the terms functions, physics and attribute space were the keywords of these presentations. The step-by-step refinement is also linked to certain logical considerations and logical inferences such as deduction, abduction, induction and methods such as circumscription are used to link these entities. What I had not yet heard was whether only physical effects are taken into account. Later on, we saw that the author also discussed other effects, such as chemical or even biological ones.

Proceeding, I understood Professor Gero to state that a main issue of his presentation was to make a distinction between theory and model, where theory is a capacity to explain, to predict, and models are the capacity to describe the products. So models are nevertheless the basis of a theory and the key elements of his presentation were similar to those of Professor Tomiyama's - like Professor Tomiyama, he spoke about functions, behavior and structure. What was interesting was that at least for behavior, a distinction must be made between expected behavior and structural behavior. This is reflected in the words describing the product "as designed" on the one hand and "as built" on the other hand. According to Professor Gero, the design process itself is unpredictable due to the fact that synthesis is an abduction process and because of the second argument, the situatedness. Summarizing, the situatedness is that the design result is a function of the designer himself. Nevertheless, Professor Gero very clearly specified eight design steps, where the step of analysis is 85% taught in universities but is still only one of eight steps. Here, the argument of emergence also came up, which can be described as taking the design environment into account. If one therefore takes the design environment and not only the product itself into account, the emergence aspects can also be handled. Reformulation was an important keyword. Reformulation steps are related to functions, behavior and structures. This also leads to the open issue we discussed: That the status "as designed" and "as built" must also be put into relation to functions and to structures as it was put into relation to behavior.

Professor Andreasen also accepts this distinction between functions, behavior and structure. However, he then states that he is convinced that we have to go one step further into detail. This can be discipline- or branch-oriented, i.e. one must discuss which functions are actually implemented in mechanical engineering , in architecture etc., what kind of behavior is described in mechanical engineering, in chemistry, in other disciplines, and which types of structures really exist in the different disciplines. And here, his remark is surely true that schools of design differ from each other at least on this second level of detail. Therefore, a designer should at least be able to formally express an artefact. This leads us to the requirement that something of the nature of a design language is necessary and must be agreed upon.

Last, but not least, some remarks to Professor Jin. He analyzed very thoroughly analyzed different approaches used in design theories. We have already discussed existing approaches, such as Quality Function Deployment (QFD), Structured Design Method (SDM, according to Pahl/Beitz), Axiomatic Design Method (ADM, according to Suh) and others. But the more important thing is that it became apparent that certain key elements need to be taken into consideration, i.e. the problem needs to be put into the context of the product, creativity and evaluation. Creativity brings with it a complexity problem, and evaluation leads to a problem of uncertainty. At this point, one also needs to take a look at decision theories. The idea of key players, presented by Professor Jin, is a very essential one. The key players in this process are not only the designer himself but also the customer on the one hand and the producer on the other hand. These key players are subject to different influences, the customers are influenced by the needs, the designers by the knowledge and the producers are also influenced by the possibilities. In certain processes, these key players are interrelated. Thus, the subsequent discussion showed that perhaps a forth player should be included. This might be someone from the field of material science. I don't know if producers also take this into consideration, but I had the impression that material science might be a further major issue to be considered.

Session Two

Design Theories in Special Areas of Engineering and Natural Sciences

Systematic Software Construction

Gerhard Goos, Uwe Aßmann

Institute for Program Structures and Data Organization

University of Karlsruhe, Germany

Keywords: software construction, software design, architecture, frameworks, design pattern, specification

ABSTRACT

Since programs are both different and similar to other technical components software construction must obey particular constraints but also shares principles with other disciplines. First these specific aspects of software and its construction are exemplified.
Then several construction principles are introduced which are analogous to other disciplines (decomposition, architectural styles, frameworks, design patterns, and specifications). These principles could form the basis of a Universal Design Theory.

INTRODUCTION AND OVERVIEW

The goal of software construction is a program which may be executed together with other programs by a computer or a network of computers. A computer in this context is any piece of hardware capable of changing state under control of the instructions of a program; this includes microprocessors as well as other electronic devices such as certain types of gate arrays, ASICs etc. We term the computer and other necessary software -- such as an operating system -- the *environment* of the software in question.
Programs are both different and similar to other technical components of such an environment [*Klir-69, Klir-91, Ropo-75*]. In consequence, software construction must fulfil some particular requirements but also shares design principles with other disciplines. In the following, several specific aspects of software are exemplified. Then software design principles analogous to other disciplines are introduced. These could form the basis of a universal design theory.

WHAT IS SOFTWARE?

Software by itself is a piece of information consisting of instructions for changing the state of a computer. Only when it is used in a suitable environment it becomes (part of) a technical product. Thus, many properties of technical products such as ease of use, durability, reliability, fault tolerance etc. can only be ascribed to the combination of the software and an environment. Actually, we must distinguish between embedded applications where the software is so tightly coupled to its environment that it cannot be judged by itself and normal applications which can be used in an arbitrary environment obeying certain standards, e.g. on a

PC with a POSIX-conforming operating system. We restrict the discussion mostly to the latter case. Except for timing relations the behaviour of the environment with respect to the given software can often be emulated by the human brain; with respect to this execution model we may consider software as a technical product and ascribe technical properties to it independently of its environment.

Each piece of software provides for interfaces to its environment and to the users of the system. These interfaces are mostly quite large and complicated and must be documented. Actually, the term software is used to comprise both, the program and its documentation. We often refer to the interface documentation as the specification of the software.

SOFTWARE FAILURES

Generally speaking, there are five possible kinds of malfunction of software:
- *Implementation failure*: the software does not react as specified by its documentation.
- *Specification failure*: the software reacts as specified but not as intended by the customer.
- *User error*: the system is fed with inappropriate (erroneous, inconsistent or incomplete) data or commands.
- *Resource failure*: the system is overloaded and fails because the processor, memory or other resources are not available with the required capacity.
- *Environment failure*: the software fails because the environment does not react as specified.

Mostly, software cannot be kept responsible for environment failures; exceptions are cases in which varying reactions of the environment are predictable and must be dealt with according to the specification of the software. Similar remarks apply to resource failures and user errors. In all cases one would, however, expect sensible reactions comprising at least a report of the symptoms of the malfunction (graceful degradation of performance).

Specification failures are errors in requirements engineering. According to [*Endr-75*], who did an extensive study during the development of an operating system by IBM, 50-70% of all failures are specification failures whereas implementation failures, the real programming mistakes, and other failure causes account for less than half of the problems; these findings have been repeatedly confirmed by other authors.

Of course, the frequency and weight of user errors and of resource and environment failures may be influenced by appropriate design decisions during software construction. This is similar to the sitatuation in constructing other technical systems. Due to the fast changing requirements there is no stable experience base of how to reduce the amount of such errors by appropriate design measures.

A particularly subtle form of failures are timing errors in embedded systems and more generally in multiprogramming, multiprocessing and networked systems. Superficially timing errors are resource failures: several branches of the program are not executed with suitable relative speeds. Since these speeds are not under control of the software designer the only systematic measure for eliminating them consists in requiring that no timing errors occur with arbitrary speeds of the program branches. We then say that the branches are executed in virtual time; timing errors become logical errors under this condition.

Usually an implementation is called correct if it obeys its specification. The process of showing the correctness is called *verification*. *Validation* denotes the process of showing that the specification fulfills the requirements. Verification thus does not imply reliability since the latter would also make statements about the frequency of the other kinds of malfunctions.

Software is digital information. There is no tear and wear of software. All implementation and specification failures are reproducible with 100% probability if they have been seen once.

Statistical methods cannot be used for measuring the frequency of implementation or specification failures; statistical data only make statements about the stability of user and environment behaviour. Because of the discrete nature of digital data slight changes of the use and environmental conditions may have a huge impact on the validity of software. There are lots of examples that software may fail even after several years of satisfactory use and after thousands or even millions of copies have been installed; such behaviour is caused by changes in user behaviour or by changes in computing speeds, etc. The failures are due to the fact that the number of possible execution paths is astronomically high even in software systems of moderate size; therefore only an extremely small percentage of all possible execution paths will ever be executed during testing and even during the whole lifetime of the system.

This implies that testing can show the absence of software failures under certain user and environmental conditions; but even a huge number of tests cannot positively ascertain the validity of a software design. Software can only be validated during the design process, not afterwards; this is one of the major differences of software construction compared to the design of other technical systems. It is also a major source of trouble in industry since engineers who have been retrained as programmers or software managers often lack this insight.

The detection of the infamous bug in the division routine of the Pentium processor after delivery of several million processors illustrates the point: The mistake was a wrongly programmed table of numbers (stored in ROM) which was used by the division routine. But chip testing after manufacture had no chance to ever reveal this logical error since it only checked the reliable physical performance of the chip compared to an equally buggy reference.

SOFTWARE CONSTRUCTION

There are four major differences between the construction of software and other technical systems:
- Software represents a logical system. It implements logical assertions of the form $\{P\}S\{Q\}$: If a precondition P is true in the beginning then after executing program S the postcondition Q will be true. Physical laws do not play any role for validating software; they may only be important with respect to environmental conditions. As a consequence, software construction is using combinatorics, mathematical logic and algebra but not analysis as its mathematical foundation. Software behaviour is discontinuous by nature.
- Specification languages and programming languages serve as CAD-tools. These languages are basically formal systems even when the designer does not conceive them as such.
- The validation process is different as discussed in the foregoing section.
- Software is easily copied and distributed. Thus the construction process does not lead to a prototype which then must be manufactured; instead it delivers the final product immediately. Especially there is no preparation for a separate production process as for physical systems.

Since there is no tear and wear there is also no repair of software in the sense of reestablishing the original properties of the product. Maintenance of software is concerned with correcting malfunctions or with adapting the software to changing user needs or environmental conditions; in both cases the result of the original design is changed and the validation and test process must be repeated.

It is often believed that it is easy to modify software. It is indeed easy to modify a line in a program. But to find out which lines to change and to ascertain that the change has no other implications then the desired ones is as difficult as modifying the design of any other technical product.

The usual measure for the size and complexity of software is the number of lines of code (loc) excluding comments although this measure shows some disadvantages. Large software systems comprise more than a million loc. The maximum performance which we observe in diploma theses is 30 - 40,000 loc per manyear. For difficult tasks such as the kernel of an operating system or a communication system this performance may go down to 2,000 loc/manyear. Error rates of one error per 1,000 loc are considered excellent.

These figures indicate that during the design phase large systems must be decomposed into modules of manageable size. These modules must have stable interfaces; system production is then basically a configuration management task.

Problems in Configuration Management

There are several obstacles with this view of system construction by configuration management:
- The rapid technical progress over the last half century has not allowed for developing stable system structures with modules of universally agreed interfaces and functionality. A statement such as ``a car comprises a body, 4 wheels, a motor, power transmission, ..." has no equivalent in software construction: E.g., the usual data base models on mainframes in the 60ties and 70ties have been replaced by relational models in the 80ties but it is already clear that relational models will not carry into the multimedia age. The estimated cost of more than a trillion dollar for solving the year 2000 problem shows that many of these old modules are still around and cannot easily be replaced by new ones. As a result system design must prepare for connecting to modules with a variety of interfaces with sometimes drastically varying functionality.
- As a further result interface design is the most difficult part of software construction as well from the technical as from the managerial point of view: Misunderstandings about interfaces make up for more than 90% of all errors.
- During system construction new versions and variants of modules are developed and released at such a rapid pace that developers have extremely difficulties to maintain a consistent prototype of the system as basis for their own development work. E.g. during the early development of the Linux operating system, a design to which several hundred people were continuously contributing over the internet, Linus Torvalds at times released a new major system version every day.
- The existing tools for supporting software configuration management have their roots in academic developments and are rather suited for small to medium sized systems up to 100,000 loc.
- Industry is very slow in viewing system construction as a configuration management task. The majority of the system developers has no formal training in computer science and software engineering but is only trained on the job. These developers are often even not aware of the available tools and of the dangers in which they run without them.

The relationship of configuration management for software and for other technical systems such as airplanes or large military applications is unclear and remains to be studied.

The Problems of System and Technology Change

The software industry and its users have not developed economic models which signal when it is time for a system change. It is unclear whether they ever will. New systems are generated by modifying existing systems in an evolutionary way. If a completely new system is developed then usually there is only a qualitative but not a quantitative measure of the advantages of the new system given.

This insight applies not only to systems but also to methodologies for system construction. Therefore there is mostly no convincing assessment of the risks of new systems and technologies possible. The difficulties in assessing these risks are very often leading to half baked decisions by which the new development is given a chance but no consequent transition is made.

THE VDI PROCESS MODEL

The VDI Guidelines 2221, [*VDI 2221*], list the common stages of the development process of technical products including software. The general system life cycle

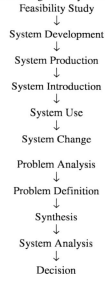

where each step is subdivided into

is the same in all cases although the meaning is slightly different for some terms. Subsequently we illustrate several basic principles of software construction with reference to this common process model. Despite of the different nature of software and its construction process, astonishing similarities to other disciplines can be identified.

PRINCIPLES OF SOFTWARE DESIGN AND CONSTRUCTION

Although computer science is a young discipline already several basic principles for system design and construction have emerged. As in other sciences, after problems are *decomposed* into parts partial solutions are composed to a complete solution. The coarse-grain structure of a system can be classified in *architectural styles*, leading to standard system structures. Beyond architectural styles, object-oriented *frameworks* supply system architectures for product families; a framework is instantiated with parameter code to deliver the final product. Frequent solutions to typical problems can be described by *design patterns* which guide implementation and teaching. Lastly, *specifications* are used to generate products from formal descriptions. The section is concluded with a comparison that identifies these construction principles in other disciplines.

This paper concentrates rather on design principles than management of software construction. The latter subdiscipline of informatics is called *software process management* and deals with problems such as the automation of the construction process and consistent variant selection in systems.

DECOMPOSITION AND COMPOSITION

Beginning with the smallest units, the following units and principles for decomposing system tasks and for integrating their solutions are used in software construction:

- *Algorithms*: The computation of a function $y = f(x)$ is achieved by an integrated piece of program. An extensive body of knowledge about algebraic, combinatorial, geometric and other types of algorithms has been developed over the last 25 years. There is also a number of design principles for algorithms such as divide-and-conquer, greedy algorithms, or dynamic programming. which mostly originated in operations research.

  In practice the use of algorithms requires to abstract from the given problem until a suitable problem representation is reached. Then this representation is compared to the alternative solutions taken from textbooks or from software libraries of reusable components. Whereas a similar approach in mechanical engineering is routinely taught to students its systematic application in software construction is still in its infancy. Studies reveal that often more than 50% of the time required for a new design is necessary for evaluating the applicability of a given algorithm drawn from a library.

- *Data Structures*: Data structures are composed from data representations and a number of functions for accessing, testing and changing the data elements. The data representation is usually called a *(data) type*; its combination with the access functions is called a *module*. In object-oriented programming the data type is usually identified with the module name; it is then called the *class* of the data structure. An important property of each module or class is that it is designed to hide the actual details of the data representation to the user. This follows the well-known black box principle of other engineering disciplines.

  As for algorithms there exists a vast body of knowledge about sophisticated data structures for various purposes. The same remarks as for algorithms apply.

- *Modularization*: Basically a module is a black box with internal state. It may be viewed as a data structure, the state variables represent the data elements. From a systematic point of view all properties ascribed to data structures can be found in modules and vice versa. Compared to VLSI-design the construction of an algorithm or a highly integrated data structure corresponds to custom design of a chip whereas modularization corresponds to cell-based designs.

- *Top-down design*: This is the standard decomposition method in software construction: The problem P is decomposed into subproblems P_1, P_2, ... and the solution of P is composed from the solutions of the subproblems. Depending on their nature the P_i correspond to algorithms, to access functions to an underlying data structure or to modules. In the latter case we also speak of a hierarchic or layered design. All known methods for software verification have been originally developed for top-down-design; other kinds of decomposition are usually much more difficult to verify.

  Technically speaking, a top-down design is a way of representing the results of a system

decomposition. Based on experience or other insights the decomposition may have been originally achieved by completely different means.
- *State-transition systems* and other forms of sequential or parallel decomposition: These decomposition methods do not consider the static structure of a system but immediately consider the properties of its dynamic execution.
- *Object-oriented decomposition*: The system is considered as a (dynamically varying) collection of cooperating objects. Objects are modules with access functions which can work independently and parallel to each other (*active objects*). Calling a function of an object is often referred to as sending the object a message. Each object is autonomous in deciding how to answer a message; this is the main difference to composing a system from modules: in the latter case the sender of the message has to predict the change of the overall system state caused by the message.
- There are a number of very specific decomposition methods underlying the functional, the logical programming paradigm and various paradigms of artificial intelligence. Although they are used in large specialised systems they cannot be considered on equal footing in general software construction with the foregoing methods.

Many of these decomposition methods are also known and used in other areas of engineering. As a general property we may note that design methods in software engineering do not make assumptions about the physical or logical representation of the modules, objects or states involved and their components. E.g., in a state transition system it does not matter whether the components of the state and the state transitions are given by mechanical components and their interactions or by electronic elements. We do, however, assume that all interactions occur in discrete time. Continuous processes are simulated by discrete processes with small step sizes.

SOFTWARE ARCHITECTURE AND ARCHITECTURAL STYLES

In mechanical engineering, a design may lead to a variety of products depending on selections made during production, e.g. a car of a certain make may have different motors, automatic or mechanical gear, and various extras. The possible selections are mostly prescribed by the design and lead to a *family of products*. In software engineering, there is no separate production process; variations are achieved by starting from a common design, the *architecture* of the system, and then allowing for modifications by specialized modules, objects or subsystems during later design phases.

In software specifications, it is beneficial to separate architectural structure of software systems from its application-specific functionality. This separation of concerns improves the reuse of both architectures and application-specific components: components can be plugged into different architectures and architectures can be parametrized by different application components. The separation leads to scalable systems, since the architecture can be adapted easily to the environment while the application functionality stays the same. It also diminishes construction costs since reusing application-specific components is facilitated.

Architectures of software systems can be classified in styles [Shaw-96]. A style fixes the coarse-grain structure of a system, i.e. how it is composed from subcomponents, as well as a basic communication structure between them. In a style, composition is hierarchically organized. Communication structures are abstracted by *connectors* that link *ports* of subcomponents. Connectors appear in different types; the used connector type forms the architectural style. A typical style is the *pipe-filter style*, in which software components exchange data via *pipeline* connectors. Components *filter* the data from an input pipeline and put it out on an output pipeline again. This style resembles factory assembly lines; data flows through a set of components.

Architectural styles facilitate documentation, testing, and maintainance. A style describes and documents the coarse-grain structure of a system abstractly, often in graphical form. This information helps testers, maintainers, and end users to understand the system. Tests of the architecture can be performed with dummy components so that parts of the system can be validated before the complete system is finished. Due to improved documentation, maintenance is facilitated since the time to analyze a system is decreased. Finally, end users are enabled to reason about the system on an abstract level.

OBJECT-ORIENTED FRAMEWORKS AND PRODUCT FAMILIES

Besides an architectural style, a software product may have a more application-oriented architecture that can be reused in another product of a family. Such a family-oriented architecture can exist in form of architectural code which is not executable by itself; an executable system is obtained only by addition of certain variant-specific components. In object-oriented systems, such architectural code is called a *framework;* users have to complete it with application classes to build a final product. Hence, a framework is a template for software product families.

The required additions may occur at design time or by dynamically linking the new components with the already running base system. In either case, certain modifications may be necessary for preparing the system for the intended extensions. A systematics for such modification steps is currently developed under the name design patterns for object-oriented systems, cf. [*Gamm-94*].

An analogue to frameworks are skeletons of concrete in houses; in order to arrive at a complete house, this skeleton has to be fleshed with walls, windows, and doors.

DESIGN PATTERNS

Design patterns describe standard solution schemes to typical design problems. They have been invented in housing architecture where they describe typical situations in houses [Alexander 77]. For instance, a *window seat* is an architectural pattern that describes seats in window corners: why people like them, which variants exist, which rules an architect should follow when he designs them, which consequences for architecture of the house result. Most importantly, patterns give the architectural solution a name so that it can be referenced by people, used for education, and documentation.

In computer science, design patterns penetrate the field of software engineering [*Gamm-94*]. Patterns describe structure, interaction and cooperation of components, and variants. Patterns specify micro-architecture; they do not impose a uniform style for the entire software system. It turns out that naming typical design situations improves communication in teams, education of students, and documentation of software systems. Additionally, patterns provide guidelines along which code can be implemented easily, adapting the design pattern to the specific application.

As an example consider the so-called client-server pattern: the server is a software system, usually comprising a data base system, which provides for a number of services. These services are function calls, e.g. initiating a data base transaction, which may be issued by clients, i.e. other software systems. The server is prepared to connect to an arbitrary number of clients; clients may search the network for finding servers providing required services. The links may be established in an ad-hoc fashion, by using proprietory mechanisms of Microsoft or other companies, or by using the industry-standard CORBA-interfaces.

Design patterns can further be classified into structural and behavioral patterns. Structural patterns describe architectural situations and interconnections among components. Behavioral patterns characterize collaboration schemes, protocols, or coordination aspects. In car construction, an example of structural patterns are physical coupling elements (cable junctions, screws, rivets). A behavioral pattern could be an engine type (2-stroke, 4-stroke).

SPECIFICATIONS AND LIBRARIES

Components of software systems may be specified using a formal language. Such a specification also describes the realization (or implementation) of the component. There exists a multitude of higher-level specification methods which describe the interfaces of components and make assertions about their input-output behaviour; [Leeu-90] and [Rech-97] describe some of them.

The most elementary one is the algebraic specification technique. It describes data structures but not their implementation as abstract data types. We illustrate it by the commonly used example of a stack which may contain values of an arbitrary type T:

STACK(T) is

    createStack: $\rightarrow$ *Stack(T)*
    push:*Stack(T)* × *T* $\rightarrow$ *Stack(T)*
    pop: *Stack(T)* $\rightarrow$ *Stack(T)*
    top: *Stack(T)* $\rightarrow$ *T*
    empty: *Stack(T)* $\rightarrow$ *Boolean*

with

    pop(push(k,t)) = *k*
    top(push(k,t)) = *t*
    empty(createStack) = *true*
    empty(push(k,t)) = *false*

end

The first part of this specification lists the signatures of the functions at the interface, i.e. their parameter and result types; the first function *createStack* has no parameters. The second part lists the axioms which these functions must obey. E.g., the rule "*pop(push(k,t))* = *k*" requires that a stack *k* is back in its former state when we first push an item *t* on it and then remove it again by help of the *pop*-function. Certain combinations, e.g. *pop(createStack)* are not covered by the axioms. This indicates error-situations and a complete specification must also deal with such problems. The specification allows for an unlimited number of *push*-operations; the stack is unbounded and cannot be implemented in full generality in a memory of limited size.

As in this example it is characteristic that specifications depend on the types T of other components which may be specified in the same manner. This kind of parametrization is called generic parametrization in contrast to the kinds of parametrizations caused by inheritance in object-oriented designs.

Specifications of this kind together with their implementations may be combined into libraries such as the standard template library for the programming language C++ [Muss-96]. The aforementioned frameworks are subsets of such libraries together with a prescription how to generate systems from them. From a systematic point of view, the present state of library design leaves many questions unanswered:

- The number of possible abstract data types and their signatures is unlimited. There are no criteria of what constitutes a consistent and primitive subset. This question has no answer

from the theoretical point of view; but also practical answers which then must make additional assumptions based on experience are not within reach.
- The obvious requirement that all specified components must be easily combinable is difficult to achieve. Each newly added component may invalidate this consistency condition. Rules for avoiding such clashes are present research topics.
- The naming conventions required for retrieving a component with required properties from a library are not universally agreed. Component retrieval is thus a difficult job especially for the uninitiated.
- Besides its functional properties a component must fulfill a number of non-functional requirements concerning its reliability, efficiency, memory consumption, etc. when it is selected for a certain purpose. Present libraries do usually not provide components with all the required properties in a certain application nor is it possible to easily adapt the given components to the given purpose. This limits the usefulness of such libraries.

COMPARISON

For all of these construction principles, astonishing correspondences can be found in other scientific disciplines. This hints at the point that major principles of design are interdisciplinary and general. The next table summarizes some correspondences (examples are given in brackets).

	Software engineering	Car construction	Architecture
Decomposition		is everywhere	
Architectural styles	Software architecture (pipe-filter)	Car style (sport, family, business)	Architectural style (gothic, baroque)
Frameworks	Object-or. frameworks (AWT)	Car series (Mercedes A-series)	Building series (prefabricated house)
Design patterns - structural - behavioral	- Software structure - Protocols	- Elements (junctions) - Engine types (2-stroke, 4-stroke)	- Standard connections - Heating methods (wood, oil)
Specifications	(Abstract data types)	Plans	Plans and maps

Decomposition, architectural styles, frameworks, design patterns, and specifications are general design principles that could form the basic elements of a universal construction theory for the engineering disciplines. Common principles could carry same names in each discipline in order to facilitate the interdisciplinary communication. Common styles and design patterns could be assembled in standardized catalogues so that they form a common terminology. Such standardization would provide excellent means for documentation of systems and education of design knowledge.

Finally, computer science deals with design of abstract objects, i.e. models of objects in the real world. Since laws for abstract objects also hold for concrete objects - otherwise the modelling is not correct - design principles which have been applied successfully in informatics, could probably also be applied in other disciplines.

CONCLUSIONS

Despite striking similarities, there are huge differences between software and mechanical system construction, based on the absence of the (re)production problem, the base in logical instead of physical laws, the absence of tear and wear, the lack of stable construction principles due to rapid change of technology and the lack of an established methodology for measuring properties of systems, their components and their construction process.

On the other hand, every discrete system may be simulated by software and can be viewed as an implementation of a system originally specified as software. Therefore all the construction principles for technical systems can be viewed also as design principles for software systems. Additionally, the behaviour of software follows logical laws for abstract objects. Since concrete objects are instances of abstract objects, the design principles of decomposition, architectural styles, frameworks, design patterns, and specifications form the basic elements of a universal design theory which is useful also for disciplines constructing concrete objects of the real world.

REFERENCES

Alex-77	Alexander, C.; Ishakawa, S.; Silverstein, M.: A Pattern Language. Oxford University Press, New York, 1977.
Endr-75	Endres, A.: An Analysis of Errors and Their Causes in System Programs. SIGPLAN Notices 10(1975), No. 6. 327--336.
Gamm-94	Gamma, E.; Helm, R.; Johnson, R.; Vlissides, J.: Design Patterns: Elements of Reusable Software Components. Addison-Wesley, 1994.
Goos-94	Goos, G.: Programmiertechnik zwischen Wissenschaft und industrieller Praxis. Informatik-Spektrum 17(1994), 11-20.
Klir-69	Klir, G. J.: An Approach to General Systems Theory. Van Nostrand Reinhold Company, New York 1969
Klir-91	Klir, G. J.: Facets of Systems Science. IFSR International Series on Systems Science and Engineering, vol. 7. Plenum Press, New York and London 1991.
Leeu-90	van Leeuwen, J. (ed.): Formal Models and Semantics. Vol. B of Handbook of Theoretical Computer Science. Elsevier, 1990.
Muss-96	Musser, D. R.; Saini, A.: C++ Programming with the Standard Template Library. Addison-Wesley, 1996
Rech-97	Rechenberg, P.; Pomberger, G. (eds.): Handbuch der Informatik. Hanser-Verlag, München, 1997.
Ropo-75	Ropohl, G.: Systemtechnik - Grundlagen und Anwendung. Hanser-Verlag, München, 1975.
Shaw-96	Shaw, M.; Garlan, D.: Software Architecture. Prentice-Hall, 1996.
VDI 2221	VDI-Richtlinie 2221: Methodik zum Entwickeln und Konstruieren technischer Systeme und Produkte. VDI, Mai 1993.
VDI 2222	VDI-Richtlinie 2222 Blatt 1: Methodisches Entwickeln von Lösungsprinzipien. VDI, Juni 1997.

Discussion

Question - K. Ehrlenspiel

I think it was interesting to see what is the structure of your software-objects, but you have not described or very poor delivered a description, what is the structure of the software-making process. You had just said: problem definition, decomposition and composition. And I think it is a parallelism to the design-methodolgy in the sixties. We also started in Germany with design methodolgy to look for the structure of design-objects, of design-elements and now since ten years we are making empirical research, what designers do during design.

Answer - U. Aßmann

Well, there is a branch of software engineering, called software-process-management where computer scientists look at process-management, but we did not present it here (due to the lack of time). So if you are interested, we should talk. We tried to concentrate on design itself, and how you describe the design process.

Question - J. Gero

Let me ask what may become my standard question: How do you create new patterns?

Answer - U. Aßmann

Well, it happens all the time. One year ago, I started to learn about design patterns more in detail and last week I looked again on the web; this time I found about three times the material I found last year. So, defining patterns happens all the time. There are new books coming out with patterns; patterns are grouped in pattern sets, (called pattern languages) and you can define pattern languages for your application domains specifically. For example, telecommunication has other patterns than car construction or CAD-design. So there is no final set of patterns. There may be some basic ones, which you can utilize everywhere, but there is no limit.

Question - J. Gero

So this is a theory of the use of existing patterns going to the web to get new ones.

Answer - U. Aßmann

Yes, you want to describe things which are already known and you want to formalize them in a standard form.

Question - R. Mackay

I find it difficult to see the convergence which I would expect to be taking place between the different disciplines, since most complex products today consist of traditional engineering components which are designed in the way which we had described this morning and a large part of the functionality is performed by information technology components which are developed in an entirely different way. And at present, I can see no indication of convergence between these two processes and I find this quite alarming because in addition to the embedded information technology in complex products, the second dimension of this whole discussion, I think should be related to the way in which information technology tools are designed to support the design process. Perhaps you can comment on both aspects of that very briefly.

Answer - U. Aßmann

Computer science lacks this convergence also. I think, computer sciencists are a bit on an island and they should read other material like these guide-lines I presented. Doing this, you discover a lot of things you can utilize for yourself and actually that happened with design patterns which were discovered in architecture. We should come together and I appriciate such a workshop like this. We should also teach students together. If we had a design-theory it could be used for first semester courses everywhere, common to all disciplines, mathematics, informatics, engineering. And I think by teaching, you get the common understanding of design to the people.

Efficient Design Methods in Microelectronic/Mechatronic Systems

Manfred Glesner, Jürgen Becker

Institute of Microelectronic Systems

Darmstadt University of Technology

Karlstr. 15, D-64283 Darmstadt

Keywords: design method, specification, high-level synthesis, hardware/software co-design, Systems on Silicon (SOC), IP-based design, standardisation, reusability, Set-Top-Box (STB), mechatronic systems, microsystem technology

ABSTRACT

The fast increasing complexity of future microelectronic and mechatronic systems requires efficient design methods. Based on the analysis of the present situation and on future views, requirements and arising problems will be analyzed, e.g. for efficient "Systems-on-Chip" (SOCs), and potential methodical solution approaches will be discussed.

HISTORY AND TRENDS

50 years ago the silicon era started with the invention of the transistor. The major breakthrough was made possible with the discovery, that the current across a semiconductor can be controlled by the addition of impurities with special properties (dopants). Affiliating differently doped semiconductors, it was possible to develop the transistor, who partially replaced the vacuum tubes, but most of all made new digital technologies possible. The development of more reliable and smaller transistors had a tremendous dynamic, and soon it was possible to integrate transistors on a silicon die. The number of transistors to be integrated increased exponentially and does so up to now. So the number of integrated transistors doubles every 18 month. Today there are 7 million integrated transistors in the Motorola Power PC, 7.5 million in the Intel Pentium II and even 10 million in the 64-bit Alpha microprocessor made by Digital Equipment. Recently NEC announced the 4 gigabit DRAM chip that will be available by the end of 2000. Often the very expensive development of memory is thought of as the motor of the microelectronics industry with its huge market potential. However, the market share of application specific integrated circuits (ASICs) [*Smit-97*] and processors (ASIPs) [*Goos-95*] has grown bigger than that of memory recently. ASICs and ASIPs are most often used in telecommunication, mechatronic and products for the consumer market. Some examples for those applications are HDTV, ISDN, digital mobile communication, automotive electronics etc. Such products are burdened with short live cycles and an immensely growing

complexity. In the last 10 years this design complexity has grown two orders of magnitude, while the live cycle has been shorted form 5 years to 18 month. This implies that new design methodologies like "rapid prototyping", "hardware/software co-design" [*Mich-95, Buch-94*] "IP-based design" [*Ajlu-98, Zori-97, Tuck-97*], and underlying technologies like "field programmable gate arrays" (FPGAs) [*Vill-97, Hart-96, Luk-97*], "mechatronic and microsystem technologies" (see section DESIGN OF MECHATRONIC AND MICROMECHANIC SYSTEMS), as well as "system-level integration strategies" [*Ajlu-98, Zori-97, Tuck-97*] dominate more and more the design of new products.

One important role plays the pressure described with "time to market", because delays in the product introduction mean lost profit and higher development budgets. One example for this trend is the development of the SPARC processor: the first version was implemented with a gate array technology in order to come in to the market very quickly, while the second version was realised with a more cost efficient technology based on standard cells [*Bohm-91*].

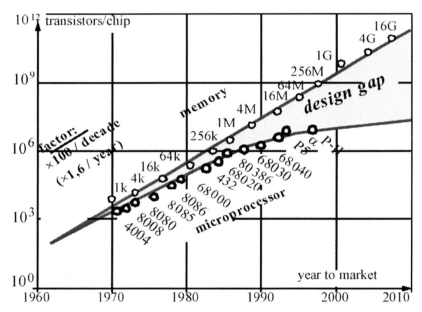

Figure 1 *Transistors / Chip growing rates: processors vs. memories (Gordon/Moore curve)*

A further problem besides the short live cycle is the tremendous growths of the complexity. The time to develop a 25K gate design took in 1991 8.1 month, whereas in 1993 the development time of a 50K gate design was only 5.5 month. One possibility to compensate the complexity growth is to increase the number of designers, but this cannot counterbalance the growth completely. The needed productivity enhancement can only be accomplished with the massive use of CAD tools and new efficient design methodologies. Figure 1 shows the Gordon/Moore curve illustrating the design gap (in transistors / chip) between regular structured memories and complex microprocessors. The lack of efficient design methods in today's commercial design tools resulting in architectural cleverness is responsible for such a gap in microprocessor design complexity.

Figure 2 shows the development of processor design from another perspective [*Henn-96*] . Starting from the first years of the microprocessor (CISC architecture) the average growth rate of performance was 35%. The performance is measured here by SPECint rating which uses the execution speed of a standard suite of C-programs (e.g. small program fragments, medium and large size applications). With the introduction of new architectural features in the 80's the pace of performance improvement increased to 58%. This was mainly driven by architectural innovations. The first performance gain did come from the introduction of Reduced Instruction Set Computer (RISC; e.g. MIPS R2000) and pipelines. In the last years the development was accelerated by features like superscalar (more than one execution unit), branch prediction, instruction reordering, out-of-order execution, and very long instruction word processors. These architectural improvements added to technology development a high performance growth, which was not caused by additional transistor usage.

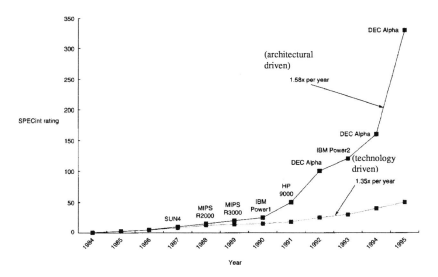

Figure 2   Performance/processor growing rates ("SPECint" benchmark): architectural-driven vs. Technolgy-driven solutions

Thus, there still exists an enourmous potential for efficient design methods explaining both:
- architectural possibilities, as well as
- technology borders.

DESIGN METHODS

ELECTRONIC DESIGN TECHNIQUES AND TRANSFORMATIONS

Electronic design is carried out in many ways by various designers for a wide range of purposes. Therefore, it is impossible to describe one methodology that applies in all cases. Instead Figure 3 shows the different phases of a general electronic design method, which is adopted with variations by several designers. Such a top-down design method divides the design process into phases: specification, functional design, logic design, circuit design, and physical design., as described in detail in [*Prea-88*].

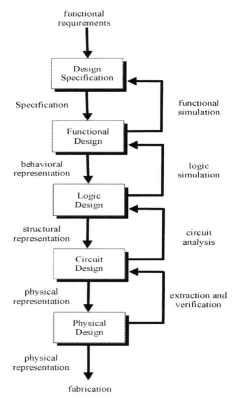

Figure 3   Phases of general electronic system design for a top-down design methodology

Each design phase is characterized by synthesis, analysis and verification steps. Synthesis derives a new or improved representation based on the representation in the previous phase. At lower levels of abstraction, synthesis is typically automatic, whereas at higher levels it is a topic of intensive research. Analysis follows synthesis in each design phase and generally has two goals: the given design has to be emulated against its requirements (e.g. size, performance, power consumption etc.), and the design has to be analyzed for behavioural, structural, and physical correctness and completeness. Verification is the final step within each design phase and proves that the synthesized representation is equivalent under all conditions of interest to another representation.

The different design phases of such a high-level top-down electronic design process and corresponding transformations are discussed in the next section.

HIGH-LEVEL DESIGN TECHNIQUES AND TRANSFORMATIONS

Silicon Compilation is a first approach for rising the level of abstraction when specifying integrated circuits. Silicon Compilation transfers exact specified structural circuit descriptions, specified at a high level of abstraction (e.g. on register-transfer-level with complex functional operators as multiplication, addition etc.), to mask layout data. For this conversion parameterizable macrocell generators are used, which split operators into bit slices or decompose them into functional subblocks. Moreover, they generate the layout of the desired blocks (e.g. RAM, ROM, multiplier) directly out of the block parameters (wordlength, address range etc.) [*Gajs-88*]. Basic building blocks, as for example gates in random-logic circuits, are included from libraries and are directly integrated into the layout.

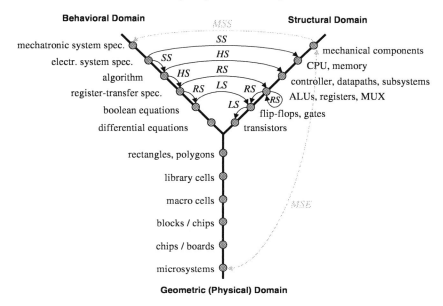

Figure 4   Synthesis steps in the extended Y-chart

Synthesis, as mentioned in section ELECTRONIC DESIGN TECHNIQUES AND TRANSFORMATIONS, is a second approach for obtaining an implementation of an integrated device out of a high level circuit specification as depicted in Figure 4, which shows an extended Y-chart with additionally integrated mechatronic and microsystem technology components.

Synthesis is a combined process of domain transformation (from behavioural to structural domain), and of refinement including several optimizing steps within the different levels of abstraction. Thus, synthesis can be performed on different abstraction levels. Dependent on the viewed level, it is called:
- system synthesis (**SS**),
- high-level synthesis (**HS**),
- register-transfer synthesis (**RS**), and
- logic synthesis (**LS**), as well as
- microsystem synthesis (**MSS**), if mechanical system components are used.

The integration of the different synthesis types mentioned above is illustrated in Figure 4.

The *logic synthesis* is performed within the lowest level of abstraction, where boolean equations are mapped onto single gates of a technology-dependent library, or onto reconfigurable hardware structures (PLA, FPGA etc.). This includes also optimizations of area and execution times, whereas only combinational circuit parts are considered. Thus, the storing elements are not viewed during the optimization on this abstraction level. In general, the logic synthesis consists of two stages:
- a library-independent logic optimization step, where the number of product terms and literals are minimized, as well as of a
- technology-dependent technology mapping step.

The *register-transfer synthesis* is located one abstraction level above the logic synthesis, and is often called sequential synthesis. This active area of research allows optimizations of the timing behaviour, which becomes more and more important, compared to chip-area optimizations. The register-transfer synthesis can be divided into two categories:
- The classical optimization of finite-state machines (FSMs), where decomposition in smaller FSMs and optimized state coding play an important role. This synthesis category includes mainly controller applications with control-flow oriented specifications.
- Resynthesis methods for data-flow oriented applications, where retiming and operator-select techniques are combined in order to derive optimized (pipelined) circuits.

One abstraction level above register-transfer synthesis is *high-level synthesis* located, where an algorithmic system specification is mapped onto data paths, controllers and subsystems.

The so-called *system synthesis* is above the high-level synthesis, whereas the it differs in the specification granularity of timing behaviour. Time is modelled within high-level synthesis on the level of clock cycles, and within system synthesis on the task level in form of relations, e.g. task A *after* task B, or task A *parallel to* task B etc..

For more details about high-level and system synthesis see section HIGH-LEVEL SYNTHESIS, and for a detailed discussion of all above mentioned synthesis types see [*Wehn-95*].

Due to the high complexity of today's microelectronic circuits and the steady progress in technological developments, nobody can design larger systems per hand in an acceptable amount of time. The demand of new complex integrated circuits as soon as possible, and the increased relevance of microelectronic in daily life requires the application of nearly complete computer-aided design environments. Therefore, developed design methods allow a fast and safe design of complex circuits and systems. Moreover, these methods realize ASIC design on a large scale.

The design flow of digital circuits, which represent the biggest part of today's integrated circuits, is illustrated in Figure 5: starting point for a design can be on one hand a hardware description language (HDL, e.g. VHDL or Verilog), and on the other hand a symbolic (graphical) input of available larger blocks, e.g. IP-based (Intellectual Property) cells (see also section **IP-based design methods**). Hardware description language specifications can either represent the behavioural level, or can contain structural information. Such HDL-models are transformed by logic synthesis steps into a netlist of single logical cells. In contrast, graphical input of functional blocks (schematic entry) realizes this synthesis step per hand.

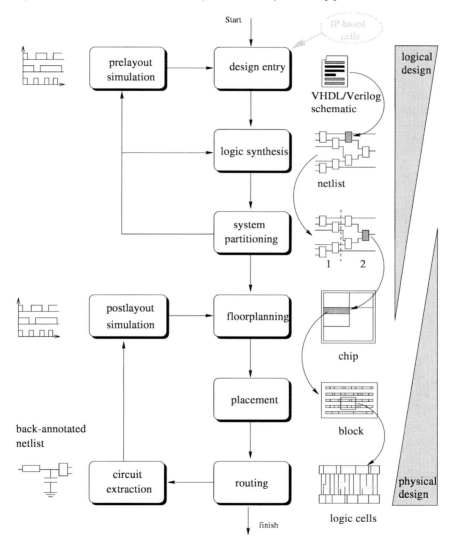

Figure 5   ASIC design flow incl. optional IP-based design cells

After a possible system partitioning and the placement of all functional blocks on the chip, a (partly) simulation of the complete system can be done. During the placement and routing phase of each single functional block parasitic capacitances and resistors arise, which have to be extracted from the layout. Thereafter, another simulation step has to be done, in order to verify the design by considering the extracted values of these factors. The final point in this design flow is the layout, which contains all necessary information for fabricating the circuit.

For reducing the general design costs and development times, existing system-level modules has to be used more and more. This process is called "reuseability" or "design-reuse" and results in new requirements for modern design methods, e.g. in designing complete "systems-on-a-chip" (SOCs). The user acceptance of this strategy depends strongly on the documentation of the reuseable modules. In the moment hardware description languages as VHDL or Verilog are used successfully here for specification and documentation. On the other side, the component developer has the authorized interest to protect his design work. In the IP-based development of microelectronic systems it is attempted to overcome this discrepancy (see section **IP-based design methods**).

STANDARDS IN TODAY'S DESIGN METHODS

Since an ASIC designer spends an increasing amount of time forcing different tools to communicate, standards have to be created for exchanging information between tools (e.g. data formats, tools etc.). An overview on the current most important public domain and proprietry data formats is given in Table 1, according to [*Zori-97*].

Public domain	Proprietary (vendor)
C, VHDL, Verilog	Synthesizable subsets
ASCII	EDA tool scripts, Verilog Change Dump (Cadence)
	Waveform Graphical Language (TSSI)
EDIF, SPICE	Design Exchange Format (Cadence) Standard Parasitic Extended Format (Cadence)
	Interpolated table lookup cell-level timing (Mentor)
	Non-linear Delay Model (Synopsys)
	Table Lookup Format (Cadence)
	Motive Modeling Format (Viewlogic)
SPICE	Layout Exchange Format (Cadence)
	SPEF, GDSII (Cadence),
	CIF

Table 1   Public domain and proprietary data formats currently in use

For the description of a layout several representation forms (like programming languages) are available. The mainly used representations are the CIF (Caltech Intermediate Form), which is a readable polygon description of the masks, and the binary GDSII-format, which also includes topological facilities which makes it a bit more powerful. The EDIF (Electronic Data Interchange Format) never attained this importance in the field of layout representation (and the latest releases do not support the layout representation anymore), but it is the commonly used netlist representation for automatically synthesized gate level descriptions, which can be used for further design steps in most CAD-tools.

For the important area of board and chip testing the following standardization efforts can be mentioned. In 1985 a group of European manufacturers formed the **Joint European Test Action Group (JETAG)** to study board testing. With the addition of North American Companies, JETAG became the **Joint Test Action Group (JTAG)** in 1986. The JTAG 2.0 test standard formed the basis of the **IEEE Standard 1149.1 Test Port and Boundary-Scan Architecture**, approved in February 1990 and also approved as a standard by the American National Standards Institute (ANSI) in August 1990. This scan chain on board level defined by the IEEE Standard 1149.1 can be also transferred on chip level for accessing easily internal signal values. Such a **design for testability (DFT)** is called **scan path** design and is often integrated automatically in today's design tools. This trend to include more test hardware on an ASIC is continued by another set of structured-test techniques for combinational and sequential logic, memories, multipliers, and other embedded logic blocks. The principle is called **Built-in Self-test (BIST)** and is to generate test vectors, apply them to the **circuit under test (CUT)** or **device under test (DUT),** and then check the response (**signature analysis**). For more details about test hardware and test pattern generation see [*Smit-97*].

For the new area of *IP-based design* (or core-based design, see section **IP-based design methods**) the IEEE Computer Society Test Technology Committee initiated a dedicated **Technical Activity Committee (TAC)** to look into the field of testing core-based chips. This TAC first met in conjunction with the international Test Conference in October 1995 and began to determine the need for standardization. The interest group identified a number of common industry standardization needs and received IEEE approval in June 1997 to become a standardization group (**IEEE P1 500 working group**), which should develop a core test description language, a core test control mechanism, and a core data access mechanism. Two proposals are now in progress:
- a **core test desription language**: would help to construct the peripheral access features (multiplexers, scan chains). It would also define the access path from core peripheries to on-chip or external test resources, such as BIST controllers.
- a **scalable architecture** to meet the core test control mechanism (the **Core Test Access Port, CTAP)** and the core data access mechanism. The CTAP provides a standard communication interface for all internal cores, at a minimum, and requires a minimum number of internal busing signals and IC pins. Used by itself, the CTAP supports internal and external core testing based on known test methods (scan path, boundary scan, BIST).

The standardization study group coordinates ist efforts closely with the **Virtual Socket Interface Alliance (VSIA)**. VISIA is working in six key areas to define the interface standards that will allow the plug-and-play of virtual sockets, which include:
- **test standards** and design guidelines for test structures, test methods for cores, and test infrastructure,
- standards-based solutions for **IP protection**,
- **core standards** that enable the implementation and verification of core-based systems,
- standards and guidelines for integration of **mixed-signal cores** in largely digital systems,
- **on-chip bus** specifications for design, integration, and testing of multiple functional blocks on a single chip, and
- **standards** to enable the core user to **evaluate and select various cores** within the context of the overal sytem-chip specification.

Without such standard interfaces, designers will find it very difficult to connect various IP cores within the same chip. While *VISIA* is trying to establish standards in this area, many EDA vendors are developing their own methodologies and tools for implementing SOC designs using IP from various sources.

Another industry trade association chartered with meeting the needs of companies and organizations, which develop and sell intellectual property, is the **RAPID initiative (Reusable Application-Specific Intellectual Property Developer).** *RAPID's* function is to promote and accelerate the acceptance and use of third-party intellectual property products within the electronics industry, incl. forcing the integration of EDA providers, semiconductor companies, and industry standard organizations. For more details see [*Ajlu-98, Zori-97, Tuck-97*].

HETEROGENOUS SYSTEM SYNTHESIS

Different methodical approaches for solving the complexity problem are analyzed: on the specification level exist different conceptual system models, which can be categorized in mainly 4 categories:

Process-oriented models: have their strength in describing parallel operations and their cooperation. Similar to control oriented models they are well suited for the description of control dominated applications, but offer a much more powerful concept for specifications, that is capable of dealing with very complex applications. On the other hand it is this complexity, that makes it hard or even impossible to perform automatic optimizations. In form of Petri-nets process oriented models have been well researched.

- **Process-oriented models:** have their strength in describing parallel operations and their cooperation. Similar to control oriented models they are well suited for the description of control dominated applications, but offer a much more powerful concept for specifications, that is capable of dealing with very complex applications. On the other hand it is this complexity, that makes it hard or even impossible to perform automatic optimizations. In form of Petri-nets process oriented models have been well researched.
- **Data-flow oriented models:** are well suited for the processing of data streams, when all data is processed using the same scheme. Therefore they are mainly used in the field of signal processing applications. Typical applications are(digital) filters or transformations. For this area there are several optimized Tools, e.g. SWP or COSSAP, that support the user by supplying an highly automated development environment.
- **Control-flow oriented models:** Control flow oriented models use finit state machines (FSM's) for specifying the reaction of a system for certain input data. Since finite state machines (FSMs) can be implemented and optimized very efficiently, they are well suited for specifying control dominated applications. This is the reason why FSM-based models are used in several tools, that are targetted for creating specifications, e.g. SpeCharts [*Gajs-93*] or StateMate.
- **Object-oriented models:** try to improve the handling of complexity by subdividing the whole system into smaller subsystems (objects). In contrast to other specification methods, object orientation tries to combine attributes with methods, that can be used to modify the attributes, into one class, that describes the objects [*Booc-91*]. This concept has already a wide use in the development of software and it is supported by several programming languages as for example C++, Smalltalk or Eiffel. Due to this success in the area of software development, there are now several approaches of transferring this concept to other areas as for example the development of hardware (e.g. Objective VHDL).

Shorter development times can be realized by novel methods of design automatation, as well as by a (partly) parallelization of the design flow. Here, techniques and experiences from different areas of microelectronics and mechanics can be used:

- **High-level synthesis / hardware-software co-design:** These two promising research areas are known since the late 70's resp. the early 90's. They discuss the automatic design

of systems specified on higher abstraction levels, as for example by C, VHDL, Verilog, C++ etc. The main difference lies in the timing model: high-level synthesis uses a cycle-based model, whereas hw/sw co-design uses coarser time granularities on instruction and task level, and includes also software parts into the final implementations.
- **Mechatronic systems / microsystem technology**: Here today's mixed electronic / mechanical systems (electronic and mechanical components are placed at different locations) should be integrated to autarkic systems with integrated information processing. In this case the different design flows for mechanical und electronic components have also to be integrated and parallelized.
- **Rapid-Prototyping**: Early detailed information about system behaviour avoids cost-intensive redesign-cycles, and allows overlapping working phases (e.g. software development). This would be impossible without a prototype, which is concerning functionality and timing nearly identical with the final system to be implemented.

The next subsection discusses above mentioned methods and experiences in detail.

HIGH-LEVEL SYNTHESIS

The high-level synthesis transforms a behavioral description at the algorithmic level of a digital system, e.g. a behavioral description in VHDL, in a structural description on register-transfer level. The final dichotomic structure consists of two main components (see Figure 6):
- an arithmetic data path, and
- a controller for sequencing the data path operations.

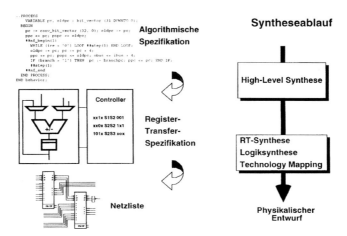

Figure 6   *Principles of transforming an algorithmic input description in high-level synthesis*

The algorithmic input description includes arithmetic and logic operations applied to variables, and control structures such as branches, loops, and procedure calls. Mapping of variables to registers or signals, operators to functional units, and control structures to actions of a controller is, in principle, a straightforward process. But there are many different ways to solve this task. Basically, three different subtasks can be distinguished:

- appropriate functional unit types have to be selected (**resource allocation**),
- operations have to be assigned to time slots (**scheduling**), and
- operations have to be mapped to single functional units (**resource assignment**).

Obviously, there is a time/area trade-off: allocating more resources allows more parallel execution of operations, thus giving higher performance at higher hardware cost (chip area). – The feasible region (subject to constraint given by the user) of the space spanned by time area is often called design space. To find an implementation corresponding to the optimum point within this space is one task of scheduling, resource allocation, and resource assignment.

For further details about efficient methods and architectural synthesis in high-level synthesis see [*Wehn-95, Mich-92*].

HARDWARE/SOFTWARE CO-DESIGN

The discipline of hw/sw co-design targets the design of heterogenous systems, which are specified in a high-level input description. Here, system partitioning in cooperating hardware and. software components is performed, whereas implemented hw/sw-systems consists typically of one or several ASICs and/or FPGAs, as well as programmable component, e.g. microprocessors or digital signal processors. The majority of digital systems are programmable, and thus consist of hardware and software components. The value of a system can be measured by some objectives that are specific to its application domain (e.g., performance, design and manufacturing cost, ease of programmability, etc.), and it depends on both hardware and software components. *Hardware/software co-design* means meeting system-level objectives by exploiting the synergism of hardware and software through their concurrent design. Since digital systems have different architectures and applications, there are several co-design problems of interest, which are associated with the classes of digital systems they arise from. Thus, the general criteria for characterizing these systems are (see Figure 7):

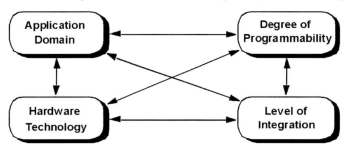

Figure 7   Distinguishing criteria of electronic systems

Application domain: In digital systems it can be distinguished between general-purpose computing systems, dedicated and control systems, and emulation/prototyping systems. General purpose computers support applications of different kinds, determined by the software developed or acquired by the user. In contrast dedicated computing and/or control systems, commonly called embedded systems, are conceived with specific target applications. Embedded system design has been growing steadily over the last few years, for example in the manufacturing industry (e.g., plant and robot control), in consumer products (e.g., intelligent home devices), in vehicles (e.g., control and maintenance of cars, planes, ships, as for example the fly by wire control system of an aircraft), in telecommunication applications, and others. Emulation and prototyping systems make a class of their own, usually based on structural programmable hardware technology by using hardware compilers, also called synthesis systems [*Mich-95*]. Such prototype systems are often used to validate a design concept.

Degree of programmability: There is a wide range in the *degree of programmability* of various digital systems, which can be programmed at the *application, instruction* or *hardware* levels. The most restrictive approach is the *application level*, where the system is running dedicated software programs that allow users to specify desired functionality options using a specialized language, e.g., setting navigation instructions in an automated steering controller of a ship. On the other side, the process of compiling application programs onto PCs or workstations results in the generation of instruction sequences for the architecture being considered, which are programmed at the *instruction level* in this case. *Hardware-level* programming means configuring the hardware (after manufacturing) in a desired way, e.g. *microprogramming*, e.g., determining the behaviour of control units by microprograms stored in memories.

Hardware technology: Digital systems rely on VLSI circuit technology, and may have components with different scale of integration (e.g., discrete and integrated component) and different fabrication technologies (e.g., bipolar and CMOS). The choice of *hardware technology* for the system components is critical, because it affects the overall performance and cost. *Re-programmable FPGAs* support multiple hardware reconfigurations on the field, i.e., after manufacturing, and open the door to novel system-level solutions, and to very interesting co-design problems. Despite the growing importance of FPGA technologies, only a small fraction of today's digital systems exploit these features of FPGA technologies, because of less performance and density of the programmable hardware parts, and FPGA chips are more expensive as compared to dedicated hardware components in standard non-programmable technologies when produced in high volumes. But the performance, density, flexibility, and arithmetic capabilities of today's FPGA devices incl. synthesis tools are increasing steadily [*Vill-97, Hart-96, Luk-97*].

Level of integration: The level of integration is a key factor in the system cost. It is usually convenient to reduce the nimber of parts of a system, by integrating as many functions as possible on a single chip *(SOC)*. Thus, chips have been manufactured that integrate digital processing, memory storage, analog functions and transducers.

Above four factors are interrelated, for example the level of integration may affect the degree of programmability of a system. Some applications require specific technologies and programmability features.

In general, system cost and performance trade-offs dictate a choice between synthesized hardware solutions or software prototypes. Cost-effective designs use often a mixture of hardware and software to accomplish their overall constraints, which is the goal of hardware/software co-design. The input specification in co-design may consist of a single or a collection of heterogenous specifications. Hardware/software co-design and its further development co-synthes is benefit from a systematic analysis of common design trade-offs, which create cost-effective systems. One way to accomplish this task would be to specify constraints on the cost and the performance of the resulting implementation, as illustrated in Figure 8. Then a systematic exploration of system's *design space* is necessary, driven by these *constraints*. The essential aspects of a typical synthesis approach for embedded systems is shown in Figure 9. A behavioural specification is captured into a system model that is partitioned for implementation into hardware and software. The partitioned model is then synthesized into interacting hardware and software components for a given target architecture.

For more details and examples about hardware/software co-design and exisisting development systems see [Mich-95, Beck-97].

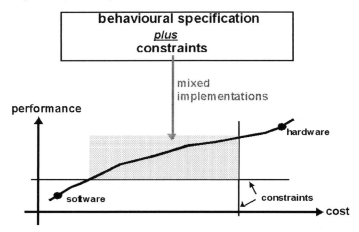

Figure 8   *Hardware/software co-design approach to system implementation*

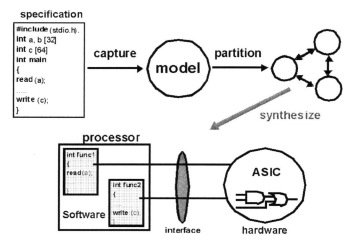

Figure 9   *Typical synthesis approach for embedded systems*

IP-based design methods

The exponentially growing number of transistors available on a single piece of silicon with the deep submicron technology poses both an opportunity and a challenge to system design teams. This current technology provides the ability to integrate the functions of many chips such as microprocessor, memories, I/O interfaces, etc. onto a single one. Therefore, the production of lower-cost chips with greater functionality than ever before have been made possible. On the flip side, the conventional ASIC design methodologies that worked on 100k gate designs are simply not sufficient to produce robust million gate chips in a reasonable amount of time. Therefore, new design strategies are needed for the System-on-a-Chip technology. A systematic *design-reuse* of proven IP-cores provides considerable promise for addressing this problem. The set-top box design is moving strongly in this direction in order to accomodate the new emerging emphasis: minimum design time, optimization at the system level, and cost-effective manufacturing. This paper shows how advances in IP-cores based System-on-a-Chip are spearheading the trend in the design of future set-top boxes (see section **Set-top boxes**).

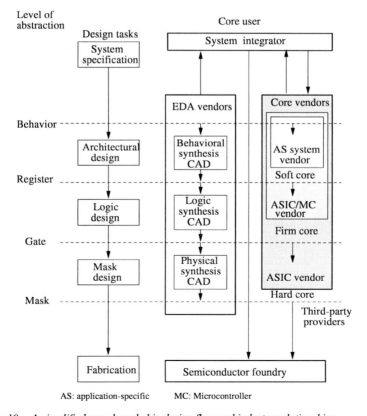

Figure 10   A simplified core-based chip design flow and industry relationships

Currently, semiconductor chip design falls into three distinct phases: behavioral synthesis, logic synthesis, and physical synthesis (see section **DESIGN METHODS**). System design has moved from a vertical process in which the design flowed through various phases in the same design house to a horizontal process in which different design houses may handle different phases. The IC design process may now move through multiple design houses and ASIC as well as electronic design automation (EDA) vendors before reaching the fabrication line (see Figure 10, according to [Zori-97]).

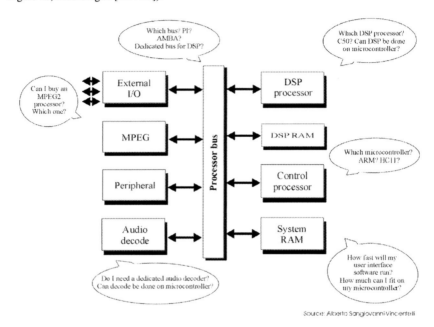

Figure 11   IP-based design of hw/sw system implementation

Based on functional and physical complexity levels, IPs currently available fall into three categories:
- **Hard Cores:** Hard cores are black boxes that have been fully implemented down to the mask-level data required to fabricate the block in silicon. They have technology-specific timing informations and a fixed physical layout that allows maximum optimization in terms of performance and density. Typically, they are targeted at providing dedicated functions, such as an Ethernet interface or an MPEG decoder. However, hard cores have the most limited vendor portability and greatest difficulty of reuse when moving to a new process technology.
- **Firm Cores:** Firm cores are technology-dependent synthesized gate-level netlist that is ready for placement and routing. They provide flexibility in the chip layout process because the core form is not fixed. A firm core's size, aspect ratio, and pin location can be changed to meet a customer's chip layout needs, and a floor planning guidelines assist the

chip designer in making tradeoffs. The technology-specific nature of a firm core makes it highly predictable in performance and area.
- **Soft Cores:** Soft cores consist of technology-independent synthesizable HDL (Hardware Description Language) descriptions. They have no physical information.

When building an IP-based hardware/software system, different decisions and trade-offs have to be viewed (see Figure 11). Alberto Sangiovanni-Vincentelli (Prof. at UC Berkeley, California) uses this diagram to illustrate such an IP-based design process of mixed hardware/software systems [*Tuck-97*].

With deep-submicron manufacturing technology moving towards 0.18-micron designs, semiconductor density reaches hundreds of millions of gates, making it possible to merge the elements of an entire system onto a single chip. Thus, an unprecedented challenge is created, because designer productivity has not kept pace. A recent sematech study noted that semiconductor densities are increasing at a 58% compound annual rate, while design productivity is advancing only 21% annually. Clearly, new approaches are required to the full advantage of multi-million-gate system chips. To bridge this gap, design methodologies have to be undergoing fundamental changes, e.g. turning to IP-cores that can be integrated into a device to provide the functionality needed for such system chips. The IP-core concept can save thousands of hours work and is spreading rapidly; in fact, a Dataquest estimate predicts that 77% of ASIC designs this year will contain some form of a core. An overview of core providers, users' experiences, and standardization efforts is available on the web (see Table 2, from [*Zori-97*].

Resource	Web URL
Core suppliers	
Hard core summary	http://www.isdmag.com/EEdesign/HardCoretables.html
Soft core summary	http://www.isdmag.com/EEdesign/SoftCoretables.html
DSP guides	http://www.bdti.com/library.html
Advanced Risc Machines (ARM)	http://www.arm.com
Altera MegaCores	http://www.altera.com/html/products/megacore.html
LogicVision	http://www.lvision.com
LSI Logic CoreWare	http://www.lsilogic.com/products/5_5g1.html
Mentor Graphics 3Soft	http://www.3soft.com
Synopsys 8051	http://www.synopsys.com/products/designware/8051_ds.html
Virtual Chips Synthesizable Cores	http://www.vchips.com/products.htm
Xilinx Core Generator	http://www.xilinx.com/products/logicore/cg_intro.htm
Core Standards Organizations	
IEEE Test Technology Tech. Commitee	http://www. computer.org/tab/tttc/index.html
IEEE P1500 working group	http://jesse.stanford.edu/coretest/
IEEE P1450 Test Interface Language group	http://stdsbbs.ieee.org/groups/1450/index.html
RAPID	http://www.rapid.org
VSIA	http://www.vsi.org
Information on cores	
Design and Reuse Inc.	http://www.design-reuse.com
Virtual Chip Design magazine	http://www.virtualchipdesign.com

Table 2   Web resources of core providers, users' experiences, and standardization efforts

Set-top boxes

"Set-top box" is a generalized term of peripheral devices that connect TVs to network-delivery system, e.g. DBS (Direct-broadcast satellite), CATV, xDSL, etc., and/or plays published media (e.g. DVD). Set-top box design is a dominant communication multimedia application. An earlier-generation set-top box architecture is a clutter of discrete ICs-such as a microcontroller, an MPEG-2 video decoder, an MPEG audio decoder, an MPEG-2 transport-layer demultiplexer, graphics hardware, a forward error correction (FEC), a channel demodulator, and peripherals. Typical peripherals in a set-top box include an NTSC/PAL encoder, an UART, a parallel port, an audio D/A, and a smart-card interface. This solution possesses serious limitations of upgradability, scalability and flexibility. For instance, the transport descrambler and graphics engine vary considerably among service providers. Due to MPEG-2 commitee's decision to leave certain requirements open, a relevant number of different transport-standard variations continue to emerge. Hence, some ICs within the discrete Set-Top box solution have to be redesigned to comply with each differing network or service provider requirement. The redesign of discrete ICs contributes even further to high costs and makes difficult the achievement of the time to market for the set-top box industry. Moreover, the discrete set-top box solution cannot efficiently handle future design requirement with an acceptable price/performance curve. It does not allows to integrate new applications quickly in response to marketplace demands.

To tackle next-generation set-top box designs successfully, system architects are taking advantage of advances in the System-on-a-Chip technology based IP-cores. IP-cores have emerged as a way to shorten time to market in the face of growing design complexity. Engineering teams are developing set-top box IP-cores and maintaining the industry-standard proven components in libraries that can be offered to system designers. If these proven components have a long life span, they must be continually updated to achieve an efficient core compatible with the technical progress of deep submicron tools and technologies, without losing any original functionality. This migration should be also done to meet the actual customers' expectations. By having set-top box core libraries available, designers turn their attention on integrating pre-designed blocks aiming at providing a single functional unit.

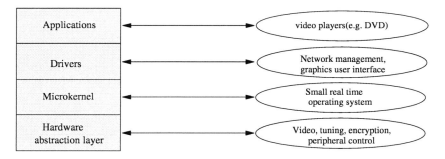

Figure 12   Software architecture of a set-top box

The development environment of a set-top box should be available with integrated software and hardware that have been quality-tested to help designers in speeding their products to market. The architecture of the test-proven set-top box software consists of four layers (see Figure 12). Powerful applications such as video on demand, pay-per-view, Web browsing, and electronic program guides should be supported. The real-time operating system should be provided with development tools for helping software engineers in quickly creating software for the supported applications. Other features include software examples (for controlling OSD functions, infrared remote controls, and program selection) and sample streams (video clips for use in testing and demonstrating applications). The provided software should include the necessary device drivers and boot code for the various integrated hardware components.

The hardware of a set-top box consists of the following subsystems (see Figure 13):
- Processing subsystem,
- Video subsystem,
- Audio subsystem,
- Peripheral control subsystem, and
- Network interface and upstream communication subsystem

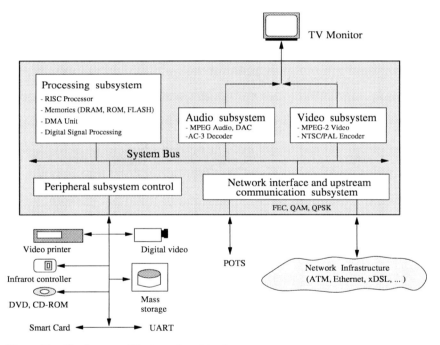

Figure 13   Hardware architecture of a set-top box

The processing subsystem includes host processors and memory units. Wide adoption has made the PowerPC processor an industry-standard platform upon which numerous third-party developers have created software tools. The components of the processing subsystem can be often found as hardcores. The video subsystem provides an NTSC/PAL encoder and typical MPEG-2 video decoder functions such as forward motion estimation, interlace functions, several frame sequence support, prediction modes, and motion compensation. The NTSC/PAL encoder converts MPEG-2 bit streams and supports NTSC/PAL displays. The audio subsystem includes an audio DAC and must support AC-3 dolby, MPEG audio layer I and II (two

channels), and linear PCM audio formats. The peripheral control subsystem consists of interface modules for standard controller units such UART, RS232, infrared unit, IEEE 1284 parallel port for connection to external PC or printer, and smart card interface. The network interface and upstream communication subsystem is becoming more and more prevalent and have to be designed according to the consumer's requirements (ATM, Ethernet or xDSL).
This subsystem performs the following functions :

- **Modulation/Demodulation**
  On the receiver side, the modulation carrier must be removed before recovering the digital transport streams. Different types of modulation are supported, quadrature amplitude modulation (QAM) for cable, quadrature phase shift keying (QPSK) for satellite, and vestigial side band modulation for over-the-air.
- **Forward Error Correction (FEC)**
  Errors introduced by the channel media (space or cable) are decoded and corrected. A soft decision Viterbi or a DVB standard Reed Solomon decoder can be used. The signal recovery from bursty noise sources will be improved through interleaving/deinterleaving.
- **DVB and DVD transport**
  Transport streams are reformatted to be MPEG-2 compliant.
- **Scrambler/Descrambler**
  The descrambler provides conditional access to information in the streams. For the upstream communication, the scrambler performs the opposite operation.
- **System level parser**
  This function disassembles streams into their original audio/data/video PES (Packetized Elementary Stream) packets and extracts programming, network, and channel information.

A set-top box core library should contain IP-cores collected according to the subsystem they belong to (see Figure 13). Many IP components are available from IP-providers (see Table 2). Before choosing an IP-core for an integrated set-top box solution, designers should take into account convenient selection criteria. Table 3 shows selection criteria for some IP-components suitable to be included in a IP-cores library of a set-top box. For more details and exapmples about set-top boxes see [*N.N.-1, N.N.-2*].

IP-core	Selection criteria
MPEG-2 Video/Audio	- Compliancy with international standards - Conformance of testbenches to bit streams - Dolby™ Certification (AC-3) - Optional support components - High level 'C' architectural models - Firmware support - Demonstration boards - Support of linear PCM Audio - Error concealment
Modulation/Demodulation, FEC	- BER (Bit Error Rate) estimation testbenches - 'C' behavioral model and testbench for system level simulation - VHDL behavioral model, testbench, and vectors
Transport and Parser	- Dependent on the supported types of MPEG-2 streams
Scrambler/Descrambler	- Customer should incorporate their own proprietary IP

Table 3   Selection criteria for some of the IP-components viewed in Figure 13

DESIGN OF MECHATRONIC AND MICROMECHANIC SYSTEMS

Mechatronics, a term consisting of *Mech*anics and Elec*tronics*, is the functional and spatial integration of the components of a Mechatronic system consisting of intelligent sensors, actuators, the information processing unit and the mechanic system itself. The goal of this integration is a higher reliability and accuracy of the new system as well as more functionality key focus and a lower price.

The traditional design flow of mechatronics systems consisting of mechanic and (micro-) electronic components starts from the specification of the end product. Based on a first concept, the system is divided into a mechanical and an information processing part. After defining the interfaces for the communication between different system components, the development for the mechanical and electronical part (including the software) is carried out independently. At the end of the development cycle integration takes place and the system can now be validated against requirements. The main disadvantage of this approach is that the test for validation is done at the end of the development process. Errors made in earlier design stages force redesign cycles which are expensive in terms of *time-to-market* and *design-for-cost*.

New design methods pay attention to a parallel development of the mechanic and electronic components (see Figure 14). It is usually very difficult to build exact mathematical models for complex mechanical processes including sensors, actuators mechanic and electronic components. Simulation based on mathematical models either does not make any reliable statements or can not be performed.

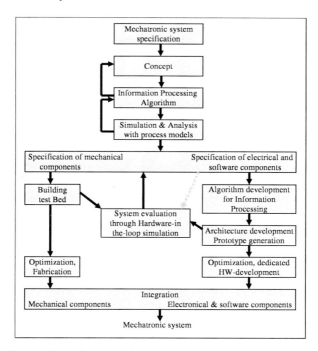

Figure 14   General design flow of mechatronic systems

Thus, the development of the mechatronic system's mechanical part concerns usually with the building of a complete test rig with sensors, actuators and the mechanical part where the prototype of the electronical part is plugged in to perform a so called *hardware-in-the-loop simulation*. With such a functional prototype it is possible to emulate a mechatronic system in real time, to change and to adapt different algorithms and to detect errors and bottlenecks in an early design stage [Le-97].

It is desirable to describe the complete mechatronic system with an "universal" description language, considering the mechanic as well as the electronic components. The main point in the implementation of the mentioned interfaces between the different levels of abstraction is the possibility to reuse existing components and, if necessary, to adapt them to user specific requirements, that means to make these components accessible to a reuse with parametric boundary conditions.

Since the middle of the eighties microsystem technology, which can be considered as continuation and integration of mechatronics, is gaining importance. Goal of the microsystem technology also is the combination of sensors actuators, passive structures and smart signal processing units to a system, but this is mainly achieved by using microelectronic technology, which allows and extreme miniaturization as well as an integration on one single chip. Following the microelectronics, the field of intelligent signal processing can already fall back on the strategies of the previously presented design methodologies. But also in the field of integrated sensor and actuator components, several design support frameworks have been developed the recent years, which support the system idea in a design process.

In order to realize such an idea, it is necessary to integrate one part of the system in the description form of the other domain. In microsystem technology this is normally done by transforming a mechanical description into the context of a microelectronic description. The main description forms in the microelectronic are on the one hand the behavioral description (e.g. with VHDL) and on the other hand the mask description in a layout which is necessary for the fabrication process itself. The determination and description of mechanical properties is mostly done on structural level using the finite element method (FEM). Thus, in order to obtain, as previously mentioned, a system representation including sensor, actuator and signal processing components, a transformation of the FEM-model into a behavioral model as well as into a corresponding layout is necessary.

As real sensors (and actuators) always show analog behavior, the use of an analog hardware description language (as superset of VHDL) like e.g. VHDL-AMS(VHDL-**A**nalog **M**ixed **S**ignal) is inevitable. First approaches to automatically generate a behavioral model from a structural description have already been realized [Hofm-97], so that by this a complete simulation of the whole system is getting possible.

As the efforts in microsystem technology are also heading for an integration of the whole system on *one* single chip, the generation of the complete layout in dependence of the used fabrication process is of crucial importance. In order to meet this goal, first approaches have also been realized, which are able to automatically generate a layout of sensor components in conformity with microelectronic using the structural model, which in return can be integrated in a complex system [Lang-98]. The chip-photo of an implemented microsystem example realized by the methods developed in [Lang-98] is given in Figure 15. This acceleration sensor has been fabricated by an extended 1.2 µm analog CMOS technology, where an anisitropic edge step has been added at the end. There are various application possibilities for this kind of acceleration sensor, e.g dynamic drive control systems, airbag control systems, speed control in virtual reality systems etc.. For more details about such acceleration sensors and its development method see [Lang-98].

Figure 15   *Chip-photo of an acceleration sensor fabricated with an extended 1.2 μm analog CMOS technology*

RAPID-PROTOTYPING

Complexity in microelectronic systems is dramatically increasing in the recent years. Considering the design methodology there are two main techniques, which have been discussed in detail in chapter HIGH-LEVEL SYNTHESIS, for managing the complexity: The consequent usage of top-down design methodologies based an hardware description languages, synthesis and static timing analysis as well as extensive component re-use (IP-based design, see subsection **IP-based** design methods).

Unfortunately, advances in *system design* have not been transferred to *system validation*, which is an emerging design task. Innovative microelectronic systems in telecommunications and multimedia for example are highly reactive, i.e. they are tightly coupled on a functional level with their real environment, and seek for a *subjective validation technique* in order to rate for example the audio- or video-quality of a multimedia system. Traditional validation techniques heavily rely on *prototypes*. A prototype in this context is a complete model of a new system or subsystem which is functional equivalent to it [*Thay-90*]. With increasing system complexity and device sizes the time to develop such a prototype was increasing dramatically and could not be tolerated any more. As a consequence, people shifted towards *simulation* as key validation technique. But long simulation cycles, the lack of appropriate simulation models as well as the disadvantage that the real system environment cannot be included into the validation phase turns simulation also to an improper validation technique for *complex system*s.

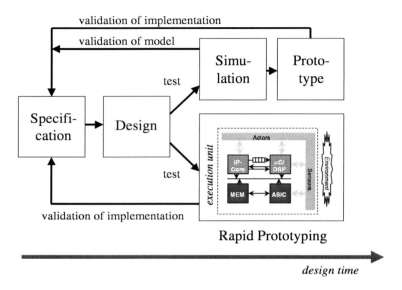

Figure 16   Design Validation using Simulation and Prototyping

Hence, functional validation becomes the bottleneck in the whole design process. Advances in the technology of reconfigurable devices as e.g. with FPGAs [*Brow-92*] and programmable interconnection devices (FPICs) [*Icub-96*] enable a new class *of reconfigurable prototyping systems* for the validation of complex microelectronic systems. During the *rapid-prototyping* step the complete system including all custom-specific circuits (ASICs/ASIPs) is mapped to the execution unit of the prototyping system (see Figure 16). The gate-level netlist representing the ASIC is emulated with an array of FPGAs. The ASIC communicates with standard IP-components as e.g. microprocessors, memory or I/O-interfaces (analog/digital) via a reconfigurable interconnection network. REPLICA [*Kirs-97, Kirs-98*] a prototyping system for the validation of embedded systems is an example platform for such a prototyping environment. Specific CAD-tools automate the time-consuming and error prone task of technology mapping and also relief the designer from the configuration of the prototyping architecture. Thus, additional costs on form of extra design time, which are introduced by this prototyping methodology are minimized. Although the architecture of the prototype may differ from that of the final product and the execution speed is approximately a magnitude lower than real-time, the prototype allows the testing of different architectures and implementation alternatives as well as the real environment may be included. Especially when third party IP-elements are used, which normally lack appropriate simulation models for system validation, the early integration of *real* core-elements into the prototype is essential and will improve design quality significantly. Additionally, when using rapid prototyping multiple design teams can work in parallel very early in the design process. For example software developments (operation systems, application software, device drivers etc.) may be tested with the prototype already before the first silicon (ASIC) has been produced, which may take some time. This will minimize optional redesign cycles and thus shorten the overall design time significantly.

CONCLUSIONS

The paper has discussed efficient design methods for today's complex microelectronic / mechatronic systems. The rapidly increasing complexity of present and future microelectronic and mechatronic systems has been sketched, as well as the major problems arising from this situation. Therefore, the historical technological development has been viewed, and trends as well as perspectives for future microelectronic/mechatronic systems have been outlined based on this.

Today's commercially available design methods and tools cannot fully exploit the potential and density of newest technologies (*design gap*). Based on the introduction of existing design principles and methods, e.g. *high-level synthesis* and *hardware/software co-design*, new promising design strategies like *IP-based* (or *core-based*) design and *set-top-boxes (STPs)* have been described. The paper tried also to stress the perspectives and possibilities of such system design strategies for handling the required complexities, as well the need of *standardization efforts* for efficiently designing such *systems-on-a-chip (SOCs)*.

Finally, a new approach for automatic design of *mechatronic systems* in *microsystem technology* has been described, which has been applied successfully during the design of an acceleration sensor. Here, the problems and importance of *rapid prototyping* for microelectronic and mechatronic systems has also been outlined, incl. the introduction of an efficient prototyping architecture for heterogenous system evaluation.

Efficient future design methods for developing complex heterogenous microelectronic / mechatronic systems have to perform fundamental new strategies, especially to exploit the potential of *deep-submicron technologies* and to manage arising problems (timing, power, fault-tolerance etc.). The paper tried to provide an overview of the first steps in this direction, and to motivate for the big challenges within this promising area.

REFERENCES

Ajlu-98 C. Ajluni: Redfining EDA In The New Age OF Intellectual Property; Electronic Design pp. 64-75, January 12, 1998.

Beck-97 J. Becker: A Partitioning Compiler for Computers with Xputer-based Accelerators; Ph.D. dissertation, University of Kaiserslautern, Germany, 1997.

Bohm-91 M. Bohm, E. Detjens, K. Keutzer, A. Sangiovanni-Vincentelli, C. Stroud, T. Willliams, J. Werner. ASIC Alternatives for System Design. IEEE Design & Test of Computers, 1991. Roundtable Discussion.

Booc-91 Grady Booch,"Object Oriented Design", The Benjamin/Cummings Publishing Company, Inc. 1991.

Brow-92 S.D. Brown, R.J. Francis, J. Rose, Z.G. Vranesic: Field-Programmable Gate Arrays. Kluwer Academic Publisher, 1992.

Buch-94 K. Buchenrieder: Hardware/Software Co-design; ITPress, Chicago, 1994.

Gajs-88 D. Gajski: Silicon Compilation. Addison-Wesley Publishing Company, Inc., 1988.

Gajs-93 D.D. Gajski, F. Vahid, S.Narayn, "SpecCharts: A VHDL front-end for embedded systems", UC Irvine Dept. of ICS, Tech. Rep. 93-31,1993.

Goos-95 G. Goossens, J. van Praet, D. Lanneer, W. Geurts, F. Thoen: Programmable Chips in Consumer Electronics and Telecommunications; NATO ASI Hardware/Software Co-Design, Tremezzo, Italy, June 19-30, 1995.

Hart-96 R. Hartenstein, M. Glesner: 6th Int. Workshop on Field-Programmable Logic: Smart Applications, New Paradigms and Compilers (FPL'96); Darmstadt, Sept. 1996.

Henn-96 Hennessy and Patterson: Computer architecture a quantitative approach; Morgan Kaufmann Pub., San Francisco, 1996

Hofm-97 Hofmann, K.: Differential model generation for microsystem components using analog hardware description languages, Dissertation TU Darmstadt, 1997.

Icub-96 I-Cube Inc.: IQX-Family Data Sheet, 1996.

Kirs-97 A. Kirschbaum, M. Glesner: Rapid Prototyping of Communication Architectures. In IEEE Workshop on Rapid System Prototyping, S.136-141, Chapel Hill, USA, 1997

Kirs-98　A. Kirschbaum, J. Becker, M. Glesner: A Reconfigurable Hardware-Monitor for Communication Analysis in Distributed Real-Time Systems; IEEE Reconfigurable Architectures Workshop, Orlando, USA, 1998.

Lang-98　Lang, M.: Entwurfsmethoden und Rapid Prototyping integrierter Mikrosysteme, Ph.D. dissertation TU Darmstadt, 1998.

Le-97　Le, T, Renner, F.-M., Glesner, M.: Hardware-in-the-loop Simulation – a Rapid Prototyping Approach dor Designing Mechatronics Systems, 8th Int. Workshop on Rapid System Prototyping, Chapel Hill, USA, 1997.

Luk-97　W. Luk, P. Y. K. Cheung, M. Glesner: 7th Int'l. Workshop on Field-Programmable Logic & Applications (FPL'97); London, Sept. 1997.

Mich-92　P. Michel, U. Lauther, P. Duzy: The Synthesis Approach to Digital System Design; Kluwer Academic Publ., 1992.

Mich-95　G. De Micheli: Hardware/Software Co-Design: Application Domains and Design Technologies; Proc. of NATO Advanced Study Institute on Hardware/Software Co-Design, Tremezzo, Italy, June 19-30, 1995.

N.N.-1　http://www.chips.ibm.com/settopbox/designkit.html

N.N.-2　http://www.mentorg.com

Prea-88　B. Preas, M. Lorenzetti: Physical Design Automation of VLSI Systems; Benjamin/Cummings Publ., 1988.

Smit-97　Smith, M.J.S.: Application Specific Integrated Circuits, Addison Wesley, 1997

Thay-90　R.H. Thayer, M. Dorfmann (Hrsg.): System and Software Requirements Engineering. Los Alamitos: IEEE Computer Society Press, 1990.

Tuck-97　B. Tuck: Integrating IP blocks to create a system-on-a-chip; Computer Design pp. 49-62, Nov. 1997.

Vill-97　J. Villasenor, W. H. Mangione-Smith: Configurable Computing; Scientific American, June 1997.

Wehn-95　N. Wehn: Verfahren zur Architektursynthese mikroelektronischer Systeme, Habilitationsschrift, TH Darmstadt, 1995.

Zori-97　Y. Zorian, R. K. Gupta: Design and Test of Core-Based Systems on Chips; IEEE Design & Test of Computers pp. 14-25, Oct. – Dec. 1997.

Discussion

Question - H.-J. Franke

You said ten years ago people thougt about an automatical method named "Silicon Compiler" and you mentioned the large gap between necessary and available designers. That leads to the question: What is the reason why the compiler was not invented and realized ?

Answer - M. Glesner

The Silicon Compiler was a tool which was operational, but was not efficient enough, because, let's say, on the upper level the exploration of the design space was not possible. And it was easier to separate the whole design task in 3 blocks as architectural design, register transfer logic design and the physical design. So that is better to handle. The automatic software we have nowadays is just for medium sized complexity, standard cell gate arrays, but even on that level you have to observe the technological result, because you have to do that extraction and check if the circuit is really properly operating. But it is an enormous complexity that is handled nowadays. Our students are regularly designing chips with complexity in the order of 100,000 gates. And imagine, the first microprocessors had some 2,000 transistors on it.

Question - K. Ehrlenspiel

Listening to your arguments, one is surprised and somewhat stunned in view of the sheer amount of technical, physical and logical aids you use in these processes - what you describe is a vastly different situation from what we are currently experiencing in the field of mechanical engineering. In mechanical engineering, we are currently working on questions such as: How does human problem-solving work? Where do creative ideas come from? What is the influence of human communication? When are what kinds of mistakes made and which corrective loops can be used to eliminate these errors? We are much more interested in the human process. This is perhaps rather logical, since 'our' technology, as opposed to yours, is not developing at such a tremendous speed anymore. But you will probably arrive at a point sooner or later when you will begin to study the human influence on your processes.

Answer - M. Glesner

It is of course a big problem that we understand how to cope with the complexity of the design process itself and there are ways, how you can approach it. The human interface in the sense: What is a real contribution, let's say, of some smart sudents? is also investigated. I admire the adventures Mr. Birkhofer is doing by inviting his students to design a certain robot and I remember some saturday afternoon where you had the final contest. And I thought, some 50 people were there and you had 500 to 1,000 people coming to that event. Of course, we have not anything similar, but what we are doing is the following, that we give students a certain design project. We give them a rough idea on the architecture of the design. Let's say, we did it with a simplified microprocessor, having only 40 instructions instead of 400 instructions. Or we gave them the functionality of a chip which can be used inside a watch. And it was always amazing for us to see how different solutions came up concerning the final architecture, concerning maximum clock speed, concerning power consumption and we really continue that work. We do it actually outside of the lecture-time during the official holiday period. We give them the task. We let them work for 4 weeks on the design projects. The best projects get some prizes. They are presented in a contest form and the best design also can go to fabrication.

Question - H.-J. Franke

Is the greater number and the more advanced quality of tools you have, relatively to our mechanical design, caused by the much higher standardization of the electronic functions ? A logical array can principally be based only on NOR or NAND-functions. Naturally that is nicer and much easier than doing the same with our different machine elements used in mechanical engineering.

Answer - M. Glesner

The role of standardisation is quite high in microelectronics, because in old times you had catalogues of devices which you can order and put on your printed circuit board. Nowadays you have also, let's say, entry points in the design process on the level of logic circuits, so that is very convenient for people outside. And also inside the tools themselves, you have a lot of - I did not discuss it, I had one transparency with me - but you have very strong interface standards.

Question - F.-L. Krause

I'm interested to get to know if within the domain of microelectronics there have been any approaches in the past to look for a kind of design theory. Or has there been always the close connection to the technical potentials. There is an important difference. Within this group we think about artefact development without thinking about the technical possibilities to realise the artefact. I've got the impression that the common way of product development within your domain is strictly technologically oriented. But, did anybody think about it before to do it in another way?

Answer - M. Glesner

I remember some form of optimization techniques to get some circuit structures synthesized, but you have some evaluation of sensitivity function to create a certain structure of a circuit corresponding to fulfill certain requirements, but that was not really operational. But I think, the creativity here in the whole business is on the level of the architecture, because on the architectural level you can come to complete different solutions. And even when you map it down on the circuit level, you can even think about completely new types of circuit realisations, e. g., when you go now for microelectronics in some 5 to 10 years to introduce nanoelectronic components, when you might create transistors with the second gate and it might be possible to have memory cells, having different logical states, so instead of binary realisations, you come to alternate solutions or other solutions, and that will give you also a lot of possibilities. But in the sense that you abstract completely the technology from the design process, that is not working correctly. Our real problem is actually, that when you are doing decisions on the architectural level that you directly know more or less the consequences later on the technology level to reflect that in your design decisions.

Question - H. Birkhofer

I believe that in certain branches of machine design we have the exact equivalent, in special machine design, for example. We can buy proper components for an entire installation there, for example from the company Bosch. The actual creativity of the developer is used to define the installation layout. This is the exact equivalent of the action described by Mr. Glesner.

Answer - M. Glesner

Also the correspondance is quite similar let's say on, when perhaps in the next talk we will see, stronger on the level of microsystems, which is quite physical and you have similar types of finite-elements-simulation as you are using it in your domain.

MICROSYSTEMS TECHNOLOGY - A NEW CHALLENGE FOR THE MECHANICAL ENGINEER

Wolfgang Menz

Institute for Microsystems Technology

Albert-Ludwigs-University Freiburg, Germany

ABSTRACT

Microsystems technology started out as the continuation of microelectronics into the field of non-electronic regimes like mechanics, optics, fluidics, acoustics and others. The first years of development were concentrated toward microstructures, i.e. the manufacture of components with dimensions into the micrometer range. The way microstructures are to be designed differ quite a lot from the conventional methods of mechanical design but contain a great potential for fabricating systems with features unknown in conventional products. Even though several microcomponents were put to industrial mass fabrication, only very few real microsystems were demonstrated. The emphasis of the next years should be put into the industrial implementation of the microsystems technology including the development of intelligent systems with novel characteristics. As a concept for low cost fabrication of small quantities of systems a modular building concept with standardized subsystems is suggested.

INTRODUCTION

Microsystems technology nowadays is one of most frequently used terms in scientific circles. In spite of the popularity of this word, the meaning of the term „microsystem" is still indistinct and despite of the great number of different interpretations in the literature, there is today no clear definition which would contain all aspects of microsystems and microsystems technology.

However, to get an idea of the initial motivation of microsystems technology one should have a closer look to the origins of microelectronics. This technology is the starting point of our considerations. No doubt, microelectronics is unique in the technical history of mankind. Unlike all other technologies which were developed in an evolutionary manner and empirically in the course of the decades, centuries or even millennia, microelectronics was designed theoretically from the beginning, meaning the origin did not evolve out of human experience via long periods of trial and error but by pure outcome of quantum mechanics. In fact we deal with a technology whose origin and technological backbone is theoretical physics.

A further unique fact of microelectronics is to be seen in the ostensible reduction of technical possibilities of microelectronics compared to classical electronics:

- The use of photolithography as a means to transmit structure information onto the substrate surface implies the renunciation of the third dimension. The imaging of the structure mask

onto the substrate surface (wafer) is basically two-dimensional, and thus microelectronics is basically two-dimensional.
- The transmitted structures consist only of rectangles and trapezoids. Again here is an obvious reduction in variety of shapes compared to the classical electronics with its abundance in different elementary structures.
- The great number of materials of conventional electronics degrades to a „mono culture" of silicon.

In spite of this apparent curtailment of the technical possibilities microelectronics experienced an unprecedented success story. The quality of electronic components were improved in the course of few decades by a factor of ten thousand, simultaneously the production costs of those components were reduced by about the same amount. Microelectronics extends nowadays in all areas of our technical and private world and is even the technological basis of our cultural shift from industrial society to information society.

The step towards microsystems technology is therefore more or less a logical consequence: Why not utilizing the same fabrication processes, the same evolution philosophy, the identical materials for other, non electronic, applications in order to achieve similar technological and economic success? Out of these considerations a technology evolved which was called originally „micromechanics", later MEMS (micro electro mechanical systems), or micromachine technology and microsystems technology. Concluding, microstructure technology demands that:

Microstructure technology must generally follow the path of microelectronics in order to be as successful.

Consequently the development philosophy must mean:
- Enabling the supply of sufficiently powerful and efficient software development tools for micro components; development, simulation and optimisation of structures on the computer,
- Transmission of these structures onto the workpiece (silicon wafer) by means of photolithography, using the advantages of a high packing density and a decrease in the size of the structure,
- Manufacture of a multitude of components with narrow manufacturing tolerances by batch processing.
- Integration of many „dull" components to an „intelligent" system.

As mentioned before microelectronics is basically two-dimensional. An integrated circuit expands in most cases laterally over several square millimetres or even centimetres, whereas the active thickness of the structure rarely exceeds a few micrometers. In micro mechanics consequently processes had to be developed whereby the production of three dimensional bodies is possible, despite the utilisation of photolithography. Over the course of the years this has lead to many variants. Three of the most important technologies are briefly described in this part.

In **silicon micro mechanics** the third dimension is generated from an etching process, with which the single crystal can be subtractively levelled to a desired shape. Special etchants such as KOH (potassium hydroxide) etches the material anisotropically from a single crystal, corresponding to the crystal morphology. The specific advantage of silicon-micro mechanics lies in the ability to be able to install on the same substrate as well as microstructure bodies (*e.g.* sensor elements or actuator elements), by a combination of etching processes and standard processes of microelectronics and also to install suitable electronic evaluation circuits. Due to the different etching rates for different crystallographic orientations the use of simulation tools is essential to predict and control the fabrication of microstructures in bulk silicon. An example of the etch progress on a microstructure is shown in Figure 1, using the program SIMODE ® [*Früh93*].

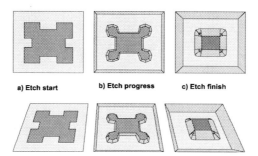

Figure 1   Simulating the etch progress at a so-called mesa structure using the SIMODE ® program. In order to get convex edges certain provisions have to be made on the mask.

Instead of etching into the bulk of the silicon crystal, a microstructure may be build up layer by layerin an additive process called **surface micromachining**. In the course of the process layers underneath may be etched selectively, called sacrificial layer technique, thus giving the designer the possibility to manufacture cavities, flexible structures and even rotating devices. The first micro motor presented in 1989 was build in this technique [*Tai-89*]. Figure 2a is showing the basic procedure, and Figure 2b the micrograph of this motor. Obviously this layered structure is inadequate for the manufacture of precise rotating devices due to the lack of supporting structural height and due to inherent tribological problems. A linearly scaling down of conventional macroscopic designs will inevitably lead to failure.

The third relevant fabrication technology is the so-called **LIGA technology** which was developed at the beginning of the 1980ís at the Nuclear Research Center Karlsruhe (today Research Center Karlsruhe) [*Beck-86*]. Here the microstructure is produced using X-ray lithography (a special variation of photolithography using the high energy radiation emitted from an electron synchrotron). This process allows the copying of the lateral absorber structures of a special mask into a thick layer of polymer, resulting in microstructures with the identical lateral shape and with straight vertical walls. By means of electroplating these plastic structures can be transferred into complementary metallic structures, which are in turn used as moulding tools for a multiplying process with hot embossing technique or injection moulding. The three major process sequences - in German words: Roentgenlithography, Galvanik, Abformung - are the components of the acronym LIGA. The parallel beam path of the X-ray source and the extreme layer thickness permit the manufacture of structures with an aspect ratio (*i.e.* the ratio of structure height to the smallest possible lateral structure) of over 100. Aspect ratios of about 1 are normal in microelectronics. By using sacrificial layer technique again, the structure can be freed from the supporting substrate, thus generating flexible or free moving components as seen schematically in Figure 3.

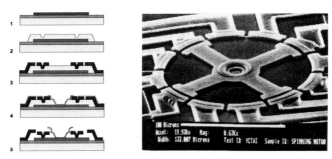

Figure 2   Left side: The basic process steps to manufacture an electrostatically driven micro motor in surface micromachining technique. 1. deposition of isolating layer and poly-silicon, 2. first sacrificial layer, 3. deposition of second poly-silicon layer and patterning, 4. patterning of second sacrificial layer and deposition of third poly-silicon (axle), 5. etching of sacrificial layers and freeing of the rotor.
Right side: SEM picture of the motor. The diameter of the rotor is about 150 µm.

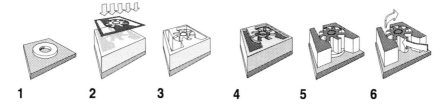

Figure 3   Process steps of a micro turbine in LIGA technique and utilizing sacrificial layers to free the rotor from the substrate. The individual steps are: 1. patterning the sacrificial layer (titanium), 2. exposure of the polymer layer on top by synchrotron radiation, 3. development of the exposed polymer, 4. electroplating of nickel to achieve the complementary structure, 5. removal of the remaining polymer, 6. selectively etching the sacrificial layer.

In all cases, silicon micro mechanics, surface micromachining, and LIGA technique as well, the third dimension is not to the free disposal of the designer, but is dependant on the technology used. In the case of the silicon micro mechanics the structure is dictated by the crystallographic morphology of the single silicon crystal, in the case of LIGA technique the third dimension is shaped by the synchrotron radiation beam, usually resulting in straight vertical walls. Since this is a serious limitation in the fabrication techniques of microstructures, a simple down-scaling of macroscopic structures is not possible, but would anyway not be advisable due to other reasons, which are discussed in the following.

With decreasing dimension the influence of the surface is dominating. This can be shown by the simple example of a cube. By shrinking a cube the surface shrinks with the square of the linear shrinking factor, but the volume shrinks with a power of three. Therefore the ratio of surface/volume is increasing linearly with the demagnification factor. If one considers physical surface properties which extend into the depth of the structure the surface/volume ratio increases even more than linearly. This results into the „surface dominance" of microstructures and has far-reaching consequences on the design of microstructures. Electrostatic forces, van der Waals forces, and surface tension, are predominant to gravitational forces.

Figure 4   SEM picture of a micro turbine fabricated in LIGA technique. The diameter of the rotor is 100 µm, the structural height 120 µm.

The before mentioned restrictions of the microstructuring technologies can be considered as a challenge to the mechanical engineer, since the microelectronics had faced similar deficiencies but has coined those to an overwhelming success story by concentrating to the remaining degrees of freedom and perfecting the processes. Taking microelectronics as a guideline the mechanical engineer has to find new concepts of designing and manufacturing microstructures and integrating these components to intelligent microsystems.

MICROSYSTEMS DESIGN

Today microsystems technology is mostly understood as the means to fabricate microcomponents. The emphasis is put preferably onto the term "micro" and the less onto the term "system". The development was concentrated on the manufacturing of components which were scaled down to micrometer dimensions or even less. A typical example for this development phase is the micro turbine, which was fabricated in LIGA technique and whose diameter corresponds to the diameter of a human hair (Figure 4) [*Wall-91*].

In the meantime microstructure technology in industry is well under way. Acceleration sensors for automotive application are fabricated worldwide by the millions. Other products are pressure sensors, gyroscopes, micro pumps and optical components.

In order to repeat the stupendous economic success of microelectronics though, the microsystems technology too has to advance toward a complete system as it is sketched out in Figure 5. This complete microsystem comprises a sensor array, an actuator array, an on-board microprocessor and several interfaces or interconnections to the outside world. These interconnections can be subdivided into interconnections for energy, information, and substances as seen in Figure 6. This figure illustrates the difficulty in fabricating complete systems: it is mainly the lack in interconnection technologies for the feeding of energy (electrical, mechanical, optical, fluidic, acoustic and so on), of information (electrical, optical, inductive, capacitive, acoustical and so on), and finally of substances (for instance in the medical field: medication, rinsing fluids, biological matter and so on) that delays the manufacture of microsystems.

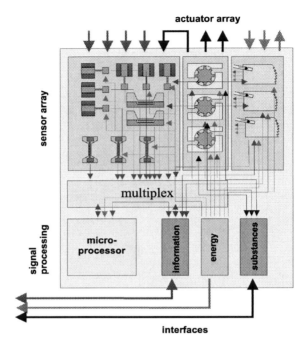

Figure 5 Schematic view of a complete microsystem comprising of sensor array, actuator array, on board signal processing, and interfaces to the outside world. Figure 5 Schematic view of a complete microsystem comprising of sensor array, actuator array, on board signal processing, and interfaces to the outside world.

A decisive advantage of microstructure technology is the potential to fabricate components (both sensors, actuators, and other structures) with micrometer dimensions in high scale integration and with economical production costs. In microsystems the integration of large sensor arrays is an essential advantage over macroscopic measurement systems. The same is true for other components which can be integrated in redundancy into the system. In case a component fails to operate it can be automatically eliminated, and a redundant component is activated instead. Therefore microsystems have the potential of self-correcting and even self-repair in case of wear or malfunction. Test routines can be performed by the microprocessor, which is integrated to the system, to check the function of safety relevant system at regular intervals or even continuously.

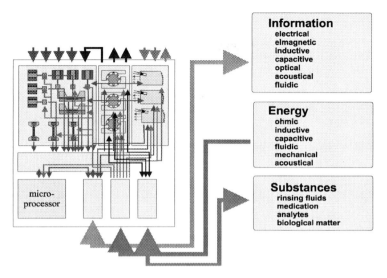

Figure 6   Interfaces on a microsystem arranged in energy, information and substances.

The assets of a microsystem can be summarized as follows:
- small dimensions and little weight,
- high measurement quality in metrology by utilizing sensor arrays,
- high reliability of the system by providing redundant components and build-in self-test routines,
- care of resources,
- high functional density,
- small power dissipation (battery operation).

It is evident from the listing above that size and weight of a microsystem are not the main assets but reliability and high measurement quality as well.

FUTURE DEVELOPMENT

What can be expected from microsystems technology in the near future? The predominant problems in systems integration can be seen in the development of a cost effective interconnection and interface technology. In opposition to microelectronics with only electrical interfaces typical microsystems contain interconnections in many different physical and chemical regimes (see Figure 5).

Future microsystems should have a much higher degree of integration compared to existing systems of today. New concepts of information processing should add more intelligence to the system to improve performance and speed. Autonomous systems with the ability to respond to unexpected situations in an intelligent manner could be designed to work in the fields of medicine, resource exploration or in dangerous environments.

Close cooperation with other scientific disciplines will certainly guide to new problem solutions and industrial applications. The aspects of artificial intelligence in combination with microsystems technology will open new fields. Polymeric chemistry can help to improve on the performance on actuators and sensors as well.

All these ideas can certainly be integrated into future microsystems. The question that arises is only, whether the market is ready to reward the scientific creativity with very high development costs for these new applications.

An important task for the near future will be the reduction of fabrication costs and the effective production of microsystems even in small quantities. One possible solution could be the development of a modular concept in microsystems. Similar to the gate-array concept of microelectronics a building block system is to be developed in microsystems technology. Standardized functional subsystems are designed in such a way, that they can be combined easily to complex systems. Exchange of some building blocks will alter the system to perform different tasks.

By means of such a modular concept the high development costs of microsystems technology can be split between different partners: the manufacturers of system modules („microsystem gate-arrays") and the manufacturers of complete systems. Certain components like power supplies, sensor arrays, optical set-ups are used in many different systems. These components can be fabricated in large quantities to moderate fabrication costs and may be bought from the supplier off the shelf. A system manufacturer may buy these system components from different suppliers, provided the individual components „speak" with each other, meaning that there is a standard hardware interconnection and a software interface as well to the components.

A module concept demonstrates a possible way to implement microsystems technology into small and medium sized enterprises with very limited development capabilities. The requirements for such a concept is:
- The availability of standardized modules for different applications,
- a technology to combine these modules to microsystems with moderate expenditure,
- software tools to simulate the systems parameters,
- and an international standardization of interfaces and interconnections.

Figure 7   *Concept of a modular system for microsystem technology. The housings carry a bumb grid array with standard connections and carry microelectronic or microstructure components in the inside.*

A collaboration between two Fraunhofer institutes (Institute for Manufacturing Engineering and Automation [IPA] and Institute for Reliability and Micro Integration [IZM]) and the

Institute for Microsystems Technology (University Freiburg) has resulted in a concept of a modular packaging design and production framework for intelligent and cost-effective microsystems (Figure 6) [*Leut-97*]. Standard packaging devices with standard interfaces allow for a simple multimode stack assembly. Utilizing these modular building block concept the implementation of microsystems in small and medium sized enterprises usually not familiar with the microstructure technologies will be emphasized considerably. Standardization of component or modular interfaces will not restrict further research and development of new components but on the contrary will speed up the industrial application of microsystems technology.

REFERENCES

Beck-86	Becker, E. W.; Ehrfeld, W.; Hagmann, P.; Maner, A.; Münchmeyer, D.: Microelectronic Engineering 4, p. 35, 1986.
Leut-97	Leutenbauer, R.; Großer, V.; Reichl, H.: In: Eder, A.; Reichl, H. (Ed.): Tagungsband SMT, ES&S, Hybrid '97. Internat. Messe u. Kongress für Systemintegration, Berlin, VDE-Verlag, pp. 77 - 84, 1997.
Tai-89	Tai, Y.-C.; Fan, L.-S.; Muller, R. S.: Proceedings IEEE Micro Electro Mechanical Systems Workshop, Salt Lake City, UT, 1-6 1989.
Wall-91	Wallrabe, U.; Himmelhaus, M.; Mohr, J.; Bley, P.; Menz, W.: VDI-Bericht 933, VDI-Verlag Düsseldorf, p. 327, 1991.

Discussion

Question - C. Weber

I can imagine that during the development process especially the integrated simulation of mechanical and electronic system components still is a considerable problem - you mentioned this in your last transparency.

Answer - W. Menz

That is correct. A few years ago everybody still was euphoric about a top-down-design. There were actual two project funded by BMBF about this matter. The idea was to copy the design concept of microelectronics, but it did not work out this way. In microsystems technology too many physical regimes are involved, as there are thermal problems, acoustical problems, optical problems and so on. The other extreme is the so-called bottom-up design. But adding components, one after the other, to a complex system is neither a successful way to achieve the optimal system. A good compromise is the meet-in-the-middle approach, which seems to be the best alternative. Here the design of a system is done by means of software tools but interactively with the human designer, who has to decide about the direction, the design will actually take. I do not see any other way at least for the next 10 years.

Question - A. Albers

You have referred to the enormous progress in microelectronics. I think, it was written in the paper, that you demand these steps in the technology for micro-systems. You have to take into consideration the complicated structure. You have shown all additional components which must be handled now, like signal and material. The structures are becoming very complex. Do you think you can make such progress in quality and cost reduction as it was done in microelectronics? Be a prophet.

Answer - W. Menz

I believe in a similar success story for microsystem technology. It may be still a long way to go. The title of my paper thus includes the word „challenge". With some products it became already reality. Take the acceleration sensor for the air-bag system. Compared with the conventional system the reliability was improved considerably. Today we have the side airbag, and in certain cars, a 3-dimensional acceleration sensor system is already reality. With this it is possible to calculate the movement of the car after an accident and to activate the right safety devices such as roll-over protection and so on. With build-in redundant components the reliability of the system is improved by orders of magnitude.

The possibility to realize sensor arrays is one of the major assets of the microsystem technology at all. With this it seems possible to build systems with continuous self-control or even with the ability for self-repair. This is vital for safety-relevant systems in the automotive business or in medical applications.

Question - D. Löhe

In micro system technology there is only a limited variety of design materials. Does this have a small or big impact on design in micro system techology?

Answer - W. Menz

It has indeed an impact on the design. Silicon can be considered as the almost ideal crystalline material with magnificent electronic and mechanical properties. On the other hand is the limitation in the patterning of a structure. Due to the technique of anisotropic etching, the microstructure has to follow the crystallographic morphology of the silicon. Microsystem technology though depends not only on silicon. Thin film technology opens up a whole new world of new possibilities in the design of structures. The engineer has to learn how to make the best out of it. On one hand there are certain limitations, but on the other hand there are great options. As I said before, the microsystem technology is a challenge for the engineer. He has to find new approaches. Only to shrink conventional machines to microsize, such as a gearbox, would lead inevitably to failure.

Question - F. Aldinger

For the production of components one has to make the basic distinction between shaping by stock removal and additive shaping. In the case of removal, one begins with materials having well defined properties and is assured that the materials properties will be retained as the size is reduced. When surfaces are built up, then a question similar to the one raised by Mr. Löhe presents itself, in other words, depending upon the nature of the shaping process, a completely new material may be generated. Should the physical dimensions be small, one enters the large field of microstructural characteristics which, in turn, affect the properties of these materials significantly. Under these circumstances, do you see a future for additive techniques? I am not talking about operations such as surface hardening in which wear-resistant layers, or the like, are generated but full components that are built up by additive techniques.

Answer - W. Menz

There is no single technology for the solution of all problems. I therefore support the hybrid approach. Competitive microcomponents have to be manufactured in different technologies to find the optimum performance for any application. Therefore, many components are hybrids fabricated by a combination of silicon bulk micromechanics, surface micro fabrication, LIGA technique, and others. We got many technologies from microelectronics for free. Now we have a huge reservoir of techniques for the manufacturing of microcomponents.

Metallic Structural Materials: Design of Microstructure

D. Löhe and O. Vöhringer

Institute of Materials Science and Engineering I

University of Karlsruhe, Germany

Keywords: Microstructure, lattice defects, strength, plastic deformation, crack initiation, crack propagation, fracture

INTRODUCTION

The most important structural materials for the design of components are metallic materials. Apart from the metallic glasses, which are not taken into consideration here, they possess a crystalline structure. In an ideal metallic element in the solid state, the positions of the atoms correspond to an ideal spatial lattice. Technical materials, however, are never in a thermodynamic equilibrium. There are lattice defects. Some of them are unwanted, for example impurities which are introduced by the manufacturing process. On the other hand, most of them are produced during the manufacturing of the material and the component to obtain desired properties. Therefore, lattice defects are elements of the microstructure which allow to design a given metallic structural material. Even though more than 10 000 metallic structural materials are available for the design of components, the types of lattice defects are rather limited and many of them - but of course not all of them - may occur in a given metallic structural material. Hence, the principles of the design of the microstructure by the utilization of lattice defects are applicable on the vast majority of metallic materials [*Ashb-80, Ashb-86, Shac-92, Smal-95, Vlac-89*].

STRUCTURAL METALLIC MATERIALS

DEMANDS

In Figure 1, it is shown schematically that for the function of a construction, mechanical, thermal, chemical, electric, magnetic, optic and other processes are used. There are demands on the construction like optimum function, high economic efficiency, little environmental pollution, little weight and so on.

These demands may, in turn, result in demands on the structural material used, for example little cost, high specific strength, high corrosion resistance and/or easy recycling. However, in a given construction, the demands on the structural material in a general sense are rather simple: There should be

- no failure at all   or
- no failure during an economic service period.

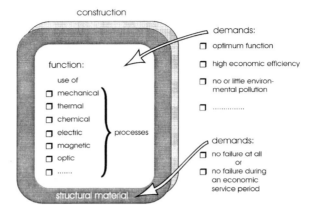

Figure 1   Demands on constructions and demands on structural materials

FAILURE MODES

The entirety of failures occuring in components made of metallic materials may be reduced to four basic failure modes, as illustrated in Figure 2: Plastic deformation, crack initiation, fracture and instability. With very few exceptions, instability does not include microstructural aspects. Therefore, it is not treated here. Plastic deformation, crack initiation and fracture can be traced down to two fundamental processes:
- generation and movement of dislocations comprising the interaction of dislocations with each other and with other lattice defects
- destruction of atomic bonding.

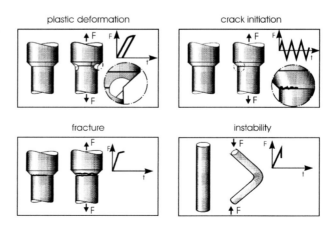

Figure 2   Failure modes of metallic structurals materials

In Figure 3 one type of dislocation, the so-called edge dislocation, is illustrated in a simple cubic lattice. In this lattice, the smallest structural unit (the unit cell) is a cube with atoms at each corner. In the left part of the Figure, a row of atoms can be seen which have no corresponding neighbouring atoms below them. This is a lattice defect which extent exceeds atomic size in one dimension. Therefore, a dislocation is a one dimensional lattice defect.

If a shear stress τ is acting on the crystal as shown in the Figure, the dislocation may move in a plane of densely packed atoms which contains the dislocation line (slip plane) in the direction of densely packed atoms (slip direction), as illustrated in Figure 3 by the sequence from left to right. During this process, repeated *reversible* breaking of atomic bonding occurs along the dislocation line (on the right hand side of the dislocation line in Figure 3) immediately followed by the re-formation of atomic bonding (on the left hand side of the dislocation line in Figure 3). By dislocation movement the crystal part above the slip plane is sheared relative to the part below the slip plane, resulting in the shown unit step of slip. This is the elementary process of plastic deformation.

On the other hand, stresses acting on a crystal may result in crack initiation and crack propagation combined with *irreversible* destruction of atomic bonding. In metallic materials, however, these processes are always combined with plastic deformation by dislocation movement.

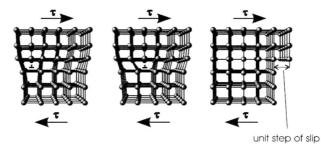

Figure 3   Dislocation movement in a simple cubic lattice

MICROSTRUCTURE

In most metallic structural materials, the fraction of one metallic element (the base element) is much higher than the fractions of all other elements. For example, in steels the fraction of alloying elements is less than 1 weight-% (carbon steels) and 5 weight-% (low alloyed steels), respectively. Similar conditions are found in many other groups of metallic materials like aluminium alloys and titanium alloys. In these cases, the choice of the base elements already determines some important properties of the structural material within relatively narrow limits: Melting point (or solidus temperature), density, elastic constants (Young´s modulus, shear modulus, Poisson's ratio), thermal expansion and thermal conductivity, as the most important ones. Hence it seems to be trivial, that a component which operates at a high temperature should be made of a material containing a base element with a high melting point (it will be shown later that just this example is not trivial at all).

On the other hand, regarding a given base element, there are many important properties which are directly related to the failure modes and hence the processes of dislocation movement and destruction of atomic bonding treated above. These properties may strongly be changed by influencing dislocation movement, crack initiation and -propagation. This is done by the generation of lattice defects which interact with dislocations, for example by thermal, mechanical, thermo-mechanical or thermo-chemical treatment. The microstructure of a metallic material is the entirety of lattice defects. Depending on the number of the dimensions of the lattice defect which exceed atomic size 4 groups are distinguished (see Figure 4):

- zero-dimensional lattice defects (point defects): For example substitutionally or interstitially solved foreign atoms, which result in solid solution hardening,
- one-dimensional lattice defects (linear defects): dislocations, which result in dislocation hardening,
- two-dimensional lattice defects (planar defects): For example grain boundaries, which result in grain boundary hardening,
- three-dimensional lattice defects (spatial defects): For example precipitations or dispersions, which result in precipitation or dispersion hardening.

To design the microstructure of a given metallic material is to produce an ensemble of lattice defects which yields the desired properties.

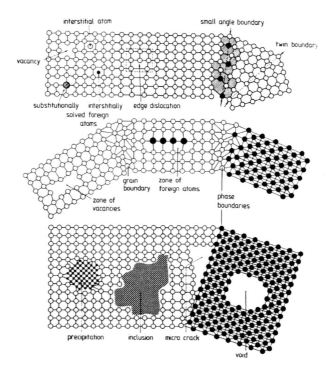

Figure 4   Lattice defects in metallic structural materials

INTERACTIONS BETWEEN DISLOCATIONS AND LATTICE DEFECTS

GENERAL SURVEY

When slip dislocations move along their slip planes, they encounter the lattice defects treated in paragraph MICROSTRUCTURE, which either confine the movement, or necessitate extra energy in order to be overcome. They may be subdivided into two categories: The first is characterized by long-range stress fields which act over several thousand lattice constants (the lattice constant is the length of the edge of the unit cell in the face centered cubic (fcc) and body centered cubic (bcc) lattice). Examples are residual stresses of dislocation pile ups on grain and phase boundaries. The other type possesses localized internal stress fields which only influence the slip dislocations over a couple of lattice constants (examples: resolved foreign atoms, small coherent precipitates). These short-range stress fields can be overcome by mobile dislocations at the cost of local lattice energy.

Hence, the flow stress required for the movement of the dislocations may correspondingly be divided into

$$\sigma = \sigma_G \text{(microstructure)} + \sigma^* (T, \dot{\varepsilon}, \text{microstructure}) . \tag{1}$$

The flow stress resulting from long-range stress field obstacles σ_G is influenced by the crystal structure of the material and the microstructure. Since the temperature influence is relatively small, σ_G is also known as the athermal flow stress component. The flow stress σ^* which emerges from the short range stress fields is extensively dependent upon the temperature T and the strain rate $\dot{\varepsilon}$ but may also be influenced by the crystal structure and the microstructure. It is also known as the thermal flow stress component.

LONG-RANGE OBSTACLES (ATHERMAL FLOW STRESS COMPONENT)

Today, one can adequately assess the structural quantities responsible for the athermal flow stress σ_G, and thus the strength of metallic materials. To illustrate this, Figure 5 shows the effects of a number of important influencing parameters on the deformation behaviour of iron and iron-based alloys. In pure iron, the flow stress increases with increasing dislocation density ρ_t (see Figure 5a). Smaller ferrite grains increase the 0.2-proof stress (Figure 5b), which is also the case when the solved silicon content is increased (Figure 5c). In the case of an FeCu alloy containing 1 wt.-% Cu, the 0.2-proof stress $R_{p0.2}$ values are influenced by the nature of the precipitations and their free particle spacing l. For coherent precipitations, $R_{p0.2}$ increases with increasing l up to a peak value at l = 0.04 µm, beyond this peak semi-coherent and incoherent precipitations cause a drop of $R_{p0.2}$ for l > 0.04 µm (cf. Figure 5d). The 0.2-proof stress of plain carbon steels increases with decreasing free ferrite length λ, as shown in Figure 5e. A reduction of λ may be achieved either by reducing the size of the cementite particles and/or by increasing the cementite content. Finally, an increase in the lamellar cementite content (or pearlite content respectively) of plain-carbon steels also increases their 0.2-proof stress (Figure 5f).

In general, the hardening of metals is governed by the elastic interaction of slip dislocations with other dislocations, grain boundaries, solved foreign atoms, particles or additional hard phases. The active mechanisms in this respect are listed in Figure 6 which includes the accompanying stress components and the surface features, which are produced by dislocations reaching the crystal surface.

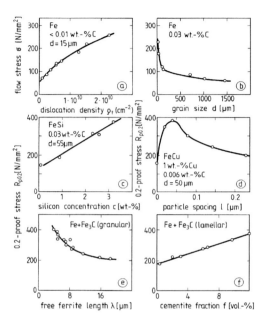

Figure 5   *Factors influencing the flow stress σ and the 0.2-proof stress $R_{p0.2}$ in iron base materials*

Dislocation hardening occurs when mobile dislocations are forced to overcome the stress fields of other dislocations on their path. This yields a stress component which increases with increasing dislocation density ρ_t, and the proportionality

$$\sigma_d \sim \sqrt{\rho_t} \qquad (2)$$

is valid.

Grain boundary hardening is caused by mobile dislocations which accumulate at the grain boundaries thus inducing slip processes in neighbouring grains. This produces the stress component

$$\sigma_g = k/\sqrt{d}, \qquad (3)$$

where k is a material dependent property. The solid solution hardening, caused by resolved foreign atoms, can be traced back to elastic interactions between the mobile dislocations and the stationary foreign atoms, situated either within the lattice slip-planes or in their direct vicinity. This leads to a stress component which increases with increasing foreign atom concentration c according to

$$\sigma_s \sim c^n. \tag{4}$$

The exponent n lies between 0.5 and 1.0. If resolved foreign atoms are able to move to new locations at adequate temperatures, a dynamic interaction can exist between mobile dislocations and diffusing foreign atoms. This is known as dynamic strain-aging which is inevitably combined with an increase of the material's flow stress.

hardening mechanisms		part of the flow stress	surface features
interaction of slip dislocations with	schematic		
1. dislocations	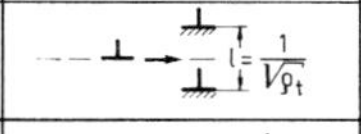	$\sigma_d = \alpha_1 Gb\sqrt{\rho_t}$	slip lines and slip bands and/or twin lamellae
2. grain boundaries	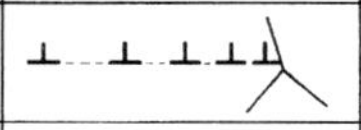	$\sigma_g = \dfrac{k}{\sqrt{d}}$	multiple slip
3. solute foreign atoms	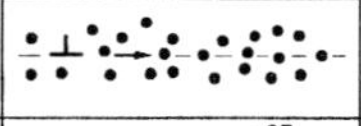	$\sigma_s = \alpha_2 G c^n$ $\quad 0.5 \le n \le 1$	increasing straightness of slip bands with decreasing stacking fault energy
4. particles a) coherent precipitates	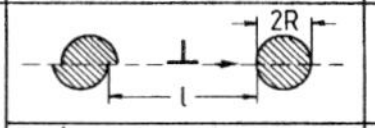	$\sigma_p^{(a)} = \alpha_3 \gamma_{eff}^m \dfrac{R^m}{1+2R}$ $m = 1$ resp. 1.5	coarse slip
b) incoherent precipitates and dispersions resp.		$\sigma_p^{(b)} = \alpha_4 \dfrac{Gb}{l} \ln\dfrac{R}{b}$	fine slip
c) granular arrangement of 2nd phases	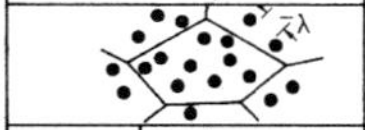	$\sigma_p^{(c)} = \dfrac{k'}{\sqrt{\lambda}}$	inhomogeneous slip
d) lamellar arrangement of 2nd phases	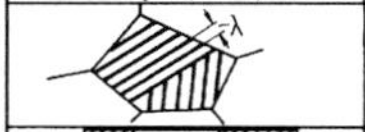	$\sigma_p^{(d)} = \dfrac{\alpha_5}{\lambda}$	
e) two phases with coarse grains	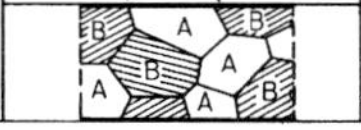	$\sigma_p^{(e)} = (\sigma_B - \sigma_A) f$	

Figure 6   Work hardening mechanisms and their contribution to the flow stress and surface topography

Particle hardening is caused by coherent, semi-coherent and incoherent precipitations or dispersions which impede the mobile dislocations. Adequately small precipitations may be cut and sheared by the mobile dislocations. For some spheric precipitations of radius R and a free distance l, eq. (5) is valid

$$\sigma_p^{(a)} \sim \frac{R^m}{1+2R}. \tag{5}$$

The exponent can take either the value 1.5 or 1.0, depending on the precipitation size and content. The reduction of the effective obstacle area of cut precipitations makes these precipitations more susceptible to subsequent shear by mobile dislocations in the same slip plane than in neighbouring ones. Hence, the plastic deformation is concentrated on a small number of glide planes which are sheared to a relatively strong extend. This leads to high, wide spaced slip steps on the grain's surface. Hence, it is refered to this as "coarse slip". Incoherent precipitations, dispersions and also large coherent precipitations cannot be cut by the mobile dislocations, but must be bypassed.

Hence the stress component

$$\sigma_p^{(b)} \sim (l)^{-1} \qquad (6)$$

is applicable, which increases with decreasing free particle distance l. This creates dislocation rings (cf. Figure 6) which surround the particles and effectively reduce their free distance l. Subsequent dislocations following the same slip path are then confronted with a higher resistance than in neighbouring planes. This leads to small and closely-spaced superficial slip steps known as "fine slip". Analogous to the grain boundary hardening (cf. eq. (3)), the presence of a granular and hard 2nd phase also hardens the material. For this,

$$\sigma_p^{(c)} = k^* / \sqrt{\lambda} \qquad (7)$$

is valid, where λ is the average free particle spacing within the phase. This relationship only is valid, however, when the particle diameter is at least several orders of magnitude larger than those causing precipitation or dispersion hardening. If the second phase is of a lamellar nature, this flow stress component is

$$\sigma_p^{(d)} \sim \frac{1}{\lambda}, \qquad (8)$$

where λ is the free inter-lamellar distance. Finally, if a coarse distribution of the second phase "B" is present within a softer phase "A", the following approximation may be applied

$$\sigma_p^{(e)} = (\sigma_B - \sigma_A) f, \qquad (9)$$

where f is the volume fraction of the second phase. Should the granular, lamellar or coarse phase particles happen to be less deformable than the matrix itself, then an inhomogeneous slip feature distribution is to be expected on the free surface of the matrix's grains.

When different types of hardening mechanisms are effective simultaneously, in many cases, the flow stress can be estimated by applying the principle of flow stress component additivity. For example, provided that only one stress component actively contributes to the particle hardening mechanisms, and that texture influences may be regarded negligible, eq. (10) is valid

$$\sigma = \sigma_G = \sigma_d + \sigma_g + \sigma_s + \sigma_p^{(i)}. \qquad (10)$$

The flow stress components combined in this relationship are practically temperature independent (i.e. athermal). σ therefore is identical to the athermal stress component σ_G.

SHORT-RANGE OBSTACLES (THERMAL FLOW STRESS COMPONENT)

The flow stress of metals is influenced by strain rate and to an essentially larger degree, by temperature. The quantitative influences of the deformation temperature and strain rate on the 0.2-proof stress are plotted in Figure 7 for a low carbon iron. For temperatures below 300 K, $R_{p0.2}$ increases sharply with decreasing temperature (cf. Figure 7a). Increasing strain rates also raise $R_{p0.2}$ values, the strain rate sensitivity of the 0.2-proof stress also being temperature dependent.

The relationship between the thermal flow stress component σ^*, and the temperature T is illustrated graphically at a constant strain rate $\dot{\varepsilon}$ in Figure 8. Here, the so-called force-distance curves of mobile dislocations in the vicinity of a specific short-range obstacle are plotted for different temperatures. In each case, the local force which must be exerted in order to move the dislocation and which is proportional to the thermal flow stress σ* has been plotted schematically as a function of the local distance x. At T = 0 K, the force F0 (or flow stress σ0) must be applied in order to overcome the obstacle, since at absolute zero no thermal fluctuations exist. At T_1 and T_2 finite quantities of the free enthalpy ΔG_1 and ΔG_2 are at disposal, whereby since $T_2 > T_1$ ΔG_2 is therefore also larger than ΔG_1. These "enthalpies" are represented by the shaded areas. It is evident that in order to overcome the same obstacle at lower temperatures, larger forces F* (or flow stresses σ*) will be necessary than at higher temperatures. Furthermore it can be observed that upon reaching the temperature T_0, the total work required to overcome the short-range obstacle is applied thermally. At $T = T_0$ F* and hence σ* both are zero. The enthalpy ΔG_0 which is available at $T = T_0$, is a characteristic parameter for a specific short-range obstacle.

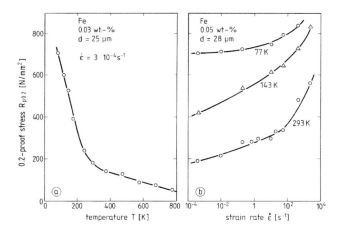

Figure 7   *Influence of a) temperature and b) strain rate on the 0.2-proof stress of low carbon steel*

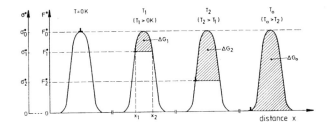

Figure 8   *Force-distance curve during interaction of a dislocation and a lattice defect at different temperatures*

The strain rate influence on σ^* can also be demonstrated in an analogous manner using the above force-distance diagrams. Increasing strain rate $\dot{\varepsilon}$ at a given temperature decreases the probability that within a specific time-interval, adequate local thermal fluctuations will occur at the short-range obstacles. The share of thermal enthalpy ΔG therefore decreases, and F^* (i.e. σ^*) increases.

For $T \leq T_0$, the thermal flow stress component may be described for a variety of metallic materials by an exponential law in the form

$$\sigma^* = \sigma_0^* \{1 - (\frac{T}{T_0})^n\}^m.$$
(11)

In this case, σ_0^* is the thermal flow stress share at absolute zero, and T_0 is the threshold temperature above which $\sigma^* = 0$. Theoretical analyses and experimental findings have verified that T_0 increases with increasing strain rate. As a result of different types of obstacles, the exponents are, for example, $m = n = 1$ for pure aluminium and a number of hexagonal metals; $m = 1$ and $n = 1/2$ for titanium alloys; $m = 2$ and $n = 1$ for pure iron; $m = 1,75$ and $n = 0,5$ for carbon steels and $m = 2$ and $n = 2/3$ for homogeneous copper alloys.

T AND $\dot{\varepsilon}$ FLOW STRESS DEPENDENCIES

The sum of the athermal and thermal flow stress components, i.e. the combination of eqs. (1), (10) and (11) delivers

$$\sigma = \sigma_G(\text{microstructure}) + \sigma_0^*(\text{microstructure})[1 - (\frac{T}{T_0})^n]^m, \quad T \leq T_0.$$
(12)

Upon calculation the flow stress σ for various strain rates in dependence on the temperature, the curve trend schematically illustrated in Figure 9 emerges. The flow stresses fall continuously with decreasing temperature, reaching the σ_G level at lower temperatures with decreasing strain rate. For metals, σ_0^* and σ_G are influenced by lattice structure, lattice constituents and structure of the faults therein.

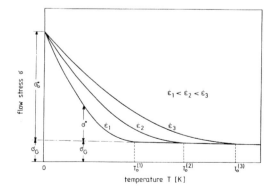

Figure 9   *Flow stress as a function of the temperature at different strain rates (schematically).*

INFLUENCE OF TEMPERATURE, LOADING RATE AND ENVIRONMENT

LOW TEMPERATURES AND HIGH LOADING RATES

As shown in paragraph SHORT-RANGE OBSTACLES, dislocation movement is hindered by decreasing temperature and increasing loading rate. Plastic deformation is also impeded by high hydrostatic stresses in triaxial stress states which may exist in the vicinity of notches or flaws. Consequently, the stress necessary for plastic deformation increases dramatically in the important metals with body centered cubic (bcc) lattice (where no close-packed slip planes are available) or hexagonal close packed (hcp) lattice (where spatial deformation also requires slip on non-close-packed planes). This involves the danger of a transition from plastic deformation by slip to brittle fracture by crack initiation and crack propagation. On the other hand, the face centered cubic (fcc) metals with a sufficient number of close-packed slip planes with different spatial orientation are not prone to brittle fracture.

In Figure 10, a schematic comparison between the behaviour of fcc metals and bcc metals is made by plotting the energy which is required for fast fracture of notched specimens as a function of the test temperature. The fracture of fcc metals is always accompanied by extensive plastic deformation, resulting in relatively high notch impact energies. On the other hand, bcc metals show a transition to britte behaviour at low temperatures. This basic behaviour reflects a property of the bcc lattice which cannot be eliminated by changing the design of the microstructure. However, the upper shelf of the notch impact energy at high temperatures and the transition temperature itself may strongly be influenced by lattice defects. It turns out that almost all effective strengthening mechanism have a detrimental effect on the ductile-to-brittle transition with one important exception: Decreasing grain size, which results in grain boundary hardening, lowers the transition temperature. This is the main reason for the unique importance of this strengthening mechanism.

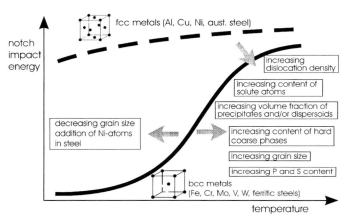

Figure 10   Influence of the temperature on the notch impact energy of the face centered cubic (fcc) and body centered cubic (bcc) metals

HIGH TEMPERATURES

Dislocation movement is enhanced at elevated temperatures, resulting in decreasing strength (see paragraph SHORT-RANGE OBSTACLES). At temperatures exceeding 40% of the absolute melting point (or of the absolute solidus temperature) plastic deformation is more and more assisted by the thermally induced movement of individual atoms in the lattice. This movement process is called diffusion and may be quantified by the diffusion coefficient

$$D = D_0 \bullet \exp\left[-\frac{\Delta H}{kT}\right], \tag{13}$$

which strongly depends on temperature and activation enthalpy ΔH. D_0 is a constant for diffusion of a given element in a given lattice and k is Boltzmann´s constant. ΔH is a temperature independent property for a given metal. However, ΔH increases with increasing strength of the atomic bonding and therefore with increasing melting temperature. Comparing fcc and bcc pure metals with similar melting temperatures, ΔH is lower (and consequently D larger) in bcc metals, because movement of atoms is easier in non-close-packed lattices.

The existence of diffusion processes always results in time-dependent plastic deformation (creep deformation) of mechanically loaded materials, the rate of which depends strongly on temperature and stress. Comparing materials with similar melting temperatures and similar microstructures, but different lattices, the creep rate at given temperature and stress is larger in a non-close-packed lattice like the bcc lattice than in densely packed lattices like the fcc lattice. This is shown schematically in Figure 11, where relationships between plastic strain and time (creep curves) of a bcc metal (for example α-iron) and a fcc metal (for example nickel) are plotted. With the exception of pure diffusional creep, which occurs at very high temperatures and low stresses, creep deformation involves dislocation movement, which may be hindered by the processes discussed in paragraph LONG-RANGE OBSTACLES.

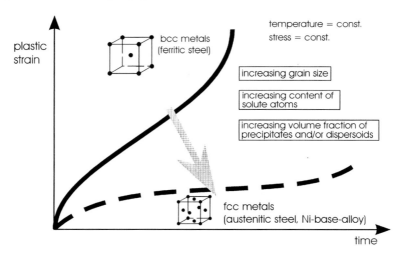

Figure 11   Creep curves of face centered cubic (fcc) and body centered cubic (bcc) metals

Therefore, as indicated in Figure 11, it is possible to drastically increase the creep strength of a given metal by using strengthening mechanisms like solute solution hardening, precipitation hardening and so on. There is, however, one important exception. Grain boundary hardening is not effective at high temperatures, because diffusion is enhanced along grain boundaries. Therefore, sliding of adjacent grains occurs along grain boundaries, where sufficiently high shear stresses are acting. Furthermore, damage like the formation of pores and microcracks may preferentially occur at grain boundaries, where high normal stresses are acting. Hence, the designing of the microstructure of a structural material which operates at high temperatures involves the utilization of various strengthening mechanism, but also the optimization of the grain size and the grain shape. This is outlined regarding an important example in section **CASE STUDY**.

ENVIRONMENT

Environmental attack may result in oxidation and/or corrosion. In some cases where oxidation or corrosion are evenly distributed, the progressive loss of structural material may increase the stress and eventually result in failure. Much more dangerous, however, is localized environmental attack enhancing crack initiation and crack propagation. For example, the fraction of fatigue life until crack initiation may be reduced and the crack propagation rate may be increased by orders of magnitude by corrosion. Since all metallic structural materials are prone to oxidation and corrosion they need a protection which reduces oxidation and corrosion to an acceptable level. A protective surface layer may be produced by the oxidation of the metal itself. Well-known examples are chromium and aluminium. If oxidation of the base metal does not produce a suitable oxide layer, it must be assured by alloying chromium and/or aluminium in sufficient amounts that a chromium oxide layer or an aluminium oxide layer is formed. Important examples are ferritic and austenitic steels, which need a chromium content of 12 % or more to gain a protective chromium oxide surface layer. Another possibility is to protect the structural metal by a coating. Sometimes, there is a constraint on the design of the microstructure by the necessity for adequate protection against oxidation and corrosion. For example, in the case of nickel-base superalloys treatet in the next section, protection against oxidation and corrosion requires high chromium contents, whereas the improvement of the mechanical properties require low chromium contents. This demand is made because high chromium contents promote the formation of brittle phases in the microstructure (sometimes after thousands of service hours) which are detrimental to ductility and strength of the alloy.

CASE STUDY:
STRUCTURAL MATERIALS FOR THE BLADES OF GAS TURBINES

DEMANDS

The efficiency of a gas turbine for power generation or aircrafts essentially depends on the temperature and the pressure of the combustion gas at the inlet of the high pressure turbine. The function of the first row of blades, which is taken into consideration here, is to draw energy from the fluid and to convert it into kinetic energy of the rotating parts of the turbine. This process implies high circumferential velocities, resulting in high centrifugal forces and thus in creep loading of the blade as indicated schematically in Figure 12. The turbulent flow of the combustion gas excites vibrations of the blade, resulting in high cycle fatigue. Especially during start-up and shut-down of the gas turbine, temperature gradients develop in the blade, resulting in thermally induced mechanical strains and stresses. Depending on the type and the operating method of a gas turbine, these procedures occur some 100 up to some 1000 times during the life of the blade, resulting in thermally induced low cycle fatigue. Finally, the combustion gas may cause oxidation and high temperature corrosion. This leads to the demands on structural materials as

- ❒ high specific strength up to very high temperatures,
    - → low density,
    - → high melting- and solidus temperature, respectively,
    - → high creep strength,
    - → high thermal fatigue strength (low cycle fatigue strength),
    - → high isothermal fatigue strength (high cycle fatigue strength),

- ❒ sufficient resistance against oxidation and corrosion,
    - → development of dense, non-volatile and thermodynamically stable oxide layers,
    - → availability of procedures and materials for suitable coatings.

Figure 12   *Gas turbine blade with centrifugal force F (black), cooling gas (dark grey arrows9 and combustion gas (light grey arrows*

SUITABLE MATERIALS

With respect to the demands „low density" and „high melting temperature", Figure 13 shows a map, where both quantities are given for a number of metallic elements and ceramic materials. Suitable materials should be found at the lower right corner of the map, and there are solely ceramic materials. There is no doubt, that carbon reinforced carbon would be an ideal structural material for the blades of gas turbines, if suitable coatings could be made available protecting the material from the severe oxidation occuring at the service temperature of a gas turbine. On the other hand, the other ceramic materials shown in the map still suffer from a lack of ductility. This is a preclusive criterion regarding the impact of a foreign body, which may cause catastrophic failure of brittle materials.

Consequently, the blades of actual gas turbines are still made of metallic materials. Figure 13 shows that there are metals with high melting temperatures like tungsten and tantalum. However, these materials have high densities which would result in very high centrifugal forces. Moreover, they do not develop suitable oxide layers and, because of their bcc lattice and their strong atomic bonding, they show a transition to brittle behaviour at temperatures well above ambient temperature (see Figure 10). Similar considerations are valid for molybdenum, niobium, chromium and vanadium. Regarding the remaining metals, one would expect from Figure 13 that titanium with a low density is more suited than nickel or iron for use as base element for blade materials. In fact, titanium alloys are used in the compressor of gas turbines at temperatures up to 550°C. However, the temperature of the combustion gas at the inlet of the high pressure turbine is close to the melting temperature of metals like nickel, cobalt, iron and titanium. Therefore, extensive cooling of the blades is necessary, resulting in a reduction of the efficiency of the gas turbine. Even in a cooled blade, it is not possible to design a microstructure in titanium alloys, resulting in sufficient strength at temperatures present in gas turbines. The unique importance of nickel-base superalloys results from the possibility to design a microstructure which provides an adequate strength at temperatures approaching 85 % of the absolute melting temperature.

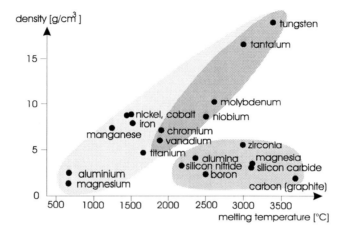

Figure 13   Density and melting temperature of some important materials

DESIGN OF THE MICROSTRUCTURE OF NICKEL-BASE SUPERALLOYS
Nickel has a fcc lattice shown in Figure 14. As discussed in paragraph HIGH TEMPERATURES, regarding diffusion in densely- and non-densely-packed lattices, this is a prerequisite for good creep strength. On the other hand, dislocations easily move in pure nickel, because densely-packed slip planes (see Figure 15) are available.

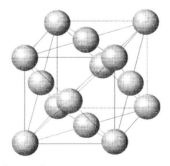

Figure 14   Face-centered unit cell

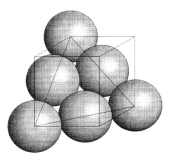

Figure 15   Densely-packed (111)-plane of the face-centered cubic lattice

To impede dislocation movement, strengthening mechanisms treated in paragraph **INTERACTIONS BETWEEN DISLOCATIONS AND LATTICE DEFECTS** are used. A rather high fraction of the nickel atoms is substituted by foreign atoms like chromium, iron, cobalt, tungsten, tantalum and vanadium as shown in Figure 16. This results not only in solid solution hardening, but also in the impediment of diffusion enhanced processes, because diffusion of substitutionally solved atoms like tantalum or tungsten is slower than the diffusion of nickel atoms („self-diffusion").

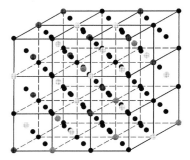

Figure 16   Nickel-lattice with substitutionally solved foreign atoms

In Figure 16 it is assumed that the foreign atoms are distributed statistically in the lattice. This phase is called γ-phase. However, if sufficient amounts of aluminium atoms are present in the lattice, the formation of a so-called γ'-phase with an ordered arrangement of nickel and aluminium atoms is observed at suitable temperatures. As shown in Figure 17, the nickel atoms occupy face centered sites and the aluminium atoms corner sites in the unit cell of the lattice. A limited fraction of the aluminium atoms may be substituted by titanium atoms. Since 1/2 of each face-centered site and 1/8 of each corner site are included to the unit cell under consideration, the composition of the phase is $Ni_3(Al,Ti)$. The γ'-phase is separated from the γ-phase as shown schematically in Figure 18. Since the lattice planes and lattice directions of the γ-phase continue into the γ'-phase, the boundary between both phases is called a coherent phase boundary. However, there are distorsions along the phase boundary, because the lattice constants do not match perfectly. These distorsions are combined with so-called coherency-stresses, which impede dislocation movement in the γ-phase. On the other hand, dislocations are only able to slip into and to shear the ordered structure of the γ'-phase at rather high stresses, which are hardly relevant concerning stresses acting in a turbine blade. Altogether, the formation of the γ'-phase is a very effective strengthening mechanism in nickel-base superalloys.

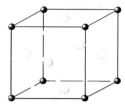

Figure 17   Unit cell of the $Ni_3(Al,Ti)$ ordered structure (γ'-phase)

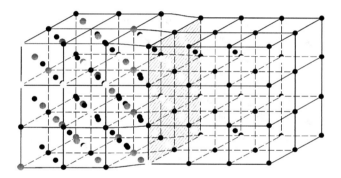

Figure 18   Coherent phase boundary between γ- and γ'-phase

In practice, an alloy is annealed at a high temperature, e.g. 1200°C, in order to statistically distribute the solved foreign atoms in the lattice corresponding to maximum entropy. By quenching to ambient temperature, this structure is frozen. After heating to elevated temperatures in the range 600°C to 900°C, the formation of the γ'-phase is initiated. Alternatively, the γ'-phase may be formed during controlled cooling from solution temperature. Combinations of both procedures are possible, too. The γ'-phase is formed as small particles. Their shape (spherical or cubical) mainly depends on the volume fraction of the γ'-phase. The dimensions of the particles range from some 10 nanometers to some 100 nanometers depending on the temperature or the temperature range of the γ'-formation. In high strength nickel-base superalloys, the γ'-volume fraction approaches 70 % or more, resulting in the microstructure shown in Figure 19. Dislocation movement is restricted to small channels of γ-phase. This microstructural design is the main feature that leads to the high strength of nickel-base superalloys at high temperatures.

In addition to solid solution and γ'-precipitation hardening, nickel-base superalloys are also, to some extend, strengthened by the presence of carbides. Furthermore, in the case of powdermetallurgically (PM) processed alloys, dispersions like Y_2O_3 may be used to obtain an adequate strength at very high temperatures, where strengthening by γ'-particles vanishes.

Figure 19   Microstructure of a nickel-base superalloy with high volume fraction of γ'-particles (light grey) in a γ-matrix (dark channels)

However, until now dispersions-strengthened PM-materials did not gain acceptance as blade materials, because the maximum level of strength is significantly lower than that of alloys with high volume fractions of γ'-phase.

Because of their high strength at high temperatures, the latter cannot be forged. Therefore, turbine blades must be produced by investment casting. Conventionally cast, a blade with a grain structure shown in the left part of Figure 20 may be obtained.

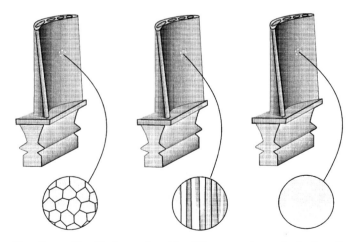

Figure 20   Gas turbine blades produced by different casting technologies: Conventionally cast with globular grains (left), directionally solidified with columnar grains (middle) and single crystal (right)

As already mentioned in paragraph HIGH TEMPERATURES, grain boundaries under high shear stresses are prone to grain boundary sliding. Grain boundaries under high tensile stresses are prone to creep damage i.e. formation of pores or microcracks. This is the case for grain boundaries which form angles of 45° and 90°, respectively, with the blade axis. Directional solidification during the casting process prevents the formation of these grain boundaries to a large extend, as shown in the middle part of Figure 20. By this procedure, columnar grains are produced which extend almost along the entire blade. The last step in this development is the complete elimination of grain boundaries by casting a turbine blade as one single crystal, as shown in the right part of Figure 20. This is done by using a seed crystal which acts as the only nucleus for the solidification.

Since every crystal has orientation-dependent mechanical and thermal-mechanical properties, a single crystal turbine blade possesses this orientation dependence on a macro-scale. Therefore, an optimization of the properties of the turbine blade is obtained by the realization of an optimum relationship between the orientation of the crystal lattice and the axis of the component. In actual single crystal turbine blades this orientation relationship is controlled to a few degrees.

REFERENCES

Ashb-80   Ashby, M. F.; Jones, D. R. H.: Engineering Materials, Int. Series on Materials Science and Technology, Vol. 34, Pergamon Press, Oxford, 1980.

Ashb-86   Ashby M. F.; Jones, D. R. H.: Engineering Materials 2, Int. Series on Materials Science and Technology, Vol. 39, Pergamon Press, Oxford, 1986.

Shac-92   Shackelford, J. F.: Introduction to Materials Science for Engineers, 3rd. Edition, Macmillan, New York, 1992.

Smal-95   Smallman, R. E.; Bishop, R. J.: Metals and Materials, Butterworth-Heinemann, Oxford, 1995.

Vlac-89   VanVlack, L. H.: Elements of Materials Science and Engineering, 6th. Edition, Addison-Wesley, Reading/Mass, 1989.

Discussion

Question - A. Albers

Mr. Löhe, you have reported about texture design and elements of a closed design theory referring to materials. It is a good example for variety. You mentioned 10.000 to 20.000 kinds of metal we use. They all have their specific applications. Are there any starting points for a predefinition of a design theory in material research today which is part of the design process.

Answer - D. Löhe

I think that these structural features, I have been talking about can be applied to all metallic materials. The question whether these features can exist in all materials is a different story. But in the design process the interaction between the features governing plastic deformation and the features governing failure and specific design elements are well known. The problem arising is of a different nature. Considering a nickel-base alloy as a 12-element or 15-element system i.e. to evaluate all phases that may exist within this system at different temperatures a 12-dimensional or 15-dimensional space would be necessary. This represents a challenge. And I think that in constitutional research there has taken place a tremendous upheaval by the aid computer aided simulation. Concerning the task to produce an alloy based on first-principles after recognition of the design elements, I think presently a lot of improvement is made to provide the tools. The design elements themselves are rather well known in the field of metallic materials. I have only been talking about metallic materials. We will learn in the following presentation that different boundary conditions might be governing in different systems.

The challenge is to find an optimized combination of these design elements, economically as always. Additionally, I have mentioned oxidation and corrosion - of course there is always a greater variety of requirements than in this simple description.

Question - J. Gero

I am trying to get an understanding of the design process. It looks to me like a combination of 3 types of processes. One is trial-and-experimentation, that is, you try some substitutions and experiments to see if the results are satisfying since you are dealing with the physical material. There appears to be analogy in the sense that this particular metal inserted at the molecular level worked here, maybe it will also work with in other combination. And then the third one is, if it works here, maybe if we increase the density it will work better. Is this the reasonable characterisation of some of the design processes that you are employed?

Answer - D. Löhe

First of all, the 30 years have been used for the development of a appropiate technology. They knew the detrimental effect of the grain-boundaries, of course, at the very beginning of this development. But it took such a long time to set it into practise, to realize the single-crystal turbine-blade. So that is that, what I said. You know the elements, you know the potential of the elements, but the question is, how to realize the elements in a given system. So in this case, the nickel-based-alloys are very, very important simply because they allow to produce this structure. And this structure is relatively stable. Imagine, the alloy is far away from thermodynamic equilibrium, so the question always arises: what happens with this structure during service? At high temperature and thousands of hours? So this structure will change, of course. It is not thermodynamically stable, but it is relatively stable. And this is a further demand on the design of the microstructure. It makes no sense to have it just for a few hours, you have to have it for 1,000 eventually 100,000 hours. But you are right, we have the elements and the question is how to realize it in other systems, e. g. in a system with a lower density which would be a huge progress in the turbine-blade technology. And I think, that is a problem which we are facing now.

Question - S. Rude

You talked about a rather large set of materials growing infinitely. You also spoke of a set of attributes (e. g. density or melting temperature). What kind of experiences have you had in material science: Is the set of attributes designers are asking for also growing, or is the set of attributes finite? And is there a known structure of attributes in material science?

Answer - D. Löhe

It might happen that some requirements are added e.g. if a component is in service under corroding conditions that were not existing in the past. If - returning to the very first figure - a framework is existing, a process that was not there before, if a function is utilised yielding new boundary conditions for the structural material, this might give rise to a new demand. But in reality the requirements are clear and have stayed easy to survey in the course of the existence of components. The origin of this tremendous variety is not hard to understand. E. g. if a specific microstructure has to be created, say such a hardened microstructure, perhaps it can be easily achieved using a very simple and low cost material in a thin structure of 3 mm. But if you want to fabricate a 1m diameter shaft for a steam turbine power plant you will not succeed and you will have to choose a material completely different. The goal is exactly the same. The demands of the design engineer might be identical but only due to producability of such a structure combined with the demand of maximum profitability there are some hundred steel for hardening and tempering.

Ceramics from Elementorganic Compounds

Fritz Aldinger

Max-Planck-Institute for Metals Research and

Institute for Non-Metalic Anorganic Materials

University of Stuttgart, Germany

ABSTRACT

The condensation of preceramic compounds is done by solid state pyrolysis, chemical vapor deposition or chemical liquid deposition. The general idea of such process routes is that the precursor molecules contain already structural units of the residual inorganics, thus providing novel paths of controlling the composition, atomic array and microstructure of materials. For the manufacture of engine components the production of covalent bonded inorganics on the basis of Si, B, C and N by solid state pyrolysis is of special interest. Using proper precursor polymers either amorphous or crystalline materials can be produced. Due to the lack of grain boundaries and a low atomic diffusivity in covalent materials their amorphous states provide thermal stability, oxidation resistance and very attractive mechanical properties at rather high temperatures. The crystalline materials also create substantial interest. Depending on the type of precursor, devitrification of the amorphous state into the nanocrystalline state occurs at very high temperatures providing the basis for thermally quite stable microstructures. Another favourable feature of the process is that for the shaping of products as parts, coatings, fibers or other components economic technics of standard polymer engineering can be applied.

INTRODUCTION

Processes of condensation of preceramic compounds are known to be proper means to build up novel materials. With these processes one starts from organoelement monomers or polymers which are chemically reacted into merely inorganic materials. Such process routes enable novel paths of controlling composition, atomic array, microstructure and shape of materials and by that their properties providing the basis for advanced component developments. These processes are especially valuable for the production of inorganics which are predominantly covalently bonded as proposed firstly [*Chan-64*], followed by others [*Verb-73, Yaji-75, Seyf-84, Peuc-90*] and reviewed recently [*Bald-97, Bill-95a, Lain-98, Ried-93*]. Since such type of ceramics cannot be densified by sintering in the pure state but only in composition with sintering additives, condensation of preceramic compounds offers the potential to transfer the beneficial thermal, chemical and mechanical properties of covalently bonded inorganic structures into materials without compromises providing a mean for the development of new classes of materials for high temperature resistant components.

CONDENSATION OF PRECERAMIC COMPOUNDS

On principle, the condensation of preceramic com-pounds can be done in solid, vapor or liquid states known as solid state pyrolysis (SSP), chemical vapor deposition (CVD) and chemical liquid deposition (CLD), respectively (Figure 1). The general idea of these processes is that the starting molecules contain already structural units of the residual inorganic material.

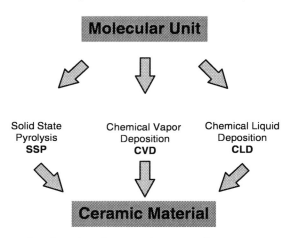

Figure 1   Types of condensation of preceramic compounds

With solid state pyrolysis typical preceramic polymers for the production of silicon-based ceramics are shown in Figure 2.

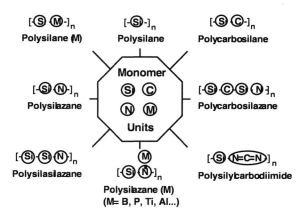

Figure 2   Typical preceramic polymers for the production of silicon-based ceramics [Bill-95a]

They all are characterised by carbon and/or nitrogen being attached directly to silicon. Other elements as e.g. boron also can be introduced into the backbone of the polymer or into its side chains.

In Figure 3 an example is shown for a polysilazane which can be used for the preparation of ternary Si-C-N ceramics.

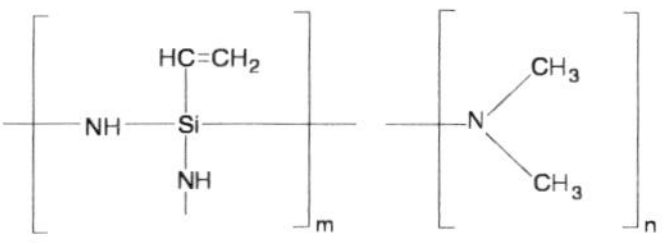

Figure 3   *Molecular structure of polyvinyl-silazane (VT50, Hoechst AG, Germany)*

The molecular structure of such polymers is characterised by silicon atoms surrounded tetrahedrally by carbon and nitrogen, which are typical structural units of silicon carbide-type and silicon nitride-type materials. This is shown schematically in Figure 4.

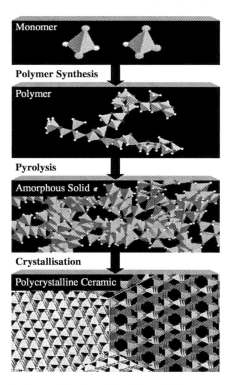

Figure 4   *Structure development during solid state pyrolysis*

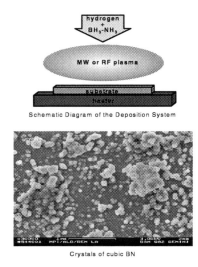

Figure 5   *Chemical vapor deposition of cubic boron nitride [Kony-97]*

The CVD process is widely used to deposit successfully inorganic compounds. Recent studies demonstrate e.g. the deposition of single phase cubic boron nitride in a hydrogen plasma using borane-ammonia-type precursors [*Kony-97*] (Figure 5).

Chemical liquid deposition offers a path to produce ceramics at ambient temperatures [*Bill-96*]. An example for the deposition of a covalent inorganic material is the formation of polsilylcarbodiimide $[Si(N=C=N)_2]_n$ by the reaction of $SiCl_4$ with bis(trimethylsilyl)-carbodiimide in a solution of toluene similar to the silica gel reaction of $SiCl_4$ with water forming SiO_2 (Figure 6).

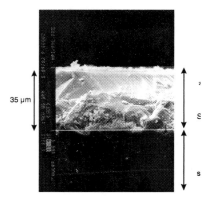

Figure 6   *SEM image of a polysilylcarbodiimide layer on a silicon substrate [Bill-96]*

Whereas CVD and CLD is only suitable for the deposition of thin-layered structures solid state pyrolysis is attractive for the production of structural components as well.

MICROSTRUCTURE DESIGN BY PRECURSOR PYROLYSIS

In detail, the process of solid state pyrolysis is outlined in Figure 7. The synthesis of monomers and polymers is followed by crosslinking the polymer chains at moderate temperatures in order to modify the rheology of the polymer and to transform it into a infusible preceramic network. The actual transformation into a real inorganic material, i.e. the ceramisation takes place by a sequence of condensation reactions during heating up the preceramic network to some 1000°C as has been described in detail elsewhere [*Bill-95a, Bill-95b, Bill-98*]. With this thermal treatment at intermediate temperatures a metastable amorphous material is formed which can be crystallized into stable phases with further increasing the temperature depending on the overall composition of the amorphous material.

Kinetically, this production process differs in principle from conventional production processes of ceramics by sintering. Due to the low diffusivity of covalently bonded atoms the sintering mechanisms need very high thermal activation, thus other processes like phase formation and grain growth will be activated, simultaneously. Since such processes affect the microstructure substantially its control is limited. In contrast with precursor pyrolysis the starting monomers and polymers reveal a rather high state of energy with respect to the final product and the activation energy for the condensation reactions, i.e. for the ceramisation is relatively small (Figure 8).

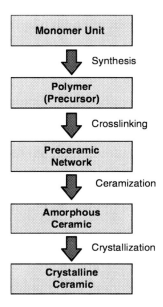

Figure 7   *Flow sheet of the production of ceramic materials by solid state pyrolysis*

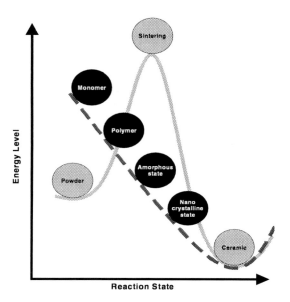

Figure 8 *Thermal activation with sintering and precursor pyrolysis (schematically)*

Thus careful heating at temperatures where chemical condensation reactions of neighboured polymer chains take place but extended diffusion is still limited enables the stabilization of states of intermediate energy levels. With such heat treatments, which can be compared to releasing and re-applying the brake during downhill driving, preceramic networks can be transformed into amorphous glass-like materials (Figure 9).

An example of grain boundary phase engineering is shown in Figure 10. Amorphous material derived from polyhydridomethylsilazane (NCP 200, Nichimen Corp. Japan) can be crystallized into composites of silicon nitride and silicon carbide with „clean" grain boundaries, whereas with small amounts of boron additions a continuous grain boundary phase can be formed consisting of a few atomic layers of turbostratic boron carbonitride [*Bill-95a, Jalo-95*].

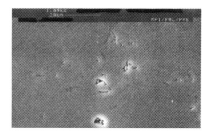

Figure 9 *Microstructure of amorphous $SiC_{1.6}N_{1.3}$ ceramic derived by pyrolysis of warm pressed polyvinylsilazane powder at 1000°C (3 vol-% porosity) [Seit-96]*

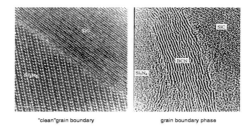

Figure 10   HRTEM image of precursor-derived Si-C-N ceramic without dopants (a) and with 1.1 wt.% B (b) [Jalo-96a, Jalo-96b]

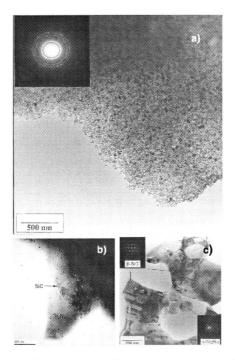

Figure 11   TEM bright field image of Si-C-N-based composites containing 1.1 wt.% B (a), no dopant (b) and 0.5 wt.% P (c) crystallized at 1800°C [Bill-95a, Jalo-95]

A further example of microstructure design is the development of nanocrystalline materials. The microstructure with a medium grain size of some 30nm as shown in Figure 11a is created by heat treating amorphous $SiC_{0.64}N_{1.06}B_{0.05}$ for 50h at 1800°C. With this treatment a grain boundary phase is formed (Figure 10b) hindering grain boundary movement thus stabilising the nanocrystalline microstructure.

Actually, the microstructure shown in Figure 11a consists of nanocrystalline silicon nitride and nanocrystalline silicon carbide, i.e. a nano/nano-type composite. Without dopants silicon nitride grains coarsen during the same heat treatment thus the microstructure is becoming a micro/nano-type composite (Figure 11b). Phosphorous dopants obviously cause grain growth of both silicon nitride and silicon carbide, thus a micro/micro-type composite is formed (Figure 11c).

PROPERTIES OF PRECURSOR-DERIVED MATERIALS

Covalently bonded inorganic materials are known for their high thermal, chemical and mechanical stability what is demonstrated by precursor-derived materials. Figure 12 shows the thermal stability of a polysilazane-derived Si-B-C-N ceramic after pyrolysis at 1050°C. In argon the material only degasses at temperatures around 2000°C, although the material contains silicon nitride which normally dissociates at much lower temperatures [*Bill-95a*]. The reason for this behavior is probably a kinetic stabilization due to the grain boundary phase surrounding the nanosized silicon nitride grains (Figure 10b), i.e. the turbostratic boron carbonitride layers form a nanosized „container" mechanically resistant against the silicon nitride dissociation pressure which is at 2000°C in the order of 20 bars.

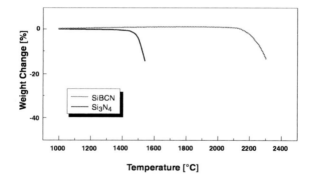

Figure 12   Thermal stability of polysilazane-derived $SiC_{1.5}N_{0.7}B_{0.3}$ ceramic and silicon nitride powder in an inert atmosphere

Since the oxidic grain boundary phases control the oxidation resistance of sintered carbide and nitride ceramics the precursor-derived materials offer a higher stability against oxygen. In air a silica layer is formed at the surface providing oxidation resistance up to 1600°C [*Ried-95*]. Thermogravimetric investigations of coatings based on amorphous $SiC_{1.5}N_{0.7}B_{0.3}$ at 1400°C in air confirm that the oxidation of substrates of carbon fiber reinforced silicon carbide (C/C-SiC) were almost totally prevented (Figure 13).

The reasons for the oxidation proof of this coating, which offers self-healing capabilities, are not yet clearly understood. However, it is obvious that the amorphous state with its very low diffusivity due to the covalent character of the material in combination with the formation of silica-based passivating surface layers play an important role.

Preliminary results indicate that amorphous precursor-derived materials have a great potential for creep resistant materials. Under compression loading, amorphous $SiC_{1.6}N_{1.3}$ ceramics derived by pyrolysis of polyvinylsilazane at 1050°C reveal after a initial strain of some 0.8 % rather small values of deformation and strain rates (Figure 14). Even after a loading of 100 MPa over some 110 h and at temperatures as high as 1550°C the deformation increases only by some 0.5% (Figure 14a). The strain rates are decreasing down to the level of as low as 10^{-8} s^{-1} (Figure14b).

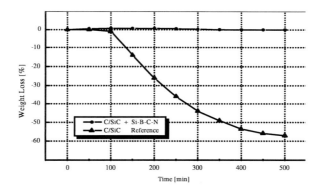

Figure 13   Oxidation behaviour of C/C-SiC in air at 1400°C with and without an oxidation protection coating of precursor-derived material [Haug-96]

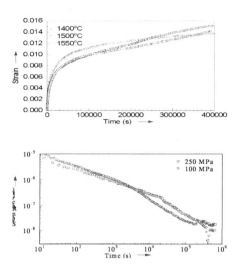

Figure 14 Strain (a) and strain rate (b) during compression creep of amorphous $SiC_{1.6}N_{1.3}$ derived from polyvinylsilazane [Bill-96, Thum-96]

The main reason for the high strain and strain rate in the initial deformation state and their decrease with loading time is probably due to a superposition of creep deformation and densification of residual porosity. These results indicate a great potential of precursor-derived materials for high temperature applications.

MATERIALS ENGINEERING

Except of their sensitivity to oxygen organoelement polymers have quite similar thermal and rheological properties as their organic equivalents. Therefore, the process engineering well-known for the production of components, extrusions, fibers, composites, coatings and infiltrated materials of standard polymers can be adapted successfully to the process engineering of polymer precursors for the production of ceramic materials. A flow sheet of processes already used for the materials engineering of organoelement polymers is shown in Figure 15.

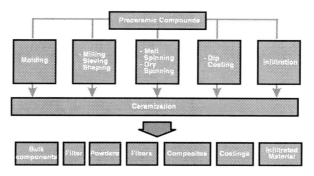

Figure 15   Materials engineering for the pro-duction of ceramics from organoelement polymers [Bill-95a]

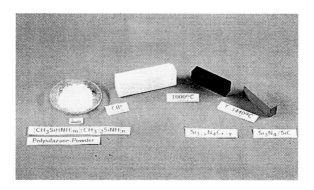

Figure 16   Demonstration of sample preparation by solid state pyrolysis of cold-compacted powder of polyhydridomethylsilazane

Of special interest for the production of components is the manufacture of bulk material (Figure 16). The polymer is thermally cross-linked and milled into a fine powder which is than compacted by cold isostatic pressing and subsequently pyrolysed at 1050°C in Ar. Finally the sample is heat treated at 1440°C for crystallisation.

With solid state pyrolysis processing the production of dense material is not trivial, since the condensation reactions during pyrolysis are combined with the evolution of gases like H_2, NH_3 and CH_4. Such species degas easily with the manufacture of fibers or coatings, i.e. with shapes which are thin in at least one dimension. In bulk material, however, such gases leave pores behind, which do not close during further pyrolysis due to its high viscosity.

In order to reduce porosity the compaction of the polymer powder was done at elevated temperature [*Bill-95b, Bill-96, Bill-98*]. Using this technique the porosity was reduced to some 3 % (Figure 9), depending on the temperature and pressure during warm compaction of the powder.

In order to make oxidation sensitive materials as carbon fiber reinforced materials usable at higher temperatures precursor-derived coatings have been developed [Haug-96] (Figure 17). The coatings have been produced by dip coating substrates of C/C-SiC into solutions of precursors and subsequent conversion into inorganics by pyrolysis. The improvement of the oxidation resistance of the substrate material is shown in Figure 13.

With respect to the development of ceramic reinforced ceramic composites the production of fibers is realized by pyrolysis (heating rate: 1.7 K/min, RT to 1100°C) of polymer „green fibers" (Figure 18).

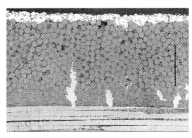

Figure 17 Cross section of a C/C-SiC substrates coated with a layer of amorphous $SiC_{1.5}N_{0.7}B_{0.3}$ [Haug-96]

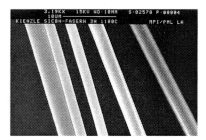

Figure 18 SEM photograph of amorphous Si-B-C-N fibers prepared by pyrolysis [Bill-95c]

The polymer fiber is produced from concentrated thf solutions of boron-containing polysilazane by dry spinning [Bill-95c]. The strength and Young's modulus of the as-received Si-B-C-N fiber are determined to be 590 MPa and 68 GPa, respectively.

CONCLUSIONS

In summary the process of solid state pyrolysis of preceramic compounds and the materials produced thereby have the following features:
Additive-free densification of covalent ceramic materials
Kinetic stabilisation of new amorphous and crystalline phases
Novel paths of microstructure design
Chemical homogeneity

High-purity materials
Moderate processing temperatures
Material and component engineering by well-known polymer processing techniques
Improved thermal, chemical and mechanical material properties
With these features precursor pyrolysis has a great potential for materials-driven advanced technologies

ACKNOWLEDGEMENT

The author likes to thank his collaborators for their close cooperation in preparing this paper.

REFERENCES

Bald-97	H.-B. Baldus and M. Jansen, Angew. Chem. 109 (1997) 338-354
Bill-95a	J. Bill and F. Aldinger, Adv. Mater. 7 (1995) 775-787
Bill-95b	J. Bill, R. Riedel and F. Aldinger, Proc. 4. Eur. Ceramics, Riccione, Italy (1995) 125-132
Bill-95c	J. Bill, A. Kienzle, M. Sasaki, R. Riedel and F. Aldinger, Advances in Science and Technology 3B, Ceramics: Charting the future, P. Vincenzini (ed.), Techna Srl, (1995) 1291-1299
Bill-96	J. Bill and F. Aldinger, Z. Metallkd. 87 (1996) 827-840
Bill-98	J. Bill, J. Seitz, G. Thurn, J. Dürr, J. Canel, B. Janos, A. Jalowiecki, D. Sauter, S. Schempp, H.-P. Lamparter, J. Mayer and F. Aldinger, phys. stat. sol. (to be published)
Chan-64	P. G. Chantrell and E. P. Popper, Special Cera-mics, Academic Press, New York (1964) 87-103
Haug-96	R. Haug, D. Heimann, J. Bill, and F. Aldinger, Verbundwerkstoffe und Werkstoffverbunde, DGM Informationsges. Verlag, Oberursel, Germany (1996) 429-432
Jalo-95	A. Jalowiecki, J. Bill, M. Frieß, J. Mayer, F. Aldinger and R. Riedel, Nanostr. Mat., 6 (1995) 279-282
Jalo-96a	A. Jalowiecki, J. Bill, F. Aldinger, J. Mayer, M. Rühle, Proc. Werkstoffwoche `96, Stuttgart, Germany (1996) 235-240
Jalo-96b	A. Jalowiecki, J. Bill, F. Aldinger and J. Mayer, Composites 27A (1996) 717-721
Kony-97	I. Konyashin, J. Loeffler, J. Bill and F. Aldinger, Thin Solid Films 308/309 (1997) 101-106
Lain-98	R. Laine, Appl. Organomet. Chem., in press
Peuc-90	M. Peuckert, T. Vaahs and M. Brück, Adv. Mater. 2 (1990) 398-404
Ried-93	R. Riedel, Nicht-oxidische Keramiken aus anorganischen Vorstufen, Materialkundl. Techn. Reihe, Gebrüder Bornträger, Berlin, 1993
Ried-95	R. Riedel, H.-J. Kleebe, H. Schönfelder and F. Aldinger, Nature 374 (1995) 526-528
Seit-96	J. Seitz and J. Bill, J. Mat. Sci. Lett. 15 (1996) 391-393
Seyf-84	D. Seyferth and G. H. Wiseman, Polym. Prepr. Am. Chem. Soc. Div. Polym. Chem. 25 (1984) 10-12
Thum-96	G. Thurn and F. Aldinger, Proc. Symp. Grain Boundary Dynamics of Precursor-Derived Ceramics, Schloß Ringberg, Germany (1996) to be published
Verb-73	W. Verbeek, German Patent 2 218 960 (1973)
Yaji-75	S.Yajima, J. Hayashi and M. Omori, Chem. Lett. (1975) 931-934

Discussion

Question - F.-L. Krause

Yes, I've got a general question. In this room the main subject is the problematic nature of Design Theory. All approaches presented her deal with the creation of something new. These are design issues. You design material. How would you describe the access to a Universal Design Theory from your point of view.

Answer - F. Aldinger

That is a difficult question that I cannot answer, to put it concisely. As indicated previously and was explained by Mr. Löhe, I would not say that matters are too complex but that, conceivably, the objections on which a design theory might be based are too diverse so that I cannot imagine how they might be tackled by such a theory. If that sounds negative, I hope that, in time, I will grow in wisdom. At the present moment, however, I cannot see things differently. That is why I am extremely curious and happy to be here where I can perhaps learn a few things. Obviously, as already intimated, we understand many criteria well, yet we are usually interested in more than a single property and want the simultaneous interaction of many properties optimized. Today every engineering material requires an optimization effort; the basic physics is understood, but the interaction of these basic principles still takes, to put it mildly, the application of trial and error methods.

Comment - H. Grabowski

If more than one theory for the design of artefacts exists, then all but one theory must be wrong, only one theory can be true. In every discipline, requirements occur and solutions are developed. The requirements in different disciplines such as material science, microelectronics, microsystem technology, or mechanical engineering may be different in detail.

But nevertheless, requirements are used in every discipline to control the elaboration of a solution. This elaboration process again may be different in each discipline. Here, the question of whether these processes are also part of a design theory or not remains to be discussed,. But every discipline again has attributes as required and attributes as built. And this set of attributes is finite (but this will be explained in my contribution later on). For now, I want to state that only one design theory can exist. Although there might conceivably be different elaboration processes in the different disciplines, it is not possible to have more than one design theory.

Question - C. Weber

Both contributions from the materials science and materials development area have made a great impression on me because - in the terminology of design science - technological problems seem to be dominating ones. On the other hand you already have a lot of knowledge concerning - again in the terminology of design science - the basic working principles, i.e. the principles of how to give materials certain required properties. The "only" remaining question seems to be how to realize these properties technologically when manufacturing components in big numbers.

Answer - F. Aldinger

Everything at the same time.

Comment - F.-L. Krause

Direct response to Prof. Grabowski. We talked about it before. In my opinion there might be an analogy to the following: world wide we have a lot of religions and all of them declare to be the only right one. Possibly it is the same in the case of theories. That means, there are a few theories but finally they have the same aims. The Theories express something similar or even alike. It can also be the situatedness talked about this morning. Is it important that we have different approaches to the term theory and that we interpret and use it in different ways. Therefore, I think this is a field to be discussed more into detail.

Question - W. Menz

As an external visitor of this workshop I suppose you are searching for the universal world formula like in physics. Would you be happy if you really had the universal design theory and could you really work with it?

Answer - H. Grabowski

I would like to give a direct answer to this question: We already discussed the subject this morning, you were not yet present at the time, and made a distinction between research and development. Research aims to increase knowledge about various issues in our world and would hypothetically end with the existence of a universal world formula. Development instead aims to use knowledge as much possible to design artefacts, non-natural objects of our world. This difference may be not extremely clear, but it is continuos. We are here a circle of researchers and should discuss this boundary.

If we tried to develop a theory for research, we would have to work for a limited time, then we would know how research works or we would know the world formula. But we intend to develop a theory for design. And this is necessary for the education and for the qualification of engineers. Even in mechanical engineering, different methods of design exist. And if you start to work as a young engineer, you fail, because what you learned is not enough and cannot be applied directly in practice in the manner of solving a differential equation. Also, in this area of mathematics, different solution procedures are known, but only one correct solution is possible in theory – and this is known before you solve the problem, because a theory exists.

Comment - K. Ehrlenspiel

I'd like to add an argument in favour of the diversity of theories. It can be assumed that reality, and thus also physical or chemical reality, is so complex, that one cannot grasp it with "one single" model. Allow me to draw your attention to the corpuscular theory and the undulatory nature of light. These two states cannot be imagined simultaneously. Normally, we simply have ideas about mechanics in our thoughts, and often they are too simple for reality. Therefore, we develop a number of rival models. This is something Mr. Hacker would have to comment on. - Thus, we live with different models which we use to grasp and deal with what is real, with reality. Since reality is so complex, why, then, should we not also work with different theories of development and design? I don't consider "one single" encompassing super-theory to be a necessary constraint in view of our limited capacity of imagination.

Comment - A. Albers

I have a similar opinion like you, Mr. Ehrlenspiel, we do not have to look for the all describing equation of the design process in every field in the first step, but I think it is very interesting and promising to lift the discussion from the narrow outer field of the machine construction to the bird's eye view and to talk about the common things. I believe we can not discuss it right now, because we do not have the time. But I think there are a lot of starting points in material science and electrical engineering for a general design-theory. It does not have to be just one.

Comment - W. Menz

In microsystem technology we have the difficulty in designing with two theoretical regimes in mind: the mechanical design and the electronic design. The difficulty may be to unite these two theories.

Comment - J. Gero
It seems as if we're using the word theory, we mean many different things. In some instances we seem to be using it to describe something, which is more like a model. We're also using it as an explanation of something. And when we make comparisons perhaps with other areas, we claim some causality of the theories, e.g. Einstein provided us the causality for gravity. How gravity actually works and what it is, what Newton did not. Newton explained the results of having gravity, but said nothing about gravity itself. And it seems that the claim - a causal theory of design - is beyond the information that we have available to us. Because a fundamental requirement for a causal theory is a detailed understanding of the phenomenon which is described and predicted and we don't have that. In fact, I would claim that we are still struggling to get the data which is the phenomenon of design itself. Once we have that, assuming we can get it in a readily representable form, then, I think, we can proceed. I think, a causal theory of design is not yet possible.

Comment - C. Weber
Only a small comment on the remark of Mr. Menz: Besides the necessity of integrating the product development processes in the mechanical and the electronic domains it seems to be of the same importance to integrate materials development on one hand and all types of product development processes on the other hand. I hear from a lot of people coming from the materials side that developers/designers have to change their attitude towards materials and materials development, not only looking up materials properties in catalogues.

Comment - W. Menz
One major difficulty in designing microsystems is the question of the material. Due to the minimizing of the structures I can not trust the established bulk material parameters any more. If the dimensions of my designs are in the order of the grain structure of the material I can not expect to find for example the Young's modulus unchanged from that of the bulk material. In the future I have to build data bases for material parameter which take into consideration among others the manufacturing history. It is a significant difference in the mechanical parameters of a microstructure if it carries a layer of nitride or oxide on the surface. The shape of the structure has an influence too on the material parameters. It is not clear yet how we achieve these multi-dimensional material tables.

From simple building blocks to complex target molecules and multifarious reactions

Henning Hopf

Institute of Organic Chemistry

Technical University of Braunschweig, Germany

email: H.Hopf@tu-bs.de

Keywords: molecular design, Aufbauprinzip (built-up principle), design of organic structures and molecules, genetic relationships, planar and three-dimensional hydrocarbon frameworks, multifarious reactions, target molecules

Hydrocarbons form the basis of organic chemistry and to a large extent of organic reactivity as well. Although not the only binary compounds in organic chemistry, they are by far the most numerous and important ones. Hydrocarbons may be "constructed" by a very simple *Aufbauprinzip* (built-up principle) that in its most elementary variant uses just three building units - the sp^3- (**1**), the sp^2- (**2**), and the sp-hybridized (**3**) carbon atom (Figure 1).

Allowing only one additional carbon atom, and considering only molecules made up of covalently bonded atoms these three basic units translate into the C-C-single (**4**), the C-C double (**5**), and the the C-C triple bond (**6**) (Figure 2).

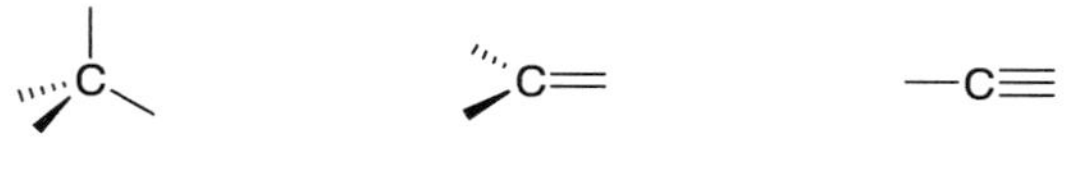

Figure 1   The building blocks of hydrocarbon chemistry.

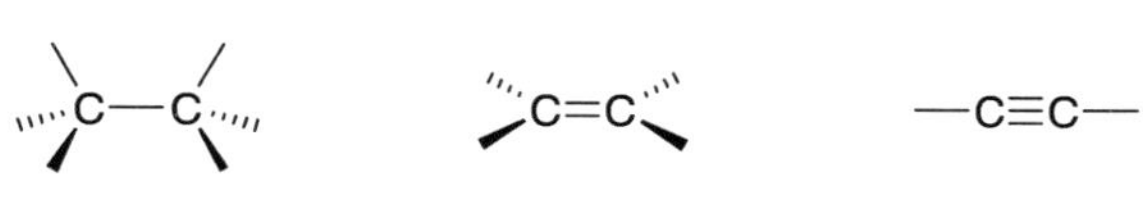

Figure 2   Single, double and triple bonds between carbon atoms - the bricks of the molecular architect.

If a chemist were given just a carbon-carbon single bond as a repetitive unit (plus the appropriate number of hydrogen atoms to "saturate" his construction at the end) - what kind and types of molecular structures could he build with it? The simple answer, given in every elementary chemistry textbook, is shown in Figure 3.

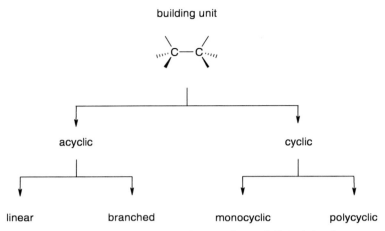

Figure 3   *Hydrocarbon structures that can be formed from C-C single bonds.*

Whereas Figure 3 does not show any structural detail, Figure 4 which presents important results of building with the "C=C-brick" does.

Beginning with the arm pointing towards the lower left it is seen that by coupling two and then three double bond units *linearly* the conjugated dienes **6** and trienes **7** are created, the first two members of the class of the *conjugated oligo-* and *polyenes.*

Preceding clockwise two and then three double bonds are joined *circularly* to provide cyclobutadiene (**8**) and benzene (**9**), the first two members of the *[n]annulene* series. Whereas the double bonds in these hydrocarbons are arranged in an *endocyclic* fashion, they are all oriented *semicyclically* in the next class of unsaturated hydrocarbons - the *radialenes*, the first two representatives, [3]radialene (**10**) and [4]radialene (**11**) of the vinylogous series being shown in Figure 4. A hybrid between the two types of double bond arrangements just discussed is shown next with the *fulvenes*, which are made up of both *endo-* and *semicyclic* double bonds. The Figure shows triafulvene (**12**) and pentafulvene (**13**) as the most simple members of this class of hydrocarbons. In contrast to the polyenes and the annulenes, the radialenes and the fulvenes are characterized by an arrangement of π-electrons which is called *cross-conjugated.* Clearly, **12** is the simplest conceivable hydrocarbon of this type. Cross-conjugation is encountered also in the so-called *dendralenes* represented at the bottom of the Figure by the first two members of this particular series, [3]dendralene (**14**) and [4] dendralene (**15**), respectively.

In all examples discussed so far, the joining of the C=C-blocks was performed by single bonds. If this task is taken over by a (then shared) carbon atom we can include the cumulenic double bond systems in our Figure as well. This is shown with the first two members, allene (**16**, propadiene) and [3]cumulene (**17**, butatriene) in the last branch of our Figure.

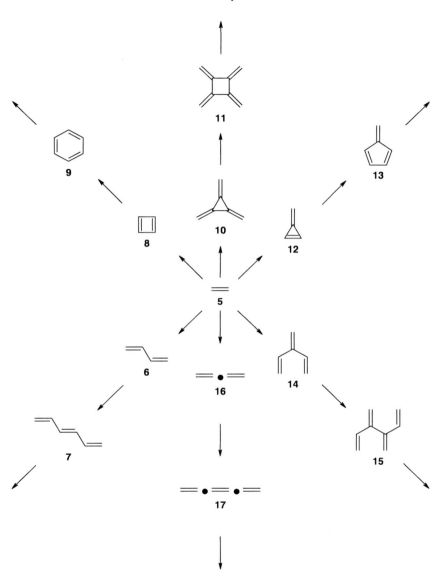

Figure 4 Molecular building with the C-C double bond: simple structures.

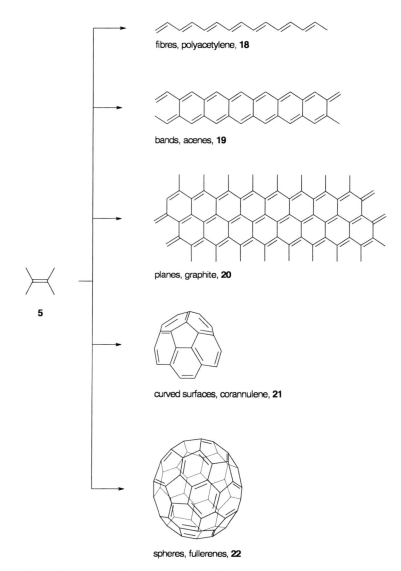

Figure 5 Molecular building with the C=C-double bond: extended structures complex "extended π-systems" as illustrated in Figure 5.

From the repetitive unit **5**, shown in the center of the diagram six branches extend.

The importance of these π-electron systems in organic chemistry differs widely. The polyolefins **6** and **7** and their higher vinylogs are subsystems of such diverse hydrocarbons as β-carotene and polyacetylene which may be regarded as archetypical examples from the *natural* and the *designed* world of organic chemistry. The overwhelming importance of aromatic hydrocarbons in both industry and fundamental research needs no further comment. Although cross-conjugation is a phenomenon often encountered in *e.g.* dyestuff chemistry, the cross-conjugated hydrocarbons - the fulvenes, radialenes, and dendralenes - are presently of far lesser importance than the first two classes of π-electron systems. This is also true for the cumulenic hydrocarbons and their derivatives and for the cross-conjugated systems.
Have we exhausted all conceivable combinations of double bonds with Figure 4 ? Not at all!
In the *next higher step of construction* we can integrate the π-systems just generated into more
As one example consider polyacetylene **18**, which we can regard as the extension of the linear combination of double bonds to infinity. We can also call **18** a fiber molecule and it has actually been likened to a "molecular wire". If two such fibers are connected by single bonds - as shown in **19** - band structures arise. These are named *acenes* and the lower members of the series have been prepared as will be described later. Continuing our *aufbau* work we can add more and more polyacetylene fibers and generate molecular π-planes, which - when polymeric - are commonly called "graphite" (**20**).

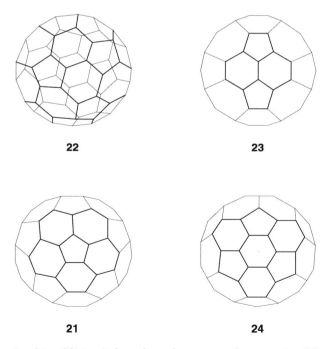

Figure 6   *Resolving C60 into hydrocarbon substructures - deconstructing C60.*

However, we must not stop here either. If other than six-membered rings are included into our growing π-planes *non-planar structures* become possible, as exemplified by the curved surface of corannulene (**21**). As the planar arrangement of double bonds reaches its ultimate realization in **20**, the curved structures lead to molecular spheres, the fullerenes, of which the presently most prominent one, C_{60} (**22**), is also shown in Figure 5. At this point, however, we have left hydrocarbon chemistry and arrived in a new field - the new forms of carbon. The close genetic relationship between these two areas also manifests itself by taking different views on C_{60} as done in Figure 6.

Depending on the "resolution" of our view we can recognize numerous hydrocarbon substructures in **22**. Relatively simple ones such as paracyclene (**23**), but also more complex ones such as corannulene (**21**) or tricyclopenta[def;jkl;pqr]triphenylene (**24,** sumanene). Many other hydrocarbon "cut-outs" are possible - as an exercise the reader is asked to deconstruct C_{60} him/herself - and a substantial number has been prepared during the last few years.

With triple bonds as the only allowed building units just one combination is possible: the molecular rod, H-(C≡C)$_n$-H. Cyclic variants are possible, but they are again new forms of carbon ("cyclocarbon"), not hydrocarbons any more.

An endless number of hybrids between the three basic systems **4** to **6** is possible. And only a minute selection will be mentioned here to show the reader what is possible.

In the alkene field, hydrocarbons formally composed out of sp^2- and sp^3-hybridized carbon-atoms are of great interest. They include highly strained olefins such as cyclopropene (**25**), the *trans*-cycloolefins **26** and the bi- or polycyclic *anti*-Bredt hydrocarbons **27**.

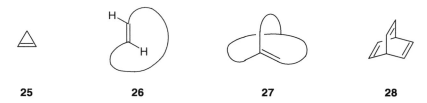

Figure 7   *A selection of alkenes built from 4 and 5 only.*

In molecules possessing more than one double bond these again must not lie in the same plane. All kinds of orientations are conceivable, the three-dimensional parallel alignment shown by barrelene (**28**) is just one (important) example.

Among molecules containing only formal single and triple carbon-carbon bonds the homologous series of the cycloalkynes is particularly noteworthy (Figure 8).

Figure 8   *The cycloalkynes: From which ring size onwards do they become isolable ? How strongly can one bend the C-C triple bond ?*

Beginning with the extremely strained cyclopropyne (**29**) - when will we reach the stability limit to allow us to actually isolate the appropriate cycloalkynes ? It is not so long ago that a deliberate attempt to generate structures such as **29** and **30** would not have been undertaken by a serious organic chemist.

So far we have applied our simple *aufbau* principle to generate *hydrocarbon structures* only. We have seen that it is quite easy to establish genetic relationships - family trees, so to speak - between many classes of hydrocarbons, thus generating order in an area which on first sight might overpower us with its huge structural diversity.

Figure 9   The Cope rearrangement - one of the fundamental processes of organic chemistry.

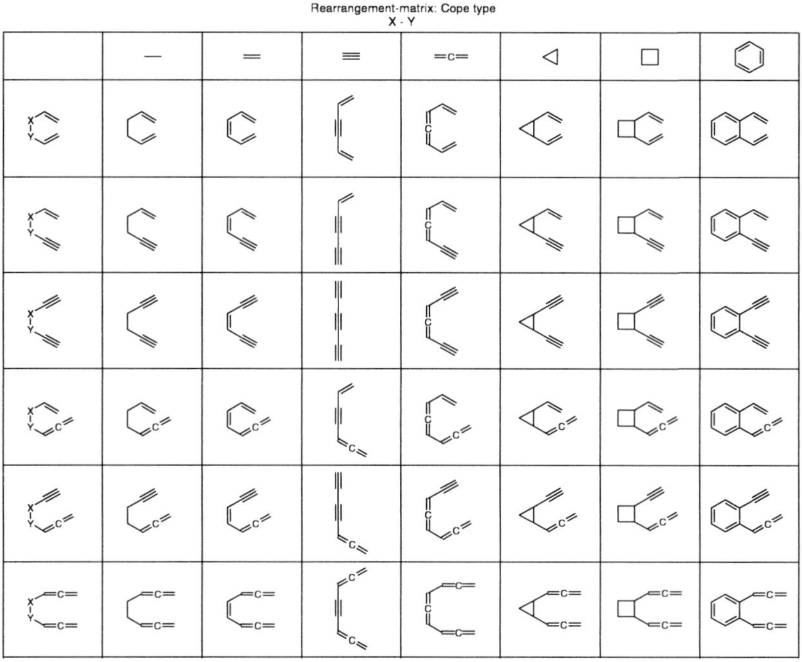

Figure 10   A selection of hydrocarbons which can undergo Cope-type rearrangements.

A very similar back-of-the envelope approach can be employed also to generate lines of heritage between *hydrocarbon reactions*. Above we have used the double bond **5** as one example to generate *structure patterns*, now we will employ just one reaction to produce *reactivity patterns* in hydrocarbon chemistry. The reaction which we select for this purpose is the *Cope* rearrangement, one of the most important and most thoroughly studied reactions in organic chemistry. In this pericyclic process one bis allyl system **33** is converted into its isomer **34** by just heating it. With R = H the process is degenerate and the reaction is an *automerization*, with R ≠ H an equilibrium between the two valence isomers **33** and **34** is established in which one side may be strongly favored.

Let us incorporate this rather simple arrangement of two double and three single carbon-carbon bonds into more complex structures! Or, to put it another way, lets us take this motive and ask what variations are possible with it. Realizing that in **33/34** all five building blocks may be replaced by structural elements of increasing degree of unsaturation and/or complexity, a *rearrangement matrix*, shown in Figure 10, results.

Without discussing every single entry of the matrix its underlying rationale is obvious: The role of the X-Y-fragment may be played by a single, double, triple etc. bond - all the way to the benzene ring, where we will deliberately stop (horizontal variation). And the double bonds of the original Cope-system can successively be replaced by triple bonds or allene groups, again ending our variations here to make the matrix not too big (vertical variations). Out of the 42 combinations shown here many have been studied and have become important in preparative organic chemistry.

But as we could continue the building process with the C-C-double bond - from Figure 4 to Figure 5 - we can incorporate the Cope process into increasingly complex - and again "three-dimensional" - hydrocarbon frameworks (Figure 11).

Beginning with divinylcyclopropane **35**, which we already encountered as one of the combinations in the initial Cope rearrangement matrix (Figure 10), we can either connect its vinyl ends by a (then bridging) methylene group or we can short-circuit these ends - the resulting hydrocarbons are homotropylidene (**36**) and norcaradiene (**38**), respectively. Homotropylidene paved the way to the most celebrated fluxional molecule of them all, bullvalene (**37**), a $C_{10}H_{10}$ molecule in which by way of repetitive Cope isomerizations any carbon and any hydrogen position can be inter-converted into any other (at about 100 °C). The most characteristic property of norcaradiene (**38**), on the other hand, is its valence isomerization to the monocyclic hydrocarbon tropylidene (**41**, 1,3,5-cycloheptatriene). When this Cope isomerization takes place in a derivative of **38** in which the bridgeheads are spanned by an additional 1,3-butadiene unit, hydrocarbon **39** results. If this experiences a norcaradiene-cycloheptatriene ring-opening the bridged aromatic hydrocarbon methano[10]annulene (**40**) results, a 10π-electron system fulfilling Hückel's rule and being aromatic on all counts. Tropylidene (**41**) itself can be used to (formally) construct higher vinylogs such as the doubly-bridged [14]annulene **42**.

Genetic relationships such as this one are typical for hydrocarbon chemistry and are a consequence of the *Aufbauprinzip* - although more often than not these connections become obvious only after extensive preparative work. For the future development of the field, *heuristic thinking and planning* in the manner presented here should be employed more often, though.

In our discussion of the significance of hydrocarbons for organic chemistry we have so far stressed structural and reactivity viewpoints and we have made the tacit assumption that in our various building exercises the bonding parameters of the building units **1** to **6** are unperturbed, *e.g.* that the angles in **1**, **2** and **3** are 109.5, 120 and 180 °, and the carbon-carbon distances in **4**, **5**, and **6** are 154, 134, and 121 pm, respectively.

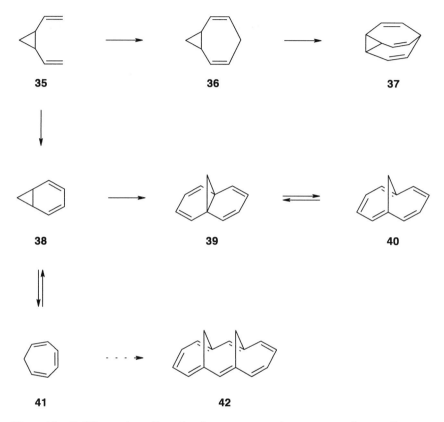

Figure 11   Building three-dimensional structures that can undergo Cope-type rearrangements.

Actually it is - and always has been since the days of *Adolf von Baeyer* and *William Perkin, Jr.* - one of the main motives of chemists engaged in hydrocarbon work to deliberately violate and even break these standard bonding situations! How this can be accomplished for **1-6** is shown graphically in Figure 12, which summarizes the most important of these bond angle and bond length deformations. Again, other molecular distortions can be conceived, and it is left to the reader to find more ways to arrange carbon and hydrogen atoms in three-dimensional space (and later prepare the "designed" hydrocarbons and demonstrate their significance for the development of organic chemistry).

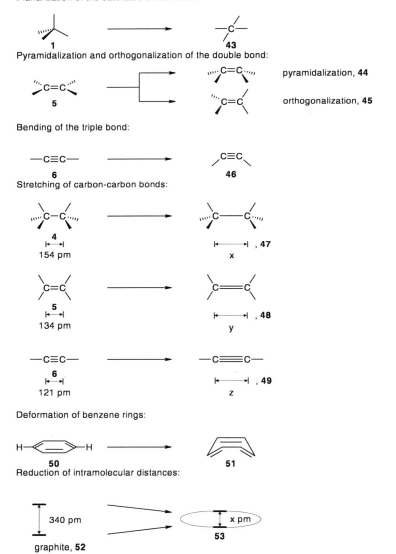

Figure 12   Ways to distort and deform hydrocarbon fragments and structures.

Consider the icon of organic chemistry, the tetrahedrally bonded carbon atom **1**: Can it be flattened to the planar carbon shown in **43** ? This would translate into planar methane if saturated with four hydrogen atoms, a molecule with anti-*van't Hoff* geometry. And the carbon-carbon double bond **5**: Under what circumstances, *i.e.* in what molecular environment, will it become pyramidalized as shown in **44** or twisted, even orthogonal as illustrated in **45** ? What one would have to do to a triple bond **6** to make it deviate strongly - as drawn in **46** - from the usual linear geometry is easier to see: its incorporation into a small or even normal ring should suffice (see above, structures **29** to **32**). Instead of deforming the bond angles we may also want to increase the bond lengths of our standard building blocks as shown in **47** - **49**. Interestingly, whereas the question of lengthening of the C-C bond has often been addressed, syematic attempts of creating particularly short bonds between carbon atoms are rare.

As far as larger organic structures are concerned, hydrocarbon chemistry offers many interesting solutions to the question of whether and how far aromatic systems, in the simplest case the benzene ring **50**, can be distorted. Is it possible to bend **50** so strongly that it forsakes its proverbial aromatic character and transforms into a 1,3,5-cyclohexa-triene (**51**) ?

Another important question concerns intramolecular distances. Starting from the 340 pm distances observed between the planes of graphites, sketched in a highly stylized way in **52**, is it possible to force these planes to a closer distances, as in **53**, thus increasing the electronic interaction between the planes ? One way to accomplish this would be by way of short molecular bridges as indicated in our diagram. It is important to realize that the construction units are not rigid moieties of a one and for all defined geometry or shape. *What their actual geometric parameters are depends on the molecular framework into which they are incorporated, i.e. it depends on the bonding situation.*

Although some of these questions may have the ring of a sports(wo)man-like competition, the wish to create a world-record for a particular molecular arrangement, they have far more serious and important implications. Firstly, to produce bonding situations that differ strongly from those encountered in usual, *e.g.* nonstrained or undeformed organic molecules as a rule requires the development of new reactions or techniques. We can thus expect an enrichment of preparative methods when we are trying to synthesize hydrocarbons which will allow us to answer the above questions. Furthermore, to define the limits of bonding (when, for example, does the contact between the carbon atoms in a stretched bond finally vanish ?) touches the very heart of chemistry, which, after all, is the science of making and breaking bonds. The study of deformed or even bizarre bonding situations can help us to understand the standard cases - as shown in **4** - **6** - better. Theoretical studies on the nature of the chemical bond have always profited strongly from molecules with unusual bond lengths and bond angles. And finally: deformed bond lengths and angles translate into unusual distributions of electrons which in turn causes surprising, often drastically enhanced chemical reactivity and hence new chemical behavior.

There is one final reason why the study of hydrocarbons attracts many chemists: *Their wish to play* is often fulfilled extremely well on this exciting playing ground of organic chemistry. Whether (the mostly male) practitioners speak of *tinker toy chemistry*, *molecular lego* or *mecano sets*, the connection to an earlier part of their lives is obvious enough.

Designing Molecules

Rainer Herges

Institute of Organic Chemistry

Technical University of Braunschweig, Germany

email: R.Herges@tu-bs.de

Keywords: Synthesis planning, molecular design, chemical structures

ABSTRACT

There are substantial differences in the design of molecules (chemistry) and artifacts (engineering). Usually the molecules that are synthesized (designed) by chemists are of the size of about 0.1 to 10 nm. For comparison: the smallest structures fabricated in chip design today are about 200 nm in width. In chemistry the components needed for construction cannot be manipulated individually. Usually a large number of molecules (about 10^{23}) are mixed and hoped to self-assemble to give the desired product. In engineering this would correspond to a massive parallel production. Using a number of heuristic rules and logical principles chemists nowadays are able to synthesize any conceivable, stable, even very complicated compound. However, there is still much room for improvement, because the strategies that are used today are still not very general and well structured.

INTRODUCTION

Among the three basic fields of natural sciences: physics, chemistry, and biology chemistry stands out by the fact that it includes a very strong synthetic element besides the analytic aspects (reductionism) which are common to all natural sciences. Probably on that account there is a chemical but not a physical or biological industry. Synthesis in chemistry is similar to construction in the sciences of engineering, however there are distinctions.

The differences between the *design of molecules* (chemistry) and *design of artifacts* (science of engineering) are as follows:

- In chemistry the components needed for construction cannot be manipulated individually. Usually a number of about 1023 pieces of each component (molecules) are mixed and hoped to self-assemble (react) to give the desired product. Imagine, an engineer would put a number of well selected components in a box shake or agitate everything and cars or other artifacts would self-assemble during this process.
- Usually there are a number of different ways the molecules can react with each other. Thus, the yield of the desired product is less than 100% and the byproducts have to be removed. The product must be identified within the reaction mixture and it has to be purified.

- The result of chemical construction (synthesis) is not an artifact with a defined function but an ensemble of particles with a statistical, macroscopic property. The macroscopic (e.g. material or biological) property is an (in principle) unknown function of the microscopic properties of the individual molecules.
- The general and paramount motivation of chemical synthesis is not the construction of a given molecular structure but the design of molecules with a well defined macroscopic property. This is often referred to as "Chemical industry is not selling molecules but properties". Unfortunately there are only empirical and fuzzy relationships between molecular structure and macroscopic properties.

SYNTHESIS PLANNING

The final goal in synthesis planning is a defined molecular structure (target). In many cases the target is a chemical compound that was isolated and identified from natural sources and for which interesting biological (pharmaceutical) properties were found. The problem that has to be solved in synthesis planning is to find a sequence of reactions that leads from readily available starting material to the target molecule.

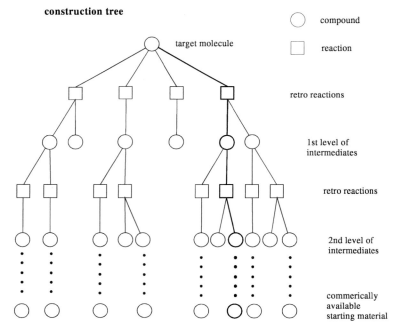

Starting from a weak empirical basis in the beginning of this century, a number of empirical rules and problem solving strategies slowly emerged. In the beginning of the 1960's an empirical level was reached that encouraged several groups to develop computer programs for synthesis planning. The attempts to squeeze the current intuitive and fuzzy chemical knowledge into algorithms in such a way that they can be programmed on a computer, forced the chemist to become aware of the heuristics of chemical thinking. Surprisingly, the main achievement of this work was not the development of useful synthesis planning programs, but the explicit expression of logical principles that up to this time were applied only intuitively. The progress in this field not withstanding, professional chemists are still able to outclass any of these computer programs. This might be due to the fact that the rules are less exactly defined than e.g. the chess rules and because synthesis includes a strong element that can be referred to as rather an "art" than pure science.

In the following chapter the general approach and some basic rules for the construction of molecules are presented. The general rules are then explained by a simple example.

Probably the most general and important strategy is to start with the target molecule and to apply retro (backward) reactions on the target molecule to generate potential precursor molecules (1st level of intermediate products in Figure 1). The process is repeated until one ends up with compounds which are commercially available. Moving down the levels of intermediates the combinatorial tree is branching increasingly because several retro-reactions can be applied to each intermediate product. This problem is not only restricted to chemistry and usually referred to as combinatorial explosion. It is advisable not follow each of the branches but finally decide in favor of one of the alternatives. The synthesis (construction) then follows the reverse direction starting from the commerically available compounds and working through the tree by applying (forward) reactions up to the target molecule.

Similar to the problems in chess playing one has to apply guiding rules to select the most promising way at each branching point (cutting the combinatorial tree). Some of the most important guiding rules are:
- short synthesis
- convergent synthesis
- strategic simplification
- use of symmetry
- protecting groups

ad *convergent synthesis*: Assuming we need 10 steps with 90% yield each to get the target molecule (TM) we can design strategies with increasing extend of convergency (overall yields are given in %):

$$A \to B \to C \to D \to E \to F \to G \to H \to I \to J \to TM \qquad 35\%$$

$$\left. \begin{array}{l} A \to B \to C \to D \to E \\ F \to G \to H \to I \to J \end{array} \right\} \to K \to TM \qquad 53\%$$

$$\left. \begin{array}{l} A \to B \to C \\ D \to E \to F \\ G \to H \to I \\ K \to L \end{array} \begin{array}{l} \to M \\ \\ \to N \end{array} \right\} \to TM \qquad 66\%$$

Advantages of a convergent synthesis strategy are

higher yields
- lower risk: If one of the steps failes, only the corresponding branch has to be modified and not the whole synthesis plan.
- logistic advantages: The work can be distributed among several teams, each working on one of the branches.

Ad *Strategic simplification*: In devising a synthesis plan retrosynthetically it is always advisable to cut bonds that simplify the molecular structure as much as possible.

The following example depicts the principle:

Target ⇐ ⇒ synthon

reagent

synthesis:

There are four bonds in the target molecule which are not symmetry equivalent: *a*, *b*, *c* and *d*. Breaking of bonds *a*, *b* and *c* each leads to a system containing three rings. Only cleavage of bond *d* simplifies the structure to a system with two rings and thus is favorable. The synthesis starts with the two-ring system, however, two groups have to be located at appropriate positions in the rings to define the points of reactivity. They are removed by suitable operations after performing the synthetic steps.

The following example is more sophisticated and includes the use of several construction rules. The target molecule *multistriatin* was isolated from the so called elm bark beetle. These insects feed on elm trees. If one of them has found a source of food (elm tree) it releases the pheromone *multistriatin* to attract more insects of its family (chemical signal transmission). Motivation to look for a chemical synthesis of *multistriatin* was to build "pheromone traps" to combat the beetles without the use of insecticides.

An experienced chemist would immediately recognize the acetal function (circled with dotted line) and apply the retro-reaction which would lead to its formation and thus generate the appropriate precursor molecule. The next step would be to look for a *strategic bond* to be broken retrosynthetically. The most suitable bond (marked with a wavy line) leads to a *symmetrical* precursor molecule. This is favourable because in the subsequent synthesis both reactive positions next to the =O group will give the same product because they are symmetry-equivalent (an unsymmetrical precursor would yield two different products which had to be separated). Moreover, the symmetric precursor is cheap and commercially available. The other component on the right-hand side can be transformed to a readily available starting product by a simple retrosynthetic operation.

Construction Plan:

multistriatin

commercially available

RCO₃H
OH⁻

commercially available

The synthesis that was derived from the retrosynthetic construction plan is self-explaining except the first step which is necessary to make the commercial starting material more reactive (to weaken the bond which has to be broken in the first step).

multistriatin

Following the empirical rules and strategies a construction plan has been developed that allows the synthesis of multistriatin in only four steps. The compound is now available in kg amounts and is used forest plantations.

Multistriatin still is a very simple example. Among the most sophisticated successfull synthesis projects are palytoxin and brevetoxin B:

Palytoxin [Klein-89]

Isolated from marine soft corals. Most poisonous substance known to date except for some polypeptides found in bacteria.

Brevetoxin B [Nico-95]

Isolated from red algea (Phytodiscus brevis Davis). Cause of the so called red tide catastrophy, (mass extinction of fish, including dolphins, whales and members of the food chain). The synthesis was performed with about 20 chemists over a period of 12 years. 123 steps were needed. The longest linear sequence was 83 steps with an average yield of 91% per step. Final yield: 4.8 mg.

CONSTRUCTION OF MOLECULAR GUESTS AND HOSTS

Recently a new kind of chemical design became more and more important. The target is not a molecule with an exactly defined structure. Only the approximate shape, the charge distribution at the surface, positions for hydrogen bonding and the flexibility of the structure are given.

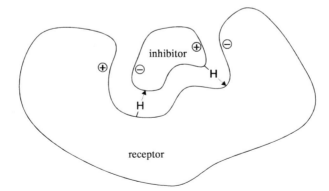

The main motivation for this kind of chemical research is the development of pharmaceutical compounds. Owing to modern analytical methods, the detailed spatial structure of the receptors of a number of physiologically important systems are known. It is the final goal of the chemists to design molecules that are complementary to the receptor in shape and charge distribution and therefore bind to it with high affinity and inhibit its function. Several protease inhibitors, that are used for the treatment of AIDS were developed this way. They inhibit the HIV-protease, which is necessary for the synthesis of the virus envelop protein that is essential for the reproduction cycle of the virus.

There are only a few very "fuzzy" rules for a rational approach to the design of guest molecules for a given host molecule or vice versa. Construction material are the 105 stable elements of the periodic system (but mainly C, H, N, O, S and P are used) each of which can adopt several valence states and hybridisations. The number of conceivable combinations increases with the size of system. The number of different molecules that can be constructed from 100 atoms is almost as large as the number of atoms in our universe.

CONCLUSION

There is no general and well defined theory of design in chemistry. Rational approaches, problem solving strategies and logical priniples exist but they are fuzzy and difficult to concretize in algorithms. The most powerful design strategy in synthetic chemistry still is to know a large number of examples and to use analogy reasoning.

REFERENCES

Nico-95 Nicolaou, K. C.; Hwang, C. - K.; Duggan, M. E.; Nugiel, D. A.; Abe, Y.; Bal Reddy, K.; DeFrees, S. A.; Reddy, D. R.; Awartani, R. A.; Conley, S. R.; Rutjes, F. P. J. T.; Theodorakis, E. A.; Sato, M.; Tiebes, J.; Xiaou, X. - Y.; Duggan, M. E.; Zang, Z.; Couladouros, E. A.; Sato, F.; Shin, J.; He, H. - M.; Bleckman, T.; Untersteller. E.: J. Am. Chem. Soc. 1995, 117, 10227-38.

Klein-89 Klein, L. L.; McWohorter, W. W. Jr.; Ko, S. S.; Pfaff, K. - P.; Kishi, Y.; Uemura, D.; Hirata, Y.; Fujioka, H.; Christ, W. J.; Cha, J. K.; Leder, J.; Finan, J. M.; Ko, S. S.; Yonaga Y.: Pure Appl. Chem. 1989, 61, 313-24.

Summary Session 2

J. Gausemeier

We have gained a very impressive insight into the development methods in the domains software engineering, microelectronics, micro systems engineering, and material science. Basic similarities with the mechanical engineering can be recognized here. This holds especially for micro systems engineering where - as well as in mechanical engineering - a system under development is being modeled by means of three essential items: substance, energy and information. Material science deals with the design of molecules and atomic structures. Faced with the shortness of time, I would like to emphasize some aspects concerning the similarities and differences among development methodologies within various domains of engineering and natural science.

In all domains we find recognized and established methods. In software engineering, for example, there are development steps like problem description, problem analysis, requirement analysis, system definition, system design, etc. In microelectronics, the required steps consist of elaboration of design specifications, functional design, and logical design. From an abstract point of view, development methods in various domains resemble themselves. Within individual steps, various methods (sub-methods) are being used for the purpose of modeling data, functions, and processes. In microelectronics petrinets, VHDL[1], and ASM[2] are some well-known examples for such methods. Many of those sub-methods may also be used for the specification of modern products of mechanical engineering, i.e. for modeling of information flows coupled with machine control processes.

The importance and moreover the necessity of a systematic approach during development in different domains is differently pronounced. In microelectronics, a strictly systematic development procedure is absolutely necessary. However, it does not hold for the mechanical engineering. A possible reason for that difference may be the way of thinking and the more haptic disposition of a mechanical engineer, who thinks less in terms of abstract, but more in terms of shape-afflicted structures. It is also well known that in the software engineering a development goal can be achieved without any rigid methodical approach. Although it is obviously not a wise development procedure, practice unfortunately clearly shows such tendencies. One reason for it surely lies in the fact that during the software development, design and manufacturing processes are highly coupled with each other, what makes it too easy to produce a completed product (e.g. code, program) without any specifications. It seems plausible that the complexity of manufacturing processes enforces a systematic development process.

This fact is also the main motivation for model synthesis and analysis. Computer models are being built in domains where real prototype building implies an expensive manufacturing procedure. Nowadays, such tendencies can also be observed in the mechanical engineering. The key word *virtual prototyping* paraphrases this objective.

In the domains microelectronics, micro systems engineering and molecular design, there is a close interdependency between design and manufacturing processes. It is (or, at least, it should be) the case in the mechanical engineering, too.

---

[1] VHDL: very high integrated circuits description language
[2] ASM: abstract state machines

A further similarity is the tendency of using solution patterns (solution elements, design patterns) during the development process. Depending on the development stage, different aspects of the system should be taken into consideration. Some examples for such aspects are behaviour, structure, or shape. Therefore, it seems appropriate to store such design patterns together with their relevant aspect models in knowledge-based systems.

Altogether becomes clear that we have remarkable opportunities within domains presented here, and that a systematic procedure is required in order to exploit them. It also does not surprise that presented methods are currently being subject of refinement and improvement efforts in the corresponding domains. Furthermore, there are many common properties among development methodologies within different domains including mechanical engineering. In my opinion, it is very important that today we already have a close collaboration of functional principles from various disciplines in products. In the future, a modern mechanical engineering product will consist of mechanical, electrical and software components, will be built out of new materials (e.g. *smart materials*), and will be subject to high miniaturization, which implies a permanent usage of newest research results at the field of micro systems engineering. Such an integration requires an utilization of appropriate integrated development methods and tools. Today, only the first steps towards that direction have been taken (e.g. hardware/software co-design). Computer-based environments for an integrated development of future industrial products is one of today's greatest challenges. The basis for overcoming these challenges has to lie in an overall design theory, which encloses all relevant domains of engineering and natural science. During this workshop, I have gained a strong conviction that such a theory is both possible and absolutely necessary. But, it requires readiness and capability for an integrated way of thinking from all the relevant domains. We will have to do it that way if we want to achieve new and innovative products, which are needed for us in order to build up our future.

Session Three

Computer Aided Development and Application of a Universal Design Theory

Universal Design Theory: Elements and Applicability to Computers

Hans Grabowski, Stefan Rude, Gunther Grein, Eike Meis, El-Fathi El-Mejbri

Institute of Applied Computer Science in Mechanical Engineering (RPK)

University of Karlsruhe, Germany

email: {gr\rude\grein\meis\mejbri}@rpk.mach.uni-karlsruhe.de

Keywords: Universal Design Theory, Design Process, Solution Patterns, Product Modeling, Requirement Modeling

ABSTRACT

Design theories can be applied by human beings or computers. This paper describes a general methodical approach to implementing the application of a design theory within the field of mechanical engineering. It serves as an informative starting point for discussing to what extent this approach can be transferred to other fields of design.

INTRODUCTION

Theories in general can be viewed as the main source of information on scientific findings. They are formulated in order to arrange and explain phenomena within the fields of human knowledge and to derive general rules regarding the existence of these phenomena. They are therefore associated with a certain *interest in cognition*. On the other hand, theories are aimed at making prognoses of the occurrence of certain phenomena within scientific fields. They are therefore also associated with a certain *interest in design*.

Theories of design [i.e. Suh-90, Yosh-87] try to explain the nature of technical products and find a procedure of general validity for the invention of heretofore unknown artefacts. In research, much effort is put into finding new facts or solution principles in a domain (new scientific findings), whereas technological development is concerned with the use of research results for finding new products, processes and methods (new development) or for improving already existing ones (further development).

A *Universal Design Theory (UDT)* is a design theory that formulates findings about design from different scientific disciplines in a consistent, coherent and compact form. It serves mainly as scientific basis for rationalizing interdisciplinary product development with respect to efficiency and reliability. A Universal Design Theory takes all the common features of different scientific domains into account in order to find statements of general validity with regard to the explanation of and the way of looking at things. In contrast to a so-called *general design theory*, a Universal Design Theory not only encompasses generic, discipline-independent knowledge, but also discipline-specific knowledge about design. It also describes the interfaces with the different design disciplines.

With regard to the establishment of a Universal Design Theory, two problems need to be solved: The problem of *universality* and the problem of *practical applicability*. These will be discussed in more detail in the following text.

THE PROBLEM OF UNIVERSALITY

Successful design requires many different skills and talents and it is undisputed that there is a high potential for the improvement of products in the application of knowledge and methods from different scientific disciplines. As a consequence, design teams should involve people with a wide variety of background knowledge, i.e. different training, experience, perspective and personalities. In order to meet this high demand for interdisciplinary product development, the following aspects must be taken into consideration:
- In various disciplines, artefacts are created without a theory of design as a scientific basis.
- Only a limited number of disciplines uses design theory approaches and these approaches are far from being a logical, homogeneous entity.

In order to enhance co-operative interdisciplinary team work, an overall framework for a Universal Design Theory must be created.

In this respect, the development of a common design language would be an important milestone. Today, everyone has their own special view of given facts and uses their own special words and expressions. Many of these overlap with similar terms from other disciplines, but people from different disciplines are incapable of understanding each other.

THE PROBLEM OF PRACTICAL APPLICABILITY

The basis for a successful product development process is the complete and correct specification of a task, represented by a certain set of requirements a design solution has to meet. The formal description of these requirements (Requirement Modeling) should be the first step in any design process. The mapping of a set of requirements onto a set of possible design parameters is shown in Figure 1.

Figure 1   *The process of design as the mapping of a set of requirements onto a set of design parameters.*

If requirements are fully defined and correctly specified, a target-oriented product development process is possible. However, we face the following problems:
- Existing design theory approaches do not completely and explicitly describe the mapping process.
- Existing design theory approaches are therefore hardly applicable by man or computer.

Statements in design theories are called *constructive* if they describe the mapping process in an explicit and complete manner. One requirement made of a UDT must be to make constructive statements in order for it to be practically applicable.

Our idea of a Universal Design Theory is as follows: First, the core of the theory is formed by a generic part which can be applied to the design of any artefact. Additionally, a specific part comprises all the domain-specific extensions. In the following text, the general design process will be reflected. We will then outline the basic assumptions of our approach and afterwards explain the elements of a Universal Design Theory, the so called Design Patterns.

DESIGN PHASES IN MECHANICAL ENGINEERING

Several German design methodologists [*Koll-94, Pahl-97, Roth-94*] have analyzed the design process in mechanical engineering. The result of their work – in which they focused on the use of computers - is a methodology subdivided into four design levels: *requirement modeling, functional modeling, effective geometry modeling* and *embodiment design* (Figure 2).

Each design process begins with the clarification of the design task. As a first approach, a *list of requirements* is given by the customer or the marketing branch. Usually, the set of requirements is incomplete, either because some requirements are missing, or because explicit requirements again imply other requirements. The completion and subdivision of all the requirements is the purpose of the *requirement modeling level*. The result of this level is a single consistent *requirement model* which is, of course, incomplete, since requirements need to be further developed during the process of design. This issue will be discussed later on.

The *functional model* describes the manner in which a technical object meets the functional requirements. PAHL and BEITZ (1997) [*Pahl-97*] define a function as follows: A function describes the *general* and *desired context* between the input and output of a system, with the goal of fulfilling a task. In a first step, the *overall function* can be derived from the requirement model. For very complex products, the overall function is divided into in *sub-functions* until a level is reached where further decomposition is not possible or useful. Thus, a *function structure* is developed. In addition to this, the subsequent integration of follow-up functions at a later stage must be taken into consideration [*Grab-96b*].

The construction of an effective geometry structure is implemented in two steps [*Roth-94*]:
- searching for and assigning of a physical effect to a function and
- assigning and adapting of the effective geometry to the physical effect and geometrical environment.

A physical effect is described by a physical law and its mathematical equation. The combination of physical effects makes up a structure of physical effects. An effective geometry consists *of effective points, effective lines, effective surfaces* and *effective spaces* and shows the schematic construction of a part. A combination of effective geometries makes up the effective geometry structure.

The concept of the morphological box was established in order to facilitate the building of physical effect structures and effective geometry structures, [*Roth-94*]. The first column of the morphological box contains the individual functions of the *function structure*. One or more *physical effects* with their *effective geometries* are assigned to each function and make up the rows. The designer connects the cells of the morphological box (the physical effects) to each other in the assumption of their technical compatibility. By these combinations of physical effects and effective geometries, several solutions are given, represented by effective physical structures and effective geometry structures. Based on this solution, the designer makes an evaluation and extracts one optimal solution in the shape of an effective geometry structure, on which he starts with the *embodiment design*.

Latest research results show that an integration of other than physical effects, e.g. chemical effects, must also be taken into consideration. In the future, effects will be classified as energy, material or information processing effects.

By the transition from effective geometry to *embodiment design*, the designer assigns material to the effective geometry and gives the effective lines and surfaces a solid design. In doing so, he must the shape-relevant requirements specified in the requirement model.

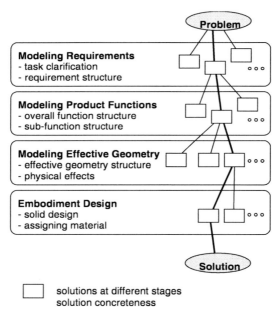

Figure 2   *Elementary solution steps in the design process*

Within these design levels, single design steps lead to individual design results. Handling of requirements during the design process is difficult, because requirements become more concrete, depending on the design results of every design level. During the course of the design process, requirements may also be refined.

SETS OF INFORMATION CREATED DURING DESIGN

The sets of information created during the different levels of design are shown in Figure 3. They can be classified as information concerning requirements (I_R), functions (I_F), physical principles (I_P) and embodiment design (I_D). These sets of information partially overlap and determine each other. In particular, every single design solution at any level of abstraction has a different effect on the subsequent phases of design.

The set of information requirements I_R fulfills a predominant task: It influences all other sets of information created during all levels of design. In this context, *design solution* means the specification of information sets I_F, I_P and I_D.

At every level of abstraction, a set of basic elements is used to describe design solutions. In this case, design solution means the result of one of the design levels after the requirement specification. Referring to three-dimensional mechanical design, the basic types of elements are:
- function
- physical effect
- action face and action face pairing

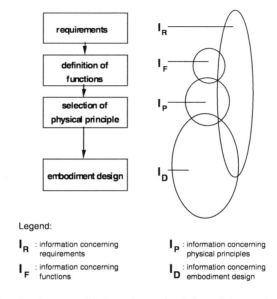

Figure 3   *Overlapping sets of information and relation of the requirement model to the other partial models*

These elements are used to describe the macrostructure of an artefact - the main intention of mechanical design. In order to describe the microstructure, the following types of elements are required:
- molecules
- atoms

However, these two types of elements will not be taken into consideration in the following text.

GENERAL ASSUMPTIONS ON THE DESIGN PROCESS

In order to work out a Universal Design Theory, some general assumptions on the process of design need to be drawn up. The set of assumptions is divided into two categories, axioms and hypotheses. Axioms are generally valid statements that do not need to be proven. Hypotheses are statements which are assumed to be true but which still need to be proven.
- The first axiom says that there is a finite number of levels of abstraction one can use to explain the process of design. With regard to the domain of mechanical engineering, these

are e.g. the above-mentioned levels of functional design, effective geometry design and embodiment design.
- The second axiom is concerned with the components on every level of abstraction. Regarding this, we can state that the set of well-known basic elements on each level of abstraction is finite at a certain point of time. Elements of a design theory therefore can only include the components currently known to us whereas the invention of new effects etc. has to be the concern of research work.
- The third axiom says that the number of transitions between the different levels of abstraction is also finite.

Based on these axioms, one can state the following hypothesis: The *invention* of a product is always a new combination of known basic elements. Here it must be pointed out that objects on the level of functional and effective geometric design must in particular be regarded as components which can also be combined. Discovery, achieved through research, is defined as the finding of new basic elements. To some extent, a design theory can help in identifying fields in which research to investigate new basic elements is necessary. A theory of design, however, cannot support the invention of new basic elements.

Here is an example that confirms our second axiom about the design level of functional modeling. VDI defines a function as [*VDI-2222*] *"the abstract general connection (relationship) between input, output and state variables of a system to fulfill a task."*

Input and output are the general quantities of design science: *Material* (M), *Energy* (E) and *Information* (I). The functions can only represent the relationship between input and output of a technical system and not its physical or chemical implementation.

In addition to general quantities, there are general functional verbs which always accompany the general quantities mentioned above. These general functional verbs are: *Store*, *transmit*, *channel*, *transform*, *change* and *connect*. Roth call these general functional verbs 'operations'. The state or state changes of the general quantities can be described using the general operations. According to this procedure, Roth associates the general quantities with the general functional verbs and derives thirty general functions with the symbols shown in Figure 4.

In order to deal with the prejudice that the application of such a Universal Design Theory prevents creative design, we would like to draw the following analogy:

Imagine the work of a composer. He is trying to find new melodies and tunes nobody has ever heard before. He has a limited amount of components at his disposal in order to achieve this. He must therefore try to assemble these components in a unique manner. First of all, he needs to consider the *mode* (e.g. major or minor), the *key* (e.g. C major, A minor, etc.) and the *notation* (e.g. treble clef or bass clef). He then has several additional tools such as the following at his disposal:
- 12 different notes in different pitches (e.g. a piano has 88 keys and thus 88 different sounds)
- the rhythm, i.e. the length of the notes (e.g. semibreve, minim, etc.)
- the dynamic range (volume, such as pp (pianissimo) or f (forte) and alteration of volume, such as crescendo and decrescendo)
- the speed of the music, e.g. adagio or allegro, and the alteration of speed, e.g. ritardando

Nobody would ever say that a composer's work - i.e. making up new songs or symphonies is not creative work although he is using only a limited number of elements. To be more precise, the creativity lies in the appropriate combination of these basic elements in order to achieve a specific atmosphere for people who hear the music. Basic elements of the effect of musical sequences on human beings are not yet known to us and could be identified as research topics as a consequence of our design theory approach.

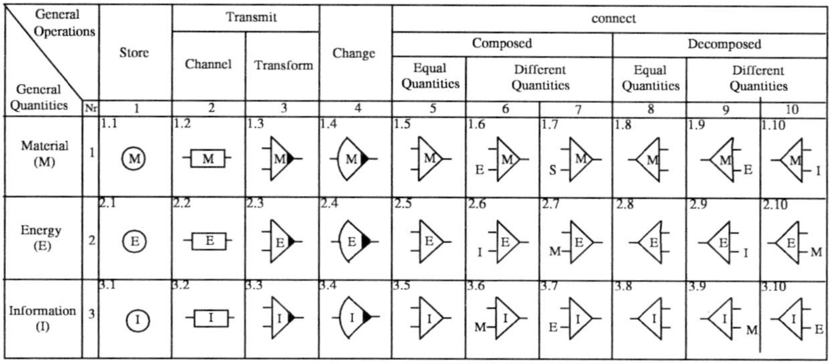

General Operations		Store	Transmit		Change	connect					
						Composed				Decomposed	
			Channel	Transform		Equal Quantities	Different Quantities		Equal Quantities	Different Quantities	
General Quantities	Nr	1	2	3	4	5	6	7	8	9	10
Material (M)	1	1.1	1.2	1.3	1.4	1.5	1.6	1.7	1.8	1.9	1.10
Energy (E)	2	2.1	2.2	2.3	2.4	2.5	2.6	2.7	2.8	2.9	2.10
Information (I)	3	3.1	3.2	3.3	3.4	3.5	3.6	3.7	3.8	3.9	3.10

Figure 4   The general functions: An association of general quantities with general operations according to [Roth-94]

From all the assumptions mentioned, we can conclude that a new design solution can be generated systematically by mechanisms of combination. These mechanisms are controlled by the set of requirements given at the outset.

The refinement of the requirement model and its interaction with other partial models on the design stages mentioned above is shown in Figure 5 [*Grab-96a*] using the development of a bicycle as an example.

Requirements - formally described in the requirement model[1] - can be classified as follows:

- Depending on their origin, we can differentiate between external requirements given by the customers and internal, enterprise-specific ones that become evident during the production process, for instance.
- We can further distinguish explicit and implicit requirements. Explicit ones are given directly at the beginning of the design task and must respond to the question „What should be generally achieved or avoided?". Implicit requirements are derived from these and usually come up during the clarification of the design task.
- We can distinguish complex requirements and elementary requirements, where elementary ones can be derived from complex ones by subdivision.
- Fixed requirements must be strictly satisfied ("need to have"), whereas desired requirements can be described as „nice to have" features of a product. If fixed requirements cannot be met, they must be either changed or rejected.
- We can further differentiate between quantitative and qualitative requirements. Quantitative ones are more operational due to their unambiguous and precise formulation. They can be measured or described by mathematic equations, for instance. In contrast to quantitative requirements, qualitative ones can only be evaluated indirectly.

---

[1]   Please note that the requirement model is a partial model of an integrated product model.

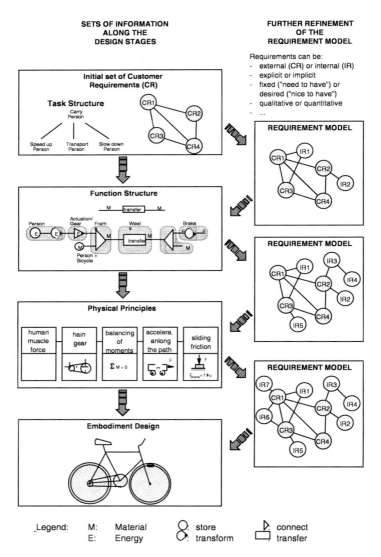

Figure 5   Further refinement of the requirement model due to confrontation with other partial models

The subdivision and nesting of requirements might be a problem if one is dealing with aesthetic judgements about the beauty of an artefact, for instance. These are typical qualitative requirements because they do not refer directly to an object and its properties. The judgement is more of a subjective expression of a viewer's reaction.

This implies the following questions:
- How can we define the qualitative requirement „beautiful"?
- How can we subdivide these types of requirements into elementary requirements?
- How can we evaluate them within the design process?

As a result of the further refinement of the requirement model during the various levels of design, single requirements can finally be mapped to certain design parameters and elements can be eliminated that are not adequate for reaching the goal.

While the set of requirements becomes increasingly complete and the single requirements become more concrete, the set of design parameters the requirements point to is reduced. Thus, a matrix can be developed that allows us to trace the transitions backwards from a certain requirement to a specific design parameter.

According to this, a set of requirements is called complete if (and only if) there is exactly one design solution that meets these requirements.

DESIGN PATTERNS

Design patterns either describe individual design solutions or types of design solutions. In the latter case, they may be represented as parameters. As shown in Figure 6, a design pattern specification is divided into three parts [Suhm-93]:
- the design solution, i.e. the actual result of the design process;
- the prerequisite, i.e. the conditions of using the design pattern;
- the environment of the design solution, i.e. a set of other design solutions that form the context of the design pattern.

A design pattern may have different properties:
- Design patterns are bound to an application context. They may differ depending on the application context.
- Design patterns are the conscious or unconscious result of a standardization process and conversely, they necessitate a standardization of a design process.
- Design patterns may be generic and individual. Individual design patterns describe individual solutions as stored in a database, for example. Generic design patterns describe types of design solutions. The description of a type of design solution can, but does not need to be parameterized.
- Design patterns may be elementary or complex.
- Design patterns may be proven or rejected. A proven design pattern has already led to a design solution released for manufacture or sale. A rejected design pattern describes a design solution that was not developed further.
- Design patterns may depend on the state of the art. In general, the view of the design process becomes increasingly differentiated over time. The design steps become smaller and the observed properties of design solutions increase.

The acquisition and the handling of design patterns require computer support. Among the major problems in this context are the complexity of the set of design patterns as well as the complexity of the information represented in a single design pattern.

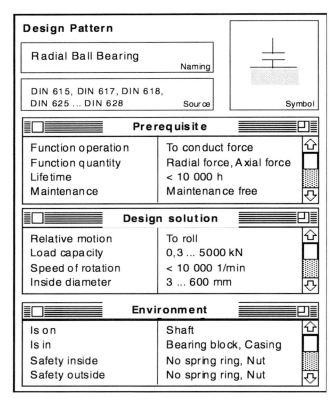

Figure 6   *Design Pattern Specification divided into Design Solution, Prerequisite and Environment*

COMPUTER SUPPORT

Based on the axioms and hypotheses mentioned in section 3, design patterns can be formally described in a manner that can be interpreted by a computer. A major problem in providing a suitable computer support is the management of the various possibilities of reasoning[2]. However, the only limitation of handling design patterns using a computer should be storage capacity and runtime performance. To solve the complexity problem, the set of design patterns must be given an order on a more intellectual or rather semantic level. Using the mechanisms of system theory [*Ulri-70*], Figure 7 gives an idea how to combine basic elements during the process of design.

---

[2]   See also Evolutionary methods in design in [MaPB-95, Gero-95].

The order may, but does not need to reflect the syntactic structure of design patterns. The order of design patterns can also be derived from the structure of reality. In particular, the order can take the following kinds of relationships into account:
- the composition relationships of design patterns,
- the implementation relationships between design patterns on different levels of abstraction,
- the functional relationships between the design patterns.

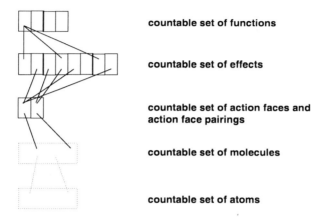

Figure 7   Combination of basic elements for the description of design solutions.

Important advantages of computer support regarding the application of design theories are:
- A theory of design becomes more applicable in practice.
- Experience-based knowledge will be made available for future design processes.
- The design process becomes easier to control since it is no longer dependent on the interpretation of a single designer.
- The design process becomes more efficient and reliable.
- The innovative design process is supported insofar as the combination of design solutions is systematically performed by computer.

OPEN ISSUES

A prerequisite for an efficient handling of design patterns is the capability of computer-aided methods and procedures to map requirements onto design solutions. However, the handling of requirements causes problems that have yet to be resolved. These problems concern the general possibility
- of formalizing the mapping process,
- of meeting given requirements,
- of completing the set of requirements,
- of reaching an optimal design solution and
- of providing knowledge about the design state transitions usually contained in methods.

REFERENCES

Gero-95	Gero, J. S.; Kazakov, V. A.: Evolving building blocks for design using genetic engineering: A formal approach. In: Advances in formal Design Methods for CAD. Proceedings of the IFIP WG 5.2 Workshop on formal design methods for Computer-Aided Design, June 1995, Chapman & Hall 1995.
Grab-96a	Grabowski, H.; Rude, S.; Gebauer, M.; Rzehorz, C.: Modeling of Requirements: The Key for Cooperative Product Development in: Flexible Automation and Intelligent Manufacturing 1996, (Proceedings of the 6th International FAIM Conference May 13-15, 1996 Atlanta, Georgia USA), S. 382-389.
Grab-96b	Grabowski, H.; Rude, S.; Langlotz, G. (1996) CAD `96 Kaiserslautern.
Koll-94	Koller R. Konstruktionslehre für den Maschinenbau Springer-Verlag, Berlin Heidelberg New York; - 3. Aufl. - 1994.
Mahe-95	Maher, M. L.; Poon, Y.; Boulanger, S.: Formalising design exploration as co-evaluation: A combined gene approach. In: Advances in formal Design Methods for CAD. Proceedings of the IFIP WG 5.2 Workshop on formal design methods for Computer-Aided Design, June 1995, Chapman & Hall 1995.
Pahl-97	Pahl G., Beitz W. (1997) Konstruktionslehre, Springer-Verlag, Berlin Heidelberg New York; -4. Aufl. - 1997.
Roth-94	Roth K. (1994) Konstruieren mit Konstruktionskatalogen, Band 1., Springer-Verlag, Berlin Heidelberg New York; Bd 1: Konstruktionslehre. -2. Aufl. - 1994.
Suh-90	Suh, Nam P.: The Principles of Design, Massachuetts Institute of Technology, Oxford University Press, 1990.
Suhm-93	Suhm, A.: Produktmodellierung in wissensbasierten Konstruktionssytemen auf der Basis von Lösungsmustern, Verlag Shaker, 1993.
Ulri-70	Ulrich, H.: Die Unternehmung als produktives soziales System - Grundlagen der allgemeinen Unternehmungslehre. Bern, Stuttgart: Verlag Haupt, 2. Auflage 1970.
VDI-2222	VDI-Gesellschaft Konstruktion und Entwicklung: Blatt 1: Erstellung und Anwendung von Konstruktionskatalogen, VDI-Verlag Düsseldorf, 1982; Blatt 2: Konzipieren technischer Produkte, VDI-Verlag Düsseldorf, 1977.
Yosh-87	Yoshikawa, H.; Tomiyama T.: Extended General Design Theory, Department of Precision Machinery Engineering, University of Tokyo, Design Theory for CAD, North-Holland, 1987.

Discussion

Question - A. Albers

Your approach is very interesting. But from the background of my practical design experience, and this must be seriously investigated, I can not accept one hypothesis from your approach: This is, that one set of requirements really leads to exactly one solution ... [Interruption by **H. Grabowski**: ... a complete set of requirements....] ... yes, I know and this is really the issue. I would like to show you with one or two examples, that there is really a matter of discussion, if the requirements are linked to specific components of a solution. But moreover, also the question must be investigated, if this hypothesis is indeed necessary.

Answer - H. Grabowski

This is a hypothesis which needs to be proven, but it is clear that in order to reach a goal, the goal must be described precisely beforehand. And if the goal is described precisely, the solution must be unique. If, however, requirements are contradictory, a goal may not be reached. But this should be investigated by a research project.

Question - T. Tomiyama

I was questioning myself and you showed us a very nice beautiful picture of decomposing requirements into functions and functions mapped onto principles and solutions. We conducted what we called design experiments and one of the findings was that designers don't design in that way. And ... [Interruption by **H. Grabowski**: They have no theory.]...Well, that's true, yes. That is the difference between theory and practice, of course. But the most serious problem is that we couldn't find any knowledge that will map functions or requirements to principles or to entities. So we think our design experiences very much influence our design activities. I'm not sure if you can explain these findings in your model, but I would appreciate your comments.

Answer - H. Grabowski

Yes, ok, due to constraints of time, I have not explained all details of our approach. There are a number of rules to be followed. One important rule is: Do not start with basic elements when complex solution patterns already exist. Complex solutions can again be divided into basic elements, if necessary. But in some cases, experience has shown that complex solutions work well. A functional or principal structure would be an example for complex solution patterns. Or a control loop. This is a special combination of basic elements for the fulfillment of special purposes, and if you know such a solution pattern, your design goal can be reached much faster. Another subject of discussion that in practice, a number of solution alternatives is developed. The reason is that the set of requirements is not complete. Most solution alternatives are not implemented. But humans memorize such ideas. And if you are using computers for the design process, computers can store these solution alternatives and in future situations, where requirements are different, these can be matched to the earlier solution alternatives: These solution alternatives can then be used. I do not see a difference between this approach and common industrial practice.

Comment - Y. Jin

It's not really a question, just a comment. I think, Professor Tomiyama just mentioned, there is no obvious mapping between the function and the principle. I also observed that in my class, I'm teaching a design class. But I think the principle is there, but is not explicit. Represented in the document maybe or other things, but when they think about specific entities, the designers still have this physical principles in their mind. And I think so, maybe one way is to make those implicit ones to be explicit, that could be a good contribution. And another thing I was thinking is, probably this principle is quite something similar to what Professor Gero said about behaviour. So there is a linkage there.

TRIZ: A Systematic Approach to Conceptual Design

Valeri Souchkov

Ideal Design Solutions, The Netherlands

INTRODUCTION

Engineering design involves the whole range of different theories and methods (e. g. [*Pahl-84*, *Suh-93*]) which aim at a mapping of given requirements and demands onto a description of physically realizable design product. However, formal design theories are only available at later design phases which are performed after a feasible design concept has been proposed and verified. Due to this, the early design phases lack sufficient computer aid and innovative design is still regarded as an art instead of science.

Altshuller's *Theory of Inventive Problem Solving* (abbreviated as *TRIZ*) is a collection of domain-independent techniques for innovative engineering design [*Alts-84*]. The TRIZ techniques have proven successful during a long-term use in various industries. Each technique consists of a number of guidelines, rules or principles which indicate how to cope with a specific problem or situation. Unlike the well-known techniques for psychological activation, for instance, brainstorming, TRIZ provides a systematic methodology for innovative engineering design.

This paper presents an overview of TRIZ - a systematic approach to conceptual engineering design. First, we discuss TRIZ background and philosophy. Then a brief introduction to three major TRIZ problem solving techniques is made. Finally, we discuss TRIZ and its role for developing CAD/CAM software.

TRIZ BACKGROUND AND PHILOSOPHY

TRIZ was originated by Russian scientist and engineer Genrich Altshuller. In the early 50th, he started massive studies of patent collections. He targeted at revealing common similarities between engineering problems and solutions that resulted in patents. More than ten years of research resulted in basic understanding of origins of inventive problems and formulation of general principles of inventive design.

Later, many researchers and practitioners of the former USSR united efforts and largely extended Altshuller's approach. It is estimated that by the end of 1984, more than 300 research and educational TRIZ centers were founded in the former USSR.

TRIZ philosophy is based on the fact that the evolution of the technology is not a random process. It correlates with the evolution of customer needs. Every field of engineering influences the evolution of the other fields. Therefore, the process of the technology evolution can and has to be studied.

Second major discovery was revealing the origin of inventive problem: a contradiction. A contradiction arises when two mutually exclusive design requirements are put on the same object or a system. For example, the walls of a space shuttle have to be light to decrease the mass of the shuttle when putting it to the orbit. However, this cannot be done by simply decreasing the thickness of the walls due to thermal impact when entering the Earth atmosphere. A contradiction results in the two conflicting design parameters: the walls have to be heavy and light at the same time.

When a designer faces a contradiction that cannot be solved by redesigning a technical system in known way, this means that he faces the inventive problem, and its solution principle resides outside a domain the technical system belongs to.

There are two ways to solve problems that contain contradictions: by finding a compromise between two conflicting parameters and by eliminating the contradiction. TRIZ is aimed at solving problems by removing the contradictions.

More than 40 years of studying patents in different areas of engineering resulted in several important discoveries which form the TRIZ philosophy:

- Every engineering system evolves according to regularities which are general for all engineering domains. These regularities can be studied and used for innovative and inventive problem solving, as well as for forecasting the further evolution of any engineering system in design terms.
- Engineering systems, like social systems, evolve through the elimination of various kinds of contradictions. The principles for eliminating the contradictions are universal for all engineering domains.
- An inventive problem can be represented as a contradiction between new requirements and an engineering system which is no longer capable of meeting the requirements. Finding an inventive solution to the problem means to eliminate the contradiction under the condition that a compromise is not allowed.
- Frequently, when searching for the inventive solution to a problem formulated as a contradiction, there is the need to use physical knowledge unknown to the domain engineer. To organize and direct the search for appropriate physical knowledge, pointers to physical phenomena should be used. In the pointers, the physical phenomena are structured according to technical functions that can be achieved on the basis of the phenomena.

Classical TRIZ which will be discussed below consists of several problem modeling and problem-solving techniques. It introduces a uniform way of modeling inventive problems by representing them in terms of contradictions and generic principles for resolving the contradictions. A comprehensive study of patent collections undertaken by TRIZ researchers and thorough tests of TRIZ within industries have proven the fact that if a new problem is represented in terms of a contradiction, then it can be solved by applying the relevant TRIZ principle. The principle must indicate how to eliminate the same kind of contradiction encountered in some engineering domain before. However, the most important achievement in TRIZ has been the formulation of general problem solving principles covering virtually all possible types of innovative and inventive problems.

In contrast to well known methods for mental activation or traditional design methods which aim at finding a specific solution to a specific problem, TRIZ organizes translation of the specific problem into abstract problem and then proposes to use a generic design guideline which is relevant to the type of the problem (Figure 1). As clear, by operating at the level of abstract (conceptual) models, the search space is significantly decreased that makes it easier to find the needed solution concept.

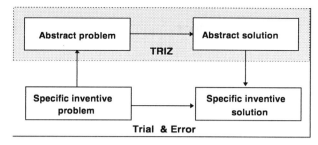

Figure 1   *Solving problems with TRIZ.*

TRIZ STRUCTURE

Modern TRIZ includes the following parts:
- **Trends of the technology evolution**. This part of TRIZ studies and formulates general trends of engineering system evolution.
- **Problem solving techniques**. The techniques aim at building a problem model and producing recommendations on how to solve the problem. Among them are:
  1. Principles for the elimination of engineering contradictions;
  2. Inventive standards for substance-field modeling to solve inventive problems through representing them in terms of substance-field interactions and applying generic patterns of interaction transformations.
  3. Pointers to physical effects. This part of TRIZ focuses on studying how to use the knowledge of natural sciences in the inventive process.
  4. Algorithm of Inventive Problem Solving - an integrated technique aimed at solving most difficult inventive problems.
- **Collections of advanced patents**. This part contains patent descriptions drawn from various engineering domains. The patents are structured according to inventive principles used to solve one or another type of contradictions. The patents can be used as analogous design cases illustrating the applicability of the principles and making a problem-solving process easier.
- **Functional and Value Analysis.** It is a modified version of traditional Value-Engineering Analysis] with the focus on functional decomposition and analysis of technical systems.

TRIZ-BASED CONCEPTUAL DESIGN

TRIZ provides a systematic support for the following phases of conceptual design:
- Analysis of ill-defined design problems by describing functions between the system components and identification of core problems by formulating contradictions.
- Generation of new solution concepts by using TRIZ problem-solving techniques: inventive principles, inventive standards and pointers to physical effects.
- Producing a technological forecast of a particular design product using TRIZ technology evolution trends.

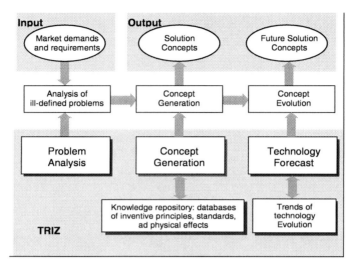

Figure 2   Conceptual Design with TRIZ

Below we will discuss each TRIZ technique in more detail.

TRENDS OF THE TECHNOLOGY EVOLUTION

The importance of the trends of evolution is that they can be used to estimate what phases of evolution a design product passed and what phases the product will pass. As a consequence, it is possible to predict what changes the design product will experience in future and to develop a strategic plan for new product development.

The most important trend of the technology evolution *is the trend of the ideality growth*. It states that during evolution over the time, any technical system tends to increase a ratio between the overall degree of performance of the system and expenses needed to provide the required degree of performance. The trend indicates a principal design requirement which every designer has to keep in mind while designing new products: a system being designed must be able to deliver every desired function with the highest degree of performance whereas the expenses required to provide the product's life-cycle should be as less as possible. The expenses in this definition are all types of energy, material and information resources required to deliver the given functionality and meet all other requirements.

Apart from the trend of the ideality growth, there are eight other laws in TRIZ:

- **Trend of system completeness:** a technical system tends to complete its material-energy structure to deliver the required function.
- **Trend of energy bypass:** a necessary condition of functioning of a system is to provide effective energy flows through all parts of the system. Accordingly the trend of ideality growth, systems tend to minimize amount of types of energy used as well as to minimize a number of energy transformations within a system.

- **Trend of irregularity of system's parts evolution**: the more complex a system becomes during the evolution the more irregularly its parts evolve. As a result, further development of the system becomes more difficult due to contradictions arising between system's parts.
- **Trend of increasing a number of material-energy interactions**: a system tends to increase the degree of interacting material-energy components to provide a higher degree of performance and controllability.
- **Trend of frequency and form adjustment.** During evolution, a system tends to adjust frequencies and forms of interacting components.
- **Trend of dynamics growth.** A system tends to replace existing designs of its movable parts or working tools with structures which have a higher degree of freedom.
- **Trend of transition to microlevel.** A system tends to replace a physical principle behind its component delivering a main function with a new physical principle which utilizes properties of more fragmented materials, particles or physical fields.
- Trend **of transition to macrolevel.** A system which has approached its limits of evolution can further evolve through merging with other systems (that produces a new function); or it can be eliminated if its function might be delivered by other systems.

The practical use of trends is possible through specific patterns. A pattern indicates a line of evolution containing particular transitions between old and new structures of a design product. Table 1 shows which phases of evolution a system phases according to the *trend of dynamics growth*.

Phase of Evolution	Description	Example
I	Solid object	Traditional mobile phone
II	Object divided into two segments	Mobile phone with a sliding part which contains a microphone
III	Two segments with a flexible link	Flip-flop phone of two parts
IV	Many segments with flexible links	Flip-flop phone of three parts
V	Flexible object	A flexible film with LCD which can be rolled in and out and stored in a plastic pen (serves also as a mobile videophone).

Table 1   The trend of dynamics growth.

We have to note that it is not easy to use the trends of the technology evolution since their formulations are too abstract to be easily interpreted in terms of a specific product. Nevertheless, the trends can be much easier learned after mastering skills with other TRIZ techniques.

INVENTIVE PRINCIPLES

A collection of inventive principles is the most known and widely used TRIZ problem solving technique. Each principle in the collection is a design guideline which recommends a certain method for solving a particular inventive problem. There are 40 inventive principles in the collection which are available in a systematic way according to a type of a contradiction that makes the problem non-solvable by the procedure of routine design. Examples of the inventive principles are:

- *Variability Principle*: Characteristics of the object (or external environment) should change such as to be optimal at each stage of operation; the object is to be divided into parts capable of movement relative to each other; if the object as a whole is immobile, to make it mobile or movable.
- Segmentation Principle: To divide the object into independent parts; to make the object such that it could be easily taken apart; to increase the degree of the object's fragmentation (segmentation).

Access to the principles is provided through a matrix which consists of 39 rows and columns. Positive effects that have to be achieved (generalized design requirements) are listed along the vertical axis while negative effects which arise when attempting to achieve the positive effects are listed along the horizontal axis (Figure 3). A selection of a pair of positive and negative effects indicates which principles should be used to solve the problem.

Parameters	*what deteriorates as a result of improvement*				
what to improve	Speed	Force	Stress	.....	Stability
Speed		13,28,15,19	6,18,38,40	.....	28,33,1
Force	13,28,15		18, 21,11	.....	35,10,21
Stress	6, 35,36	36,35,21		.....	35, 2,40
.....	.....	....	....	.....	.....
Stability	33,28	10,35,21	2,35,40	.....	

Figure 3 A matrix of principles for engineering contradiction elimination. Numbers indicate what principles have to be used: 1 - Fragmentation; 2 - Removing; 10 - Preliminary action; 13 - Other way round; etc.

Example. The weight of a short steel pipe is small enough and does not hinder the movement of the pipe inside a kiln during thermal processing. However, to process a long pipe is more difficult: its large weight makes the transportation difficult. In this situation, a contradiction arises between the parameters „length of the movable object" and „weight of movable object". One of the inventive principles suggests the use of pneumatic and hydraulic structures to eliminate this kind of contradiction. One of the known solutions to the problem is to create an air cushion in the kiln, which provides the required support and movement of long pipes.

INVENTIVE STANDARDS

Another TRIZ problem solving technique is a collection of so-called **inventive standards**. While inventive principles operate with generalized technical parameters, inventive standards are more formal and context-dependent since they operate with a specific model of a design product. This makes the inventive standards more accurate than the inventive principles.

While the inventive principles have to be used when a designer faces a contradiction, the inventive standards are used in those situations when a problem involves so-called *undesired interaction* between two or more system's components. There are for types of the undesired interactions:
- missing: some parameter of a component has to be changed during operation, but we do not know how to change it.
- harmful: a component produces harmful effect.
- excessive: an action of one component on another is too strong.
- insufficient: an action of one component on another is too weak.

To model different technical problems in uniform way, so-called *Substance-Field Modelling* is used. The basic idea behind the modeling paradigm is that any part of a design product can be represented as a system of interacting substances. Interactions between the components are represented via physical fields.

In reality, a physical field is a carrier of a function which is delivered by one of the modeled components. Examples of physical fields are mechanical, acoustic, thermal, electrical, magnetic and electromagmetic.

A basic *substance-field model* (SFM) consists of two substance components and a field between them (Figure 4). For instance, SFM of a cup of coffee can be modelled as two substance objects: coffee (fluid) and the cup. Both the components interact with each other via mechanical field.

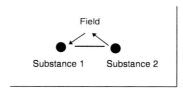

Figure 4   *A basic substance-field model: Substance 1 and Substance 2 - interacting substance objects; Field - a physical field provides the interaction.*

A level of detail is chosen by a designer on the basis of his personal point of view and may vary in each particular situation. Substance components that have complex physical structure can be generalized and modelled in a black-box manner. Boundaries of the system are usually defined by two interacting substance components and a field providing an energy flow between the components.

In real systems, there are might be many types of fields involved into the same interaction. Since SFM is aimed at problem solving, it is always recommended to limit ourselves to a single field that is a cause of a problem. As seen, every technical systems can be modeled in this way.

A problem is formulated as undesired interaction between two components. For instance, a mechanical field keeps coffee in the cup but the same field make coffee particles stick to the cap. The same interaction can be described as both desired and undesired. So it would be nice if the coffee particles could not stick to the cup.

To obtain a solution to a problem represented in terms of SFM means that the physical structure which contains the undesired interaction (source SFM) has to be transformed into a structure in which the desired interaction is achieved (target SFM). An inventive standard defines such transformation patterns.

The original term „Inventive Standard" introduced by Altshuller means that there is a common, or standard method to solve different problems which result in identical problem models. To solve a problem with nventive standards, there is no need to formulate a contradiction.

An inventive standard consists of two parts. The left part specifies conditions and a problem: what type of the source substance-field model is and what restrictions on introduction of additional components are. The right part shows a pattern of a solution.

A primary inventive standard is: *If there is an object which is not easy to change as required, and the conditions do not contain any limitations on the introduction of substances and fields, the problem is to be solved by synthesizing a **SFM**: the object is subjected to the action of a physical field which produces the necessary change in the object. The missing elements being introduced accordingly* (Figure 5).

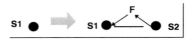

Figure 5   Basic rule of substance-field transformation.

Example. To remove air from a powdered substance, the substance is subjected to centrifugal forces.□

Another inventive standard is: *If there is a SFM which is not easy to change as required, and the conditions do not contain any limitations on the introduction of additives to given substances, the problem is to be solved by a transition (permanent or temporary) to an **internal complex SFM**, introducing additives in the present substances enhancing controllability or imparting the required properties to the SFM* (Figure 6).

Figure 6   Transition to internal complex SFM.

Example. To detect very small droplets of liquid, a luminophore is added to the liquid in advance. Then, using ultraviolet light, it is easy to detect the drops. □

Let us have a look how the inventive standards can be used to solve a problem. During arc welding, the arc does not reach every particle of a powder which is inserted into the gap between two articles. What to do? In this situation, we have two substance components: the arc and the powder. Interaction is provided via thermal field. However, the interaction is insufficient and using another powder or to increase the intensity of the arc is not allowed.

According to the described model of the problem we have to apply the following inventive standard: *If there is a SFM which is not easy to change as required, and the conditions do not contain any limitations on the introduction of additives to given substances, the problem is to be solved by a transition (permanent or temporary) to an internal complex SFM, introducing additives in the present substances enhancing controllability or imparting the required properties to the SFM.*

Solution: It is proposed to insert an exothermic substance into the gap beforehand that will produce extra heating of the powder without increasing the arc intensity (Figure 7).

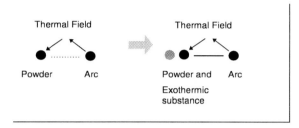

Figure 7   Solving a problem with the inventive standards.

SYSTEMATIC ACCESS TO PHYSICAL KNOWLEDGE

Another TRIZ problem-solving technique is a Pointer to physical effects. While inventive principles and inventive standards do not produce recommendations in terms of what particular physical fields and substances should be used to solve a problem, the Pointer to Physical Effects establishes links between specific physical effects and technical functions the effects are capable of delivering.

Studies of the patent collections indicated, that inventive solutions are often obtained by utilizing natural phenomena not used previously in a specific area of technology. Knowledge of natural phenomena often makes it possible to avoid the development of complex and unreliable designs.

For instance, instead of a mechanical design including many parts for precise displacement of an object for a short distance, it is possible to apply the effect of thermal expansion to control the displacement.

Finding a natural phenomenon that would be capable of meeting a new design requirement is one of the most important tasks in the early phases of design. However, it is nearly impossible to use descriptions of natural phenomena in a form as they are presented in handbooks on physics or chemistry. The descriptions of natural phenomena available there yield information on certain properties of the phenomena from a scientific point of view, and it is unclear how these properties can be used to deliver specific technical functions.

TRIZ Pointers to the effects bridges a gap between engineering and science. In TRIZ Pointers, each natural phenomenon is identified with a multitude of technical functions that might be achieved on the basis of the phenomenon.

The search for effect is possible through formulation of a problem in terms of a technical function. Each technical function indicates an operation that can be performed with respect to a physical object or field. Examples of the technical functions are „move a loose body", „change density", „generate heat field", and „accumulate energy".

A fragment of the pointer to physical effects is shown in Table 2.

Function	Effects
To separate mixtures	Electrical and magnetic separation. Centrifugal forces. Adsorption. Diffusion. Osmosis. Electroosmosis. Electrophoresis.
To stabilize object	Electrical and magnetic fields. Fixation in fluids which change their density or viscosity when subjected to magnetic or electric fields (magnetic and electrorheological liquids). Jet motion. Gyroscopic effect.

Table 2   Fragment of the pointer to physical effects.

Example. How to accurately control the distance between a magnetic head and a recording surface of a tape?
In the TRIZ pointer to physical effects, the function „to move a solid object" refers to several effects. One of the effects is the physical effect of magnetostriction: a change in the dimensions and shape of a solid body made of certain metal alloys during magnetization.
The magnetic head is fixed to a magnetostrictive rod. A solenoid generating magnetic field is placed around the rod. A change of the magnetic field's intensity is used to compress and extend the rod exactly to the required distance between the head and the recording surface.

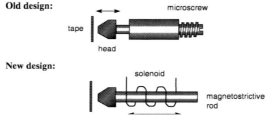

Figure 9   Solving a problem with TRIZ pointer to physical effects.

ADVANTAGES AND DISADVANTAGES OF TRIZ

As shown by numerous industrial case studies (e. g. [*Lync-97*]), TRIZ helps to shorten a new product development time and to quickly find new solution concepts. This is possible due to the following advantages of the TRIZ methodology:
- Since inventive design is a knowledge-intensive process, success of inventive design depends on how fast the needed knowledge can be found. TRIZ helps to organize the fast search for needed knowledge.

- TRIZ provides a systematic access to the previous experience of many generations of inventors. This experience is generalized and presented in a form of inventive design rules and guidelines.
- In some cases, it might be clear what function is needed but unclear what physical principle can be used to deliver the function. To organize and guide the search for proper physical principles, TRIZ pointers to natural phenomena and effects are used.
- All design products evolve over the time according to the same domain-independent trends. These trends are used for effective problem solving as well as for forecasting the further evolution of a specific design product.
- TRIZ does not replace human creativity. TRIZ restructures the thinking process of a designer and provides fast access to the needed knowledge, but it does not solve problems independently of the designer.
- No previous inventor's skills are needed to effectively solve new inventive problems.

On the other hand, there are also major disadvantages of the approach behind TRIZ:
- No formal definition of a contradiction is available in TRIZ.
- TRIZ does not provide exact recommendations on how to formulate contradictions with respect to a particular problem. As a result, a contradiction is constructed ad-hoc since no analysis of a prototypical design is performed.
- To identify an inventive principle which has to be used for solving a problem represented as a specific conflict, the conflict has to be reformulated in terms of generalized engineering parameters. However, this can only be done intuitively since no translation technique is available in TRIZ.
- Inventive principles and standards do not propose a solution to a given problem. They only recommend a method which was used to solve a similar contradiction before.

These disadvantages make the process of learning TRIZ and mastering skills with it very slow. Since TRIZ operates within vast knowledge domains, it is clear that it needs to be aided by a computer.

COMPUTER-AIDED TRIZ AND CAD/CAM SOFTWARE

Recently, a number of software packages supporting design problem solving with TRIZ have been developed. Among them are Invention Machine [*Tsou-93*] and Innovation Workbench [*Brah-95*]. Although all packages incorporate different approaches to represent TRIZ information and organize the problem solving process, they form a new category of computer-aided design tools which support a conceptual phase of engineering design.

While traditional CAD/CAM systems focus on processing and computing geometrical and material aspects of specific designs, TRIZ-based software packages structure access to previous inventive experience stored in the form of inventive principles and indexed physical principles. According to a given problem formulation, TRIZ-based packages propose information on what the generic behavior of a design solution should be rather than what form and geometry the solution should have.

In summary, TRIZ-based packages organize mapping between the function and behavior of a concept which is still to be found, whereas CAD/CAM systems map functional and geometrical specifications directly onto already known design solutions stored in the database.

In addition, CAD/CAM systems propose specific, „ready-to-manufacture" descriptions of solutions that makes such systems relatively easy to learn and use. TRIZ-based packages are well-organized interactive systems which only help with finding general recommendations on how to solve problems, or at best, indicating what physical principles to use. A designer should be able to interpret this information and translate it into a feasible solution. No sufficient computer aid has been available so far to support this step. This causes certain difficulties in the use of the software by many designers since the gap between general recommendation and specific solution can be very large. It is our belief that to be accepted by a wide audience, TRIZ-based software has to bridge this gap and be able to generate solutions in terms of specific problems instead of displaying general recommendations.

As seen from the discussions, to be successfully transferred to a computer, TRIZ needs formal background. To solve this problem, the project INDES aimed at formal modeling of TRIZ knowledge was initiated in the University of Twente, The Netherlands in 1993 [Sush-96]. The goal of the project was to restructure TRIZ knowledge and to create a sharable domain-independent ontology for modeling innovative design knowledge. The results of the project can be found in [Souc-98].

CONCLUSIONS

Since TRIZ is comprised of a large number of empirical rules, it is difficult to evaluate from the point of view of exact sciences. Unlike fundamental sciences, TRIZ is not based on the axiomatic approach and does not include formal means for problem solving and verification of results. Instead, its techniques resulted from a comprehensive study of previous engineering experience that does not guarantee that the techniques will be applicable to *every* situation that may occur when designing new products. No proof of absolute applicability is possible due to the heuristic nature of TRIZ.

On the other hand, TRIZ discovered a number of principles and introduced new concepts which, although have not been formalized yet, have proven their applicability for solving practical engineering problems and considerably accelerating the process of new product development. This fact should not be neglected when studying TRIZ. Many years of experience with using TRIZ indicated that the discovered patterns and principles can be successfully applied to solve virtually any inventive problem. For this reason, TRIZ rapidly became a part of the engineering curriculum worldwide as a general methodology for conceptual design of new products and developing new technologies.

In summary, major contributions of TRIZ to engineering are:
- TRIZ discovered a systematic nature of technology evolution and described a number of domain-independent evolution trends.
- TRIZ introduced a new classification of design solutions.
- TRIZ proposes to regard a contradiction as a cause of inventive problems and states that inventions results from eliminating contradictions.
- A set of basic principles for contradiction elimination was proposed.
- Access to the basic principles was organized in a systematic way.
- TRIZ proposed to model design products in terms of substance-field interactions and apply generic patterns to transform the physical structure of products.
- TRIZ proposed a novel way to couple physical principles and technical functions.

We believe that TRIZ concepts such as contradictions and principles for substance-field modeling might be used for developing a fundamental science of innovative engineering design.

REFERENCES

Alts-84   Altshuller, G.: Creativity As an Exact Science, Gordon and Breach Scientific Publishers NY USA, 1984.
Brah-95   Braham, J.: „Inventive ideas grow with TRIZ", Machine Design. Vol. 67 No. 18. October 12 1995.

Lync-97	Lynch, B.; Saltsman; Young, C.: „Windshield/Backlight Molding - Squeak and „Buzz" Project (Case study), The TRIZ Journal, December. (http://www.triz-journal.com/archives/97dec/97dec.htm), 1997.
Malm-93	Malmqwist, J.: „Computer-Aided Design of Energy-Transforming Technical Systems", Proceedings ICED 93, The Hague, 1993.
Pahl-84	Pahl, G.; Beitz, W.: Engineering Design: A Systematic Approach, Springer Verlag Berlin, 1984.
Souc-98	Souchkov, V.: Knowledge-based Conceptual Engineering Design. Ph.D. Thesis, University of Twente, Enschede, The Netherlands, 1998. To appear.
Suh-93	Suh, N.: Principles of Design, Oxford University Press, USA, 1993.
Sush-96	Sushkov, V.; Alberts, L. K.; Mars, N. J. I.: „Innovative Design Based on Sharable Physical Knowledge", Proceedings Artificial Intelligence in Design 96, Kluwer Academic Publishers Dodrecht, 723-742, 1996.
Tomi-95	Tomiyama, T.; Xue, D.; Yoshikawa, H.: „Developing an Intelligent CAD system", Artificial Intelligence in Optimal Design and Manufacturing, PTR Prentice Hall, Englewood Cliffs NJ USA, 83-112, 1995.
Tsou-93	Tsourikov, V. M.: „Inventive Machine: Second Generation", AI & Society, 7(1), Springer International, UK, 62-78, 1993.

Software-Repository for Universal Application within the Development Process of Artefacts

F.-L. Krause, C. Kind, R. Heimann

Production Systems and Design Technology

Fraunhofer Institute, Berlin, Germany

Keywords: artefact configuration, application system, software, product development, process, description

ABSTRACT

Artefacts can be configured by a certain amount of elements, which consists of other elements likewise and/or basic elements, which are not further divisible. The kind of the basic elements depends on the domain in which the artefact is developed and used. Against this background software to support the product development process can be configured as well by the configuration of software atoms which represent the lowest state of granulation for a specific application purpose. To support the application system generation on the basis of the description of the product development process, done by the designer himself, (globally) distributed software elements will be detected according their specific functionality, collected and configured according to the specific needs of the product developer. In this contribution a concept for the realisation of this approach will be described.

INTRODUCTION

Software becomes increasingly important to support the development process of artefacts. The suitability of software to meet the requirements of a task has a considerable impact on both artefacts and development processes concerning time, costs, quality and degree of innovation.

Up to now, software support for developers is mainly focused on the application of basic systems providing only a limited acceleration of development processes. Much more benefit can be taken from the use of task-oriented application systems [*Spur-97*]. Application software leads to a higher quality of the results of product development and an acceleration of the development process itself. Furthermore, the learning process to have complete command of the system will be shortened. In opposite to basic systems, application software fits as good as possible to the task which has to be carried out within a specific process step. For this it's necessary to get the demands on the application software deriving from the product development process itself as well as from the artefact to be developed.

An Universal Design Theory (UDT) has to cover all aspects of artefact development. Therefore, if not specified differently, in this contribution the term development refers to the development of any artificial product (artefact). There are two important areas in which software have to be considered within this research field. On the one hand the result of the development process - the artefact - can be software itself and therefore an UDT has to include the development not only of material products but of immaterial products too. Software belongs to the last. On the other hand software to support the development process has to be part of an UDT as well, since the world gets digitised and digitalisation needs software.

In this contribution both aspects are combined: the application of software to support the specific tasks within the product development process and the development process of these specific software artefacts. To deal with the requirements of an efficient software development an approach based on the definition of a software repository is introduced. The software repository will be discussed concerning its structure and the elements in it. Furthermore a methodology to derive demands on application systems from the artefact development process and the artefact itself is represented.

With respect to the universal approach of the design theory the possibilities to generalise the outlined ideas for software development are regarded and parallels to the development of material and other immaterial artefacts are taken into account.

DESIGN THEORY AND SW-DEVELOPMENT

A design theory is a theory about designs or designing and is therefore oriented towards the product as well as to the process. Within this context a theory is a set of general statements about a specific part of the world which define a system and relate variables of that system. The statements are intended to be true, and are intended to explain and predict observations in the world. A specific theory is centred around a domain, a set of phenomena, or around an activity. The observation referred to by a design theory could be of human, tool or object behaviour. Predictions, for example, might be about ranges of effectiveness of mechanisms and structures, the appropriateness of methods for different phases of a problem, or the predicted human action in a particular design situation [*Brow-95*]. On this basis an Universal Design Theory has to describe the scientific edifice of the basis, rules and principles in the whole room of development of artificial products either material or immaterial.

Corresponding to the state of the art in product development, artefacts are described digitally by the application of information technologies. Because of this an Universal Design Theory has to take the computer aided development of artefacts into consideration. Information technology comprises both computer hardware and software. As mentioned above basic systems don't provide satisfying functionality and according to this an unacceptable efficiency leading to the demand of the development of task-specific application systems, Figure 1.

Untill now the development of application software is in many cases a new development project. Requirements on the software to be developed must have been anticipated at a point of time, when many of the demands have not been specified because of the missing principal and functional decisions made during the development process. They have to be defined explicitly and cannot be derived implicitly from the development activities directly. Thereby a tremendous expenditure of time and costs is caused. This procedure leads to the effect, that more functionality is realised than necessary and that the functionality provided by the system is less goal-directed. This procedure of software development has to be improved. Corresponding to other structured methodologies applied in product development processes a method for the development of task-specific application systems has to be generated. The question, which application system is required at a specific stage within the product development process of an artefact can only be answered by analysing and describing the process as well as the product itself. Specific expressions to describe products and processes have to be defined. Consequently, a methodology to describe industrial products and development process by attributes and the standardisation of these describing terms has to be one purpose of an Universal Design Theory.

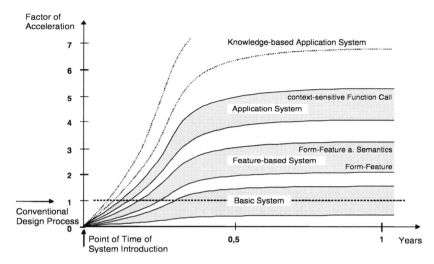

Figure 1   *Acceleration of development activities by the use of application systems [Spur-97]*

Due to the high degree of uncertainty an entire description of the hole development process will not be possible in the run-up to development activities, when most relevant decisions haven't been made. Furthermore, it is necessary to complete the process description concurrently during the execution of development tasks [*Krau-98*]. Because of this, the configuration of application systems has to be flexible and fast to realise the configuration while process runtime.

To reach high flexibility in software configuration it is essential to make the basic software elements available which are necessary to configure every required application software within a specific domain. In the following the basic software elements are called software atoms. They are structured and organised in a software repository containing all software atoms. From this repository an amount of software atoms can be picked up and combined to a software system comparable with a chemical box of bricks from which chemicals can be taken and brought together to a new chemical bond. Furthermore techniques and algorithms are required to automate the software configuration.

Thus, the definition and construction of software repository as well as the development of new techniques and algorithms for an automated software configuration has to be part of an Universal Design Theory.

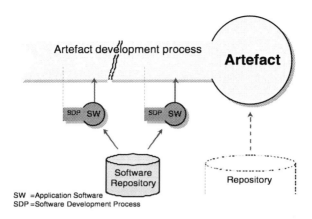

Figure 2   Integration of software and artefact development processes based on repositories

SOFTWARE REPOSITORY APPROACH

UNIT CONSTRUCTION SYSTEMS

A potential approach to meet the demands of a highly effective and efficient configuration of software can be seen in the definition of a software-repository as a part of the UDT. This repository, if necessary adjusted to specific products, classes of products or industrial sectors, contains software elements, so that every software-solution can be realised by the configuration of a specific number of these. Consequently, the repository can be compared with construction sets used by engineers for development of artefacts, where components are taken from these construction sets and brought together to a complete system.

The approach to set up an artefact by combining or configuring specific elements can be found in the field of development of material artefacts in many places and different qualities. In the field of mechanical products especially for the consumer goods (cars, white products) the platform concept and modularization of products is already or, in the case of capital goods (railway vehicles, machine tools) becomes increasingly important.

The same concept is realised in unit construction systems, where the product with its specific functions is build up by the combination of single parts and/or components, which realises different tasks. An examples is given by gearboxes with different number and kind of axles and gearwheels. A reduced approach is pursued by size ranges which follows similarity laws and realises the same function within a broad range of size [Pahl-97]. In contrast to this a more flexible system concerning scope and functional flexibility is realised by the mechanical assembly technique, which is applied for many different field of application but with quite specific tasks, such as conveyor system techniques or booth building systems.

All systems are characterised by clearly defined interfaces between the single parts or components which can be combined. The most important rationalisation effects of construction sets in the field of mechanical engineering can be found in both design and manufacturing. The product development is done once for many product variants with more effort than for a single product solution, but with very little effort for the single product realisation. In the manufacturing phase the lot sizes increase and tools, jigs and fixtures and machine tools are working to capacity [*Spur-94*].

An even more general approach can be found in the toys area. A set of different elements with a standardised interface can be used to build with very different functions and appearance such as `Lego´ or `Fisher-Technik´.

In the area of chemistry different atoms can be configured to molecules. The basis is the periodic system of elements which is surely the most powerful basis system to configure artefacts, since every material artefact is finally based on it. There are only some more than one hundred elements and the rules to combine them are defined by themselves and can be explained by the quantum theory. On the other hand it is expensive to reach specific functions realising a chemical artefact. The behaviour of chemical artefacts, which are a combination of several different elements cannot be derived from the knowledge about the behaviour of the single elements easily.

A virtual aspect of the repository approach can be found in the area of CAD technology. As well in the area of 2D drawing generation as in the room of 3D modelling the application of primitives to realise design tasks and to describe the results have been common use. Lines, circles and other two-dimensional elements have been arranged representing products on drawings and 3D elements such cubes, cylinders or cones have been applied to generate computer internal representations of the geometry of a product [*Spur-84*].

Within the CAD fields two more advanced examples of the approach to define complex structures by configuring single elements are the voxel-technology and the application of feature libraries [*Spur-97*]. The very difference between those approaches of virtual artefact configuration are the number of basic elements or atom the artefact will be configured of. The voxel technology only knows the cube as the basic element where as the feature approach has to offer many different features. In contrast the effort to define the relations between the features is smaller since the interfaces are predefined by themselves.

The aim of a software repository in general is to support the flexible configuration of application software. Furthermore, the application of the repository has to be as easy as possible since the people applying it should do that with very little or even no effort. According to the changes within a specific field of application regarding scope, requirements, and other aspects, the elements of a software repository have to be interchangeable easily according to the demands. Additionally, a software repository has to be an open system concerning the technological realisation.

STRUCTURE OF THE SW-REPOSITORY

According to the period system of elements software atoms represent the smallest unit of software systems. These software atoms can be used to build up any complex software system as a whole. Every single software atom is not further divisible. Here it's necessary to analyse, if the software atom is not further divisible corresponding to the original meaning of the word „divisible" or because of any other reasons. One of these reasons could be that a further subdivision is not ingenious. This aspect leads directly to the problem of the right degree of granulation of the software atoms concerning complexity and degree of specification. In case that the size is too small, the software atoms are quite simple but difficult to handle. Too many attributes would be needed to describe the relations between the atoms respectively the position of the atom within the context. On the other hand too complex software atoms would lead to a decreasing effectiveness. Functions of the atoms could be represented redundantly or could consist of more features than necessary to meet the requirements of a specific problem. However, obviously it is better to include unnecessary functionality than not to include functions the product developer needs.

Another question is, which criteria are suitable to describe and to determine software atoms unequivocally and entirely with all relevant attributes. Thinking about the source-code of software single words or even single letters could be possible criteria. But for sure it is unreasonable to consider a letter or a simple word as a software atom in view of the fact that the meaning of an expression derives from the combination of letters and words combined with syntax and semantic. Software in general consists of procedures, subroutines, data and other operational components. Accordingly it is probably more suitable to structure the repository by logical or functional aspects. Another classification criterion for the software atoms is given by their hardware requirements and software interfaces.

Beside the software atom repository a second repository in which all possible relations between software atoms are represented is necessary. By the application of the relations it can be described how software atoms are connected to each other. It is to be expected, that only by the determination of the required software atoms, relations and the assignment between them the demanded software system can be derived unequivocally. Probably the same atoms with different relations or the same relations in a different arrangement lead to different results.

The relation repository is in the first place a logical system. This means that in case of realisation the relation-information do not have to be described separately but they can be part of the atoms themselves as well. Compared to reactions of chemical elements the question of the relation is also important. Bringing together two or more chemical elements the way the react is not necessarily known. It is even more difficult to find the right elements looking for specific characteristics of the compound. Comparable to this it is not clear in which way two or more software atoms will work when they are brought together. It just might be a linear addition of the functionality but it is also possible that the functions interfere with each other and change the results unexpectedly.

As a basic prerequisite to realise the approach described above the hardware and the software platform have to support the integration of the software atoms. This leads to specific requirements on software technology which is used to realise the software repository. It has to offer a kind of logical software basis where the atoms are stored and functions to bring them together and integrate them in a combined system. Here, the product developer works for example on an intelligent communication layer which makes the application transparent to him so that he will not recognise the origin of the software he uses. And the software has to be capable of being plugged in without additional adaptation effort.

Up to now different technologies to support or realise this approach have been developed and are still under development. They differ in the dependence on the hardware and software platform and the flexibility to realise specific applications in different application domains. For instance the programming language C/C++ is quite independent form the computer platform if it corresponds to the ANSI standard. Leaving this standard by applying specific libraries the independence will deteriorate as well. In other programming languages such as Smalltalk or Java the availability of so-called virtual machines determine the ability to be used on a broad range of platforms.

Another requirement aspect can be derived from the physical distribution of the software repository world-wide. In this connections standards to realise distributed software systems such as CORBA or DCE has been developed [*Bake-97*]. A more advanced approach which is currently under development is aimed by mobile agent systems. Software agents execute their tasks independently and could be applied to search the software repository globally and collect world-wide distributed software atoms according to the requirements. The search of the agents can be realised using the description of the process and the product. To support the discovery of the appropriate software atoms by the agents, every software atom has to be enlarged by a declaration part which explains its features and characteristics and can be analysed by the software agent.

PROBLEMS OF A SOFTWARE REPOSITORY

Generally, in a software repository there are different software elements which have to be brought together to realise a specific task. With respect to this software repository high attention has to be paid to the following points:
- interfaces between the elements,
- application and integration of the elements,
- complexity of the elements,
- completeness of the repository, and
- openness of the system

which means the possibility to cross the border of the programming language depending on specific hardware and software interfaces.

In the field of practice a user-oriented method for software-configuration is only practical in case that all required software atoms are available. But the question is, if there is a way to make sure that the repository is complete and any software system can be derived only from software atoms available in the repository. It can be doubted that a repository can be described entirely and that there is a way to reach completeness. Therefore, a specific mechanism has to assure that the repository can be filled up in case that an atom is required which is not available. Against this background it has to be analysed, if it's possible to quantify and to describe the amount and kind of software atoms for a determined product or class of products as a sub-repository of an universal one.

Another field of problems derives from future requirements. Even if the repository is complete for today, it is certainly not valid for the future as well due to changing technologies and cultures. Furthermore, presently used atoms will not be sufficient for the software of tomorrow and have to be adopted as well. Therefore, the extendibility of the software repository has to be considered. Despite technologies and requirements of tomorrow cannot be anticipated yet the structure of the software atoms in the repository and their flexibility have to be described. These aspects have to be considered for the development of an UDT.

PROCESS DESCRIPTION FOR SOFTWARE CONFIGURATION

The requirements a software-solution has to meet arise from the artefact itself and its development process. The description of both, artefact and process, which are closely related to each other, must be done by the designer as the owner of the product and process knowledge. His process know-how has to be opened up by making available a special definition language which is similar to the developer's diction to reach high acceptance. The attributes provided by this process definition language and used by the designer to describe the development process represent the process in an explicit manner. Furthermore, implicit specifications of requirements on application software can be derived from the product description. These information are used to identify expressions to describe software systems which support the designer at specific tasks.

These expressions have to be transformed into algorithms. By these algorithms the appropriate software atoms and relations can be identified, picked up, handled and arranged correctly for a definite task. This software configuration process has to be supported by tools for software-engineering in the design field called CASE-for-CAD.

Most of the process descriptions can be made within the early phases of product development. Here the use of the language and the description realised by the language must be focused on the definition of development activities. In this case the activities are linked to each other by the flow of information generated in an early process step and required in a later one. Further descriptions are made by product oriented commitments like definition of functional structure, determination of principals and concrete solutions. These commitments contain implicit information which can be used for the specification of application systems.

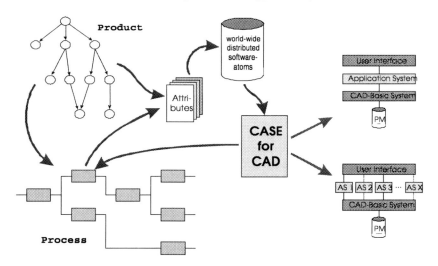

Figure 3   Configuration of application systems deriving from product description

The whole of the attributes used to describe product and development activities are defining the demand on the required application system entirely. Based on the attributes, intelligent technologies like software-agents search for appropriate software atoms within defined environment like local databases or world-wide distributed networks (Intranet/Internet). This approach is graphically represented in Figure 3.

To generate an application system with respect to the specified demands by the configuration of software atoms two procedures are possible in principle, the top-down approach and the bottom-up approach.

The basic idea of the top-down approach is to reduce the entire space of requirements down to smaller part spaces. For this, structuring rules have to be provided corresponding to which the entirety of the demands can be divided semantically. These rules provide, how and how far the spaces have to be divided considering the software atoms within the repository. This procedure will be repeated until the level of requirements on software atoms is reached.

Referring to the requirement on the application system in the bottom-up approach suitable software atoms are selected directly and combined according to possible relations. Each combination is analysed and assessed concerning the fulfilment of the postulated requirement. Here, a target-oriented and structured procedure would indirectly lead to the top-down approach.

The configuration of an application system by software atoms based on the given relations must be done by a CASE-tool, which is adapted for the development and configuration of CAD-application systems (CASE for CAD). The CASE-tool has to provide workflow management functionality to integrate the configured application system into the CAD-environment context sensitively.

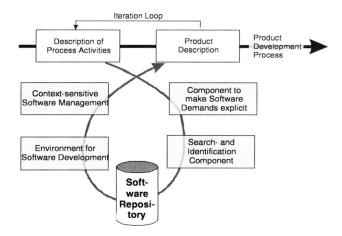

Figure 4   *Implicit product- and process information becomes explicit*

The specification of the artefact, of its development process and of the required software takes place in an iterative specification loop. In a first step, process activities are described explicitly and/or implicitly, Figure 4. Based on this information, the required application system is realised as explained above. From the following product describing activities including decisions and commitments further process describing attributes are generated, and used for the specification of the next required application system. This procedure repeats iteratively until the product development process results in the overall target artefact. The used design methodology and developer's design logic must be represented, so that the appropriate functionality is made available right in time.

CONCLUSION

In this contribution, the importance of the use of application systems to support product development activities has been illustrated. Out of this the development of a methodology for the generation of application systems as a part of the Universal Design Theory was derived.

Furthermore, parallels to the application of repository-oriented development processes of different artefacts have been presented.

To meet the requirements on an efficient development process an approach for the flexible and task-oriented configuration of application systems was introduced. This methodology is on the one hand based on a product and process description and on the other hand on the definition of a software repository. The software repository contains software components with a defined granulation, describing all possible application systems entirely within a determined domain of application. The product and process description is used to derive requirements on application software, attributed by the use of a standardised vocabulary. Additionally side there is the software repository containing software atoms described with regard to their characteristics. Based on the attribution of both sides fitting software atoms can be identified and joined to an application system fulfilling the corresponding requirements. To realise this, demands on a mechanism for the automated configuration of task-adapted application systems have been set up.

To reach the aims described, terms, syntax and semantics of a product and process description language has to be developed. It has to be analysed to what extend an Universal Design Theory makes such a language available. Both, the formulation of software requirements and the characterisation of software atoms has to be realised by a standardised attribution, which has to be extendible in case the occasion arise. Furthermore, structure and extension of the software repository, if necessary limited to a specific field of application, has to be defined. Additionally, the content has to be generated. Intelligent software technologies have to be determined respectively developed. This software has to be able to identify and to handle appropriate software atoms based on the attributed software requirements and to join them to an entire software system. Finally it is to strive for the evaluation and verification of the basic functionality of such a method by a prototypical scenario.

The idea that artefacts can be configured out of single elements is one way to approach an Universal Design Theory. For this, it could be helpful to demonstrate an specific way how to generate single basic elements, order them in a definite structure and integrate them to a specific artefact. The approach of a software repository described in this contribution represents one step into this direction.

ACKNOWLEDGEMENT

In this document, results of the project DFG Kr 785/10-1 „Computer Supported Methods for the Description and Evaluation of Design Processes and Design Results" are presented, financially supported by the Deutsche Forschungsgemeinschaft (DFG).

REFERENCES

Bake-97 Baker, S.: CORBA Distributed Objects. Addison-Wesley, New York, 1997.
Brow-95 Brown, D.C.: A Design Theory. Proceedings of Strategic Planning Workshop on Design Engineering. Tempe, Arizona, May 1995.
Krau-98 Krause, F.-L.; Heimann, R.; Raupach, Chr.: Optimisation of Product Development by Key Figures and Simulation. Proceedings of the CIRP Seminar STC Design: New Tools and Workflows for Product Development, May 1998, Fraunhofer IRB Verlag, Stuttgart, 1998.
Pahl-97 Pahl, G.; Beitz, W.: Konstruktionslehre. Springer Verlag, Berlin, Heidelberg, New York, 1997.
Spur-84 Spur, G.; Krause, F.-L.: CAD-Technik. Hanser-Verlag, München, Wien, 1984.
Spur-94 Spur, G. (Hrsg.): Handbuch der Fertigungstechnik: Fabrikbetrieb. Band 6, Hanser Verlag, München, Wien, 1994.
Spur-97 Spur, G.; Krause, F.-L.: Das Virtuelle Produkt: Management der CAD-Technik. Hanser-Verlag, München, Wien, 1997.

Discussion

Question - U. Aßmann
With the languages you mentioned, I want to ask whether you describe the process that is going on in the product or whether you describe the process of building the product, namely configuring it.

Answer - F.-L. Krause
The process I mean is the product development process. This is necessary since the person who has to describe the process should be the product developer. The question has been discussed in the field of knowledge processing for a long time: Is it necessary to employ a knowledge engineer to acquire knowledge? I think the answer is no, because otherwise there will always be a lack of information in the sense of the „Stille Post". That means somebody thinks about something and talks about it to somebody else. The second person does not understand this correctly and acts in the wrong way. Why didn't the first person acts itself right away. This is possible if appropriate tools are provided, which can be realised and applied easily.

Question - C. Weber
It is more a remark. I think, there is quite some interest out of industry especially car-industry, to get such flexible tool-systems in a distributed way for using. I think there are some parallel projects into this direction, not as big as the one you showed, but I think the direction is looked for in industry as well.

Answer - F.-L. Krause
Well, I regard this to be a little different. Industry and automotive industry in particular avoid to develop software themselves since they don't want to bind personnel resources for software development and maintenance. I imagine that this would be more attractive if it were more simple to get access to application software. Additionally, companies should understand this as a possibility to record their specific knowledge since the application software represents know-how generated within the company and reflecting the development process itself.

Question - A. Albers
You made a very interesting starting point, we have discussed it once. Do you think, there is a chance to raise the alarming low 3-D CAD share in our industry, especially in small-medium enterprises. There must be a rethink of the situation. I know from conversations that many companies do not dare to make the step into the 3-D world. They stay in the 2-D world, which has certain demands, but is not so advanced in the leading group.

Answer - F.-L. Krause
In the last year we performed the study "New Ways of Product Development" in the "Berliner Kreis". One of the results was, that about 80% of the German engineering industry apply CAD

and 80% of these apply 2D-CAD. And what is even worse: 70% of all enterprises are satisfied with the state of there information technology. This phenomenon will not lead towards the next century or millennium. The question is: Is it easy to handle the tools described to realise 3D-support. I think, little enterprises avoid to invest money for something they know not to be economical, at least in the design area. It is the crux that following phases such as process planning, manufacturing and other tasks, may profit from the application of 3D-CAD but because of the cost-centre oriented thinking it is difficult to prove this in the design area as well. That's my idea to find an appropriate way for smaller companies.

Question - C. Weber

My remark is not meant in a destructive way. But I think to get industry interested one has to present more than purely technologically oriented solutions. Even legal problems have to be considered and solved: Who is responsible for the correctness and the correctness of use of software tools provided via networks and so on? Probably these new approaches and techniques must be examined in a very broad sense.

Answer - F.–L. Krause

This is a very important question. I think this is one of the aspects of the questions how to process knowledge. Against the background that there is a release of drawings to avoid problems of product liability there is the question how to release knowledge which will become part of a database. Therefore, you may ask the question how to release software one wants to use within the system. In general I think this is part of research. We are just doing a little project which might be a starting point for more. I think we have to deal with a lot of subjects, mainly in the area of standardisation of exchange, of adaptability, and of flexibility. We can't foresee if there is a way to come by solutions fast. At first it is important to have some solutions as a basis for further discussions.

Comment - J. Gero

It is a comment. A little longer legal aspects. This has been tested in a number of court cases, where engineers have used software either they have used it incorrectly, which is the most common case by the way, or the software had bugs in it which they were not fortuned enough to discover. In both cases the engineer is liable because the contract both formal and implied is between the engineer and the client. There is no statement about what sorts of tools the engineer uses. In both cases. Once very, very large, one of a extremely large factory happen to be build for IBM in Brazil. The engineer was a Canadian engineer. The factory collapsed during construction and IBM sued the engineer. It was very clear and unequivocal, the engineer was completely responsible, no questions. And I think that applies, probably universally, it is up to the engineer to make sure the tools are the correct tools and that they are used appropriately. That is one of the reasons the engineer gets paid.

Design Methods before the Change of Paradigms? Design Research in Germany, a Short Synopsis

Hans-Joachim Franke,

Institute for Engineering Design, Machine- and High Precision Elements

Technical University of Braunschweig, Germany

Keywords: Design Theory, History of design science in Germany, Ladenburg Discussions, Formal Methods for design

ABSTRACT

The introduction will present a short background of the history of design science in the German-speaking countries.
The starting-points, topics and partial results of the Ladenburg Discussions on „Psychological and Educational Questions in connection with Systematic Design" are discussed subsequently, with special attention to cognitive psychology.
A report is given on the resultant most important topics regarding research in design science.
The study group „Bild und Begriff" (image and word) that originated from the Ladenburg Discussion is presented with their aims and some discussion topics.
An outline is given of the most important reserach projects on computer-aided design that are being carried out parallel to this in Germany .
Some theoretical attempts, partly already from the Seventies, are discussed.
Finally, the author presents his conclusions.

INTRODUCTION

In the German-speaking countries, first attempts of turning design into a science can be found as far back as in the middle of last century (Redtenbacher [*Redt-1863*], Reulaux [*Reul-1865*]). A strong revival of these endeavours - after some important publications by Wögerbauer, cf. Figure 1, in 1942 [*Fran-58*] and Rudolf Franke [*Wöge-42*], Bock, Bischoff and Hansen [*Bock-55, Bisc-53, Hans-66, Hans-74*], J. Müller [*Müll-67*] as well as F. Kesselring [*Kess-54*] and Martyrer [*Mart-60*] - did not take place before the late Sixties of this century.
This new beginning can be linked with the names of Rodenacker [*Rode-66, Rode-68*], Pahl [*Pahl-72a, Pahl-72b, Pahl-86*], Beitz [*Beit-71, Roth-69*], Koller [*Koll-71, Koll-85*] and Roth [*Roth-69, Roth-82*]. To sum up, it might be said that the emphasis of these publications aims at **providing heuristic methods for practical engineers and for the education of engineering students.**
A comparison of proposals is shown by J. Müller in [*Müll-90*], cf. Figure 2. A very good survey on design science publications up to 1990 is to be found there, too.
In a way, VDI Standard 2221 [*VDI-2222*] provides a certain conclusion of the above attempts.

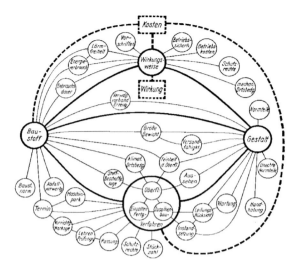

Figure 1   Wögerbauer's representation of design as a general „constraint system".

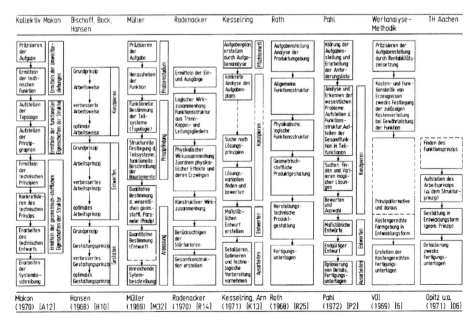

Figure 2   Survey of proposals by various authors in German-speaking countries regarding design methodology according to J. Müller [Müll-90].

THE LADENBURG DISCUSSIONS

In the beginning of the Nineties it became clear that the proposed methods were very useful for the education of engineering students, the practical application on the one hand and the systematic extended development on the other hand did, however, not turn out completely satisfactory. On J. Müller's of Chemnitz initiative, even before the political change in the former German Democratic Republic, a group of design researchers from East and West Germany had several meetings, at first in East Germany and after 1990 also in West Germany. These meetings led to the desire to discuss the further development of design science more profoundly.

Led by G. Pahl, an inter-disciplinary discussion circle was therefore established, which - with grateful acknowledgement - was supported by the the Daimler-Benz Foundation in Ladenburg and which took place on the premises of said Foundation.

As in past years a number of publications had already originated from the group of Dörner, Ehrlenspiel and Pahl on the cognitive-psychological background of design it was reasonable to invite psychologists and pedagouges to take part in the discussions. From Mai 1992 to October 1993 six 2-day discussions took place under the heading:

„**Psychological and pedagogical questions in connection with systematic design**".

Besides the permanent members, guests were invited to discuss special subjects, cf. the following list of participants.

List of participants (permanent members* and guests):

Dr. Petra Badke-Schaub, Universität Bamberg, Lehrstuhl für Psychologie 11
Prof. Dr.-Ing. Wolfgang Beitz*, TU Berlin, Institut für Maschinenkonstruktion
Prof. Dr.-Ing. Herbert Birkhofer*, TH Darmstadt, Maschinenelemente und Konstruktionslehre
Prof. Dr. Meinolf Dierkes, Wissenschaftszentrum Berlin für Sozialforschung
Prof. Dr. Dietrich Dörner*, Universität Bamberg, Lehrstuhl Psychologie II
Prof. Dr.-Ing. Dr.-Ing. E.h. Heinz Duddeck, TU Braunschweig, Institut für Statik
Prof. Dr.-Ing. Klaus Ehrlenspiel*, TU München, Lehrstuhl für Konstruktion im Maschinenbau
Prof. Dr.-Ing. Hans-Joachim Franke*, TU Braunschweig, Institut für Konstruktionslehre, Maschinen- und Feinwerkelemente
Prof. Dr. Winfried Hacker*, TU Dresden, Institut für Psychologie
Prof. Dr.-Ing. Werner Heinrich*, Büro-Organisation Roland Zeller GmbH, Dresden
Prof. Dr.-Ing. Joachim Hennig*, Institut für Konstruktionstechnik und Anlagengestaltung, Dresden
Prof. Dr-Ing. Günter Höhne* TU limenau, Institut für Maschinenelemente und Konstruktion
Prof. Dr.-Ing. Gerhard Hönisch, TU Dresden, Institut für Maschinenelemente und Maschinenkonstruktion
Dr.-Ing. Michael Jeske, Körber AG, Konstruktion-Verfahrenstechnik, Hamburg
Prof. Dr. Wolfgang König, TU Berlin, Institut für philosophie, Wissenschaftstheorie, Wissenschafts- und Technikgeschichte
Prof. Dr. Werner Krause*, Universität Jena, Institut für Psychologie
Prof. Dr. Johannes Müller* , Chemnitz
Prof. Dr.-Ing. Dr.h.c.mult. Gerhard Pahl*, TH Darmstadt, Maschinenelernente und Konstruktionslehre
Dr. Heinz-Jürgen Rothe, Humboldt-Universität Berlin, Fachbereich Psychologie
Prof. Dr-Ing.habil. Jürgen Rugenstein*, „Otto von Guericke"-Universität Magdeburg, Institut für Maschinenkonstruktion
Dr. -Ing. Andreas Rutz, Lindauer Dornier GmbH
Prof. Dr. Helga Thomas*, TU Berlin, Institut für Erziehung, Unterricht und Ausbildung
Dr. Rüdiger von der Weth, Universität Bamberg, Lehrstuhl für Psychologie

The following subjects were dealt with:
- Design methodology
- Knowledge and structurizing knowledge
- Application barriers and acceptance problems
- Guidance and team behaviour in the design process
- Development of competence
- Future goals for research.

The discussion results were documented in a book [*Roth-82*]. As it is impossible in this place to deal with all discussion topics, I will merely present some individual topics and a certain final result.

An important starting-point of the discussions used to be the observations by the group of Dörner, Ehrlenspiel and Pahl. Students, PD students and some experienced design engineers were observed, preferably by a video camera ; they had the task to develop an optical apparatus that was to be moved up and down as well as in two swivel angles and to be locked, cf. Figure 3 which is the solution by one of the test persons.

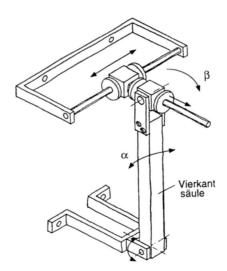

Figure 3   *Embodiment design by a test person for a swivel frame of an optical apparatus, to be mounted on a wall.*

By observing the actual activities and comparison with the general approach steps of systematic design, cf. Figure 2, it was possible to establish several hypotheses regarding the differenciating of good and less good problem solvers.

Pahl summarized this as follows: „Successful problem solvers need
- knowledge of facts
- knowledge of methodology and
- heuristical competence.

The knowledge of facts and methodology alone are obviously not sufficient. The problem-solving behaviour is - besides creativity and motivation - influenced by heuristical competence, which as an individual property can be charcterized as follows:
> Turnover of knowledge into suitable actions with new and complex situations. Ability in dealing with the system, which includes the ability for analyzing and abstracting, assessment of cross-linkage, ability of reflection and control. Decision behaviour, such as evaluation of dependencies, assessment of importance and urgency, determination and consistency, flexibility in general approach.

This results in well observable qualities of good problem solvers [*Rutz-85, Dyll-90, Pahl-92*]:
- active acting, even in cases of unclear conditions
- good pre-knowledge in systematic ideas (mental model)
- right measure between concreteness and abstraction
- adhering to aims with flexibility in general approach
- permanent control of results. "

On the part of the psychologists, the most important aspects treated were **the role of the memory** (Dörner) and **the knowledge of acting** (Hacker) [*Dörn-87, Hack-92*].

Particularly interesting in designing is the interrelationship between linguistic as well as image and motoric memory. All forms of memory are able to call upon each other. Language, however, enables us in particular to disengage ourselves from experience. Here, an important role is played by relations such as „part - whole", „abstract - concrete", as well as relations between space and time. The memory calls upon knowledge basically according to the association principle. Basic forms are the call upon similarities and the „complex supplement". A relatively demanding form of the call upon similarities are analogies in topologies and structures.

Besides the activating of knowledge components, however, the so-called protocol memory is of decisive importance, too. It serves to control and steer the procedure. In case of sufficiently complex problems it has to be supplemented by outward storing means, e.g. sketches and notes.

The knowledge or ability of acting according to Hacker comprises the following:
- knowledge of facts, e.g. material properties
- procedures, e.g. patterns of acting and abilities
- ability for organizing and planning (meta-procedures).

The activities taking place during designing are multi-stage relations, e.g.:
- subject relation,   designer - ability - adjustment
- object relation,   designer - object of design
- final relation,   measures - desired results
- means relation,   object of design - ability - tools (e.g. computer)
- local and time relations.

The activities are mostly worked off „opportunically", starting from the well-known. Only with very complex problems and with very good problem solvers the planning proportions will increase as compared with the routine proportions.

RESEARCH AIMS AS A RESULT OF THE LADENBURG DISCUSSIONS

Here, only few important research aims can be presented. Details may be seen from [*Pahl-94*].

EVALUATION AND EFFICIENCY CONTROL OF DESIGN METHODS

- Criteria, standards of value

- Measuring techniques
- Comparative analysis of alternative procedures

DESIGN METHODOLOGY
- Product quality gained by methodical approach
- Comparison between generating and variating-correcting approach
- Possibility of planning and controlling design processes
- Design laboratories
- Typical and critical problem situations
- Efficient structures of organizations

PSYCHOLOGICAL AND EDUCATIONAL QUESTIONS OF DESIGNING
- Development of heuristic competence and their conditions
- Combination of linguistic and image thinking
- Interdependency of analysis and synthesis
- Knowledge structures with beginners and experts
- Re-structurizing knowledge suitably in order to stimulate problem solving
- Aids in case of unclear conditions and criteria.

CONTINUATION OF THE LADENBURG DISCUSSIONS IN THE „BILD UND BEGRIFF" (IMAGE AND WORD) WORKING CIRCLE

Due to Klaus Ehrlenspiel's [Ehrl-83] initiative, part of the Ladenburg Discussion members - no longer supported, but connected with each other by common interest - meanwhile engaged themselves in 6 workshops from autumn, 1993, to February, 1998, dealing with the utilization of mental representations of language and image in designing.
In particular they discussed particularly the conditions and neccessities for the utilization of the various representations and for their change .

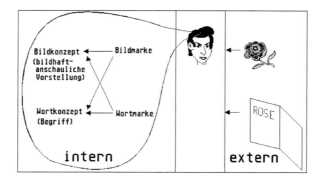

Figure 4   *Multi-mode memory model according to Engelkamp*

It may be considered a safe hypothesis that successful designers change flexibly from image-thinking into language-thinking and vice versa. It is considered a long established cognition that learning works especially successful if it is simultaneously done through several channels or modes, e.g. simultaneously watching, listening and writing (motoric channel).
Questions around the „utilization of pictures and sketches" are also pursued by Andreasen and Tjalve, who often took part in discussions in Germany, cf. [Andr-97, Tjal-75].

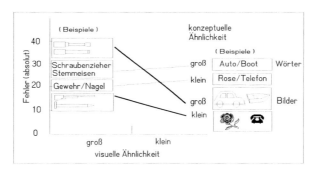

Figure 5   Different memorizing achievements (measured as number of errors) when learning image- or word-like objects (according to Nelson, Reed and Willing).

RESEACH CONCERNING COMPUTER-AIDED DESIGN

From the author's point of view, the most important topics of the last ten years were:
- Representation of knowledge and product models
    - Further development of product models and interface standardisation [Grab-86, Spur-86, Krau-88, Grab-89, Baue-91, Spur-97, Will-93]
    - Knowledge-based methods for designing [Grab-90, Adam-91, Krau-92a, Kläg-93, Spec-95, Kick-95, Fisc-96, Pete-97]
    - Computer-aided design catalogues [Rein-94, Spec-95, Schu-96]
    - Use of internet for designing [Abel-95, Birk-97, Fran-97, Meer-97a]
    - Digital mock-up, virtual reality [Spur-97, Döll-97]
- Phase-specific individual methods
    - Computer-aided listing of requirements [Kläg-93, Fran-94, Kick-95]
    - Features and gadgets for problem-adapted modelling [Krau-92b, Schu-96]
    - Parametrical design [Fisc-96]
    - Computer-aided integration of embodiment and calculation [Schu-96, Döll-97]
- Integration methods and tools
    - Integrated designers' workstations [Beit-90, Meer-91, Weig-91, Beit-92, Fisc-96, Pete-97]
        - CAD reference model, cf. Figure 6. [Abel-89, Birk-92, Diet-94]
        - Computer-aided tools for the preventive securing of quality [Webe-94, Spur-97, Pete-97]
        - Fast prototyping [Spur-97, Döll-97]

Due to the great number of papers only few have been cited. A great number of papers that are interesting with regard to design science originated from the research groups around
M.Abramovici, R.Anderl, W.Beitz, H.Birkhofer / G.Pahl, K.Ehrlenspiel / K.Lindemann, D.G.Feldmann, H.-J.Franke, J.Gausemeier, H.Grabowski, Grote, F.L.Krause, J.Klose, H.Meerkamm, H.Seifert / E.G.Welp, S.Vajna, C.Weber.
A very good and relatively present-day survey is contained in Spur/Krause „Das virtuelle Produkt" (The virtual product) [*Spur-97*].

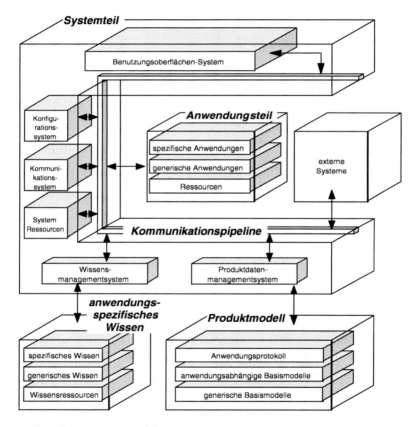

Figure 6   CAD reference model

FORMAL METHODS FOR DESIGNING

WHY FORMAL DESIGNING METHODS?

In „Objective knowledge, a revolutionary plan" [*Popp-95*] (*This title and the following quotation were re-translated*) Karl Popper compares - superficially considered - two in their essence nearly related columns, remarking that for him only the right-hand column is interesting. The problems of the meaning, and particularly of the terminology are of little interest to him. Terms are merely means for the wording or summarizing of theories, they are exchangeable.

	IDEAS i.e.
MEANINGS or TERMS or EXPRESSIONS	STATEMENTS or ASSERTIONS or THEORIES

can be expressed in

WORD	STATEMENTS.

These can be

SENSIBLE	TRUE

and their

SENSE	TRUTH

can be reduced by

DEFINITIONS	DEDUCTIONS

to that of

UNDEFINED EXPRESSIONS	BASIC STATEMENTS

Trying in this manner (not to reduce but) to establish

SENSE	TRUTH

leads to an infinite regress...

If we believe Popper (I am using the word „to believe" deliberately) we can follow that for designing consistent systems of statements on the basis of classical terms of truths and of theories must also be aims for design science in future.

What makes me believe Popper?
His concept alone of standing the test - as solution proposal for the induction problem that has been discussed since Platon - seems to be the only route that leads to stable relations between the three worlds
   {real world - mental world - formal theoretical world}.

It will, however, be necessary to solve a basic problem:
Designing must lead to a complete way back into the real world! This means, however, that besides formal theoretical organs of such a science, a
> **flexible, enlargeable image of real facts, real knowledge**

must become a very decisive component of a design theory.
Besides the knowledge of nature and technology, this includes also the
> **knowledge of economical, political (in the sense of norming on the one hand and value-relatedness on the other hand), social, organizational and mental interrelations.**

FORMAL CONTRIBUTIONS TO ENGINEERING DESIGN IN GERMANY

As world-wide a number of formal design theories are being discussed at present, as examples here only Yoshikawa [*Yosh-83*] and Suh [*Suh-95*] are mentioned, it might be of interest, in view of the extensive literature on design methodology in German, to consider a small number of less well-known formal German contributions.

Limitation of the object considered to synthesis processes

Since the beginning of the scientific penetration of engineering tasks, formal methods have played a great role - e.g. the utilization of mechanics and thermodynamics .The established engineering sciences generally start on the basis that there is a structurized system that can be analyzed, corrected and optimized.

This will, however, not be enlarged upon. Here only contributions will be discussed that are **primaryily constructive** in the sense that - starting from the desired functions and features - for the first time they **syntheticize systems** or preesent at least **tools for the qualitative discovery of topology, structure and/or embodiment.**

This idea is already based on **three requirements - we might also call them axioms :**

- The desired functions and features of technical systems can be sufficiently described formally.
- There are universal elements that can be defined as starting-points for syntheticizing processes.
- There are operators that attribute these universal elements to desired functions and features and/or derivative rules for their suitable (i.e. desired and functions-and features-guaranteeing) and consistent (non-contradictory) combination.

Examples

Rudolf Franke [*Fran-58*]

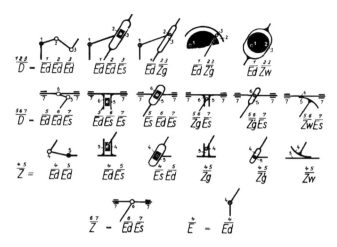

Figure 7   Definition of kinematic elementary objects

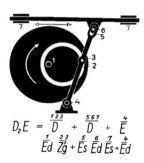

Figure 8   Combination of more complex elements from elementary objects

Waldvogel [*Wald-69*]

Waldvogel uses the example of connection means to show the possibility of comparing characteristics quantities by means of Boole's terms, and desired features with the characteristics of real solution alternatives.

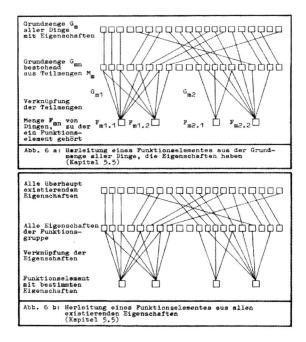

Figure 9   Quantity- theoretical formulation by Waldvogel

He considers certain classes of characteristics which he designates by letters.
A: Geometry of connecting surface
B: Seecuring the connection
C: Flexibility of the connection
D: Forces and moments
E: Sealability
F: Suitability of material
G: Assembling-ability
H: Production-ability
The individual characteristics are further sub-divided by hierarchical indices, e.g.:
A1: possible with plane connecting surface
A6: possible with spherical connecting surface.
The characteristics are then coded with a „quantity term" in the shape of a conjunctive standard form. The following example shows the term of the functional element „glued

connection" of the functional group „rigid force transmission", which for the selecting process has already been simplified by the negative members.

Glued connection: =
(A1+A2+A3+A4+A5)*A7*B12*
(D1+D12+D21+D22+D3+D32+D41+D42)
*(E1+E2+ E3+ E4)
*(F11+ F12+ F13+ F14+ F15+ F16+ F17)
*F23*G11*G12*G38
G53(H11+H12+H14+H15+H18+H111+H112)*(H31+H32+H33)

Waldvogel concluded his contribution with an astonishingly far-seeing statement:
„Nowadays it is still often stressed that talent, resourcefulness and intuition are required for designing; partly it is even talked of the „art of designing". Nevertheless, it has to be pointed out - which is to be proved by the present contribution - that designing is done on the basis of logical combinations of ideas. The matters that can be approached by a formal-mathematical treatment, which contribute to the success of designing, should be searched for, made tangible, thus to rationalize the design aactivity.

The designer is to be freed from ever-recursing similar tasks so that his thinking capacity is not claimed by those tasks. In view of the problems to be solved in future, endeavours must be made to releive the designer in his task and support him with the use of technical aids. Kybernetic machines should support the designer not only with the solution of mathematical but also design problems."

Koller [*Koll-71, Koll-85*]

Koller proposed a „concept algebra". This was to be based on 12 elementary functions and their respective inverse functions, such as „conducting", and „insulating" or „connecting" and „separating". This idea was, however, not operatively continued as far as the author knows.

Roth [*Roth-82*]

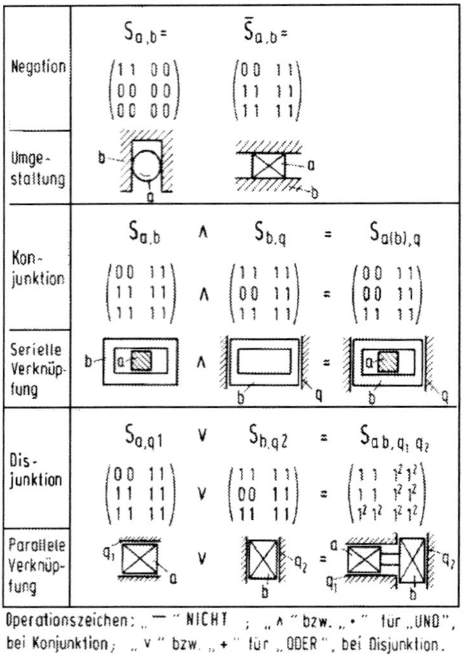

Negation	$S_{a,b} = \begin{pmatrix} 1 & 1 & 0 & 0 \\ 0 & 0 & 0 & 0 \\ 0 & 0 & 0 & 0 \end{pmatrix}$	$\bar{S}_{a,b} = \begin{pmatrix} 0 & 0 & 1 & 1 \\ 1 & 1 & 1 & 1 \\ 1 & 1 & 1 & 1 \end{pmatrix}$	
Umge-stattung			
Kon-junktion	$S_{a,b} \begin{pmatrix} 0 & 0 & 1 & 1 \\ 1 & 1 & 1 & 1 \\ 1 & 1 & 1 & 1 \end{pmatrix}$	$\wedge \quad S_{b,q} \begin{pmatrix} 1 & 1 & 1 & 1 \\ 0 & 0 & 1 & 1 \\ 1 & 1 & 1 & 1 \end{pmatrix}$	$= \quad S_{a(b),q} \begin{pmatrix} 0 & 0 & 1 & 1 \\ 0 & 0 & 1 & 1 \\ 1 & 1 & 1 & 1 \end{pmatrix}$
Serielle Verknüp-fung			
Dis-junktion	$S_{a,q1} \begin{pmatrix} 0 & 0 & 1 & 1 \\ 1 & 1 & 1 & 1 \\ 1 & 1 & 1 & 1 \end{pmatrix}$	$\vee \quad S_{b,q2} \begin{pmatrix} 1 & 1 & 1 & 1 \\ 0 & 0 & 1 & 1 \\ 1 & 1 & 1 & 1 \end{pmatrix}$	$= \quad S_{ab,q_1 q_2} \begin{pmatrix} 1 & 1 & 1^2 & 1^2 \\ 1 & 1 & 1^2 & 1^2 \\ 1^2 & 1^2 & 1^2 & 1^2 \end{pmatrix}$
Parallele Verknüp-fung			

Operationszeichen: „ − " NICHT ; „ ∧ " bzw. „ · " für „UND",
bei Konjunktion; „ ∨ " bzw. „ + " für „ODER", bei Disjunktion.

Figure 10   Formal generating of connecting characteristics by means of so-called „Schlußmatrizen" (connection matrix) according to Roth.

For the special, however constructive part problem of connecting means, Roth developed the method of the connection matrix that permits the synthetic generating or formal variating of connecting properties by means of formal steps.

Kopp [*Göde-31*]

In a very interesting contribution Kopp shows graph-theoretical methods for the structure synthesis of kinematic mechanisms.

Simonek [*Herm-71*], **Krummhauer** [*Abel-95*]

The main points of both publications were contributions how to generate, in a structural qualitative way, novel solutions from the formulations of physical effects.

Simonek developed the so-called „special functional structure", cf. [*Baue-91*].

H.-J. Franke [*Fran-76*]

Franke was concerned with the limitations of algorithmability of design processes. There he met with the problem of the countability, completeness and decidability of logical systems.
Such limits of formal systems had first been clearly worked out by Gödel [*Göde-31*]. A summary regarding problems of this kind was done by Hermes [*Herm-71*].
Further, Franke was concerned with the problem of contradictory tasks and practical problems of the combinatory explosion.
In 1976, Franke showed in [*Fran-76*] how „production systems" can be used for generating solutions:
„As already shown in the chapter on the Markow algorithm, new words can be deducted from given words by means of substitutions. As a continuation the term „Production System", ...is to be discussed, which term was coined in the theory of the formal languages.

Definition: Production system

A is a finite alphabet. A* is the quantity of all words over A. R is a finite quantity of production rules (substitutions). Every production rule is a pair of words U γ A* and V γ A*, which leads the word U into word V. A production system P is then pair (A,R).
Production systems are used in order to define formal languages. A special application now is the combination of words by means of a system of rules. Cf. the following example (table 15).

A: { E,W,S,L,(,),| }   X1 g A*
R.:   E -> EWE   X2 g A*
      E -> ESE
      E -> ELE
      E -> E(E|E)

   (X1EIX2E) -> (X1E|X2E)E

Table 15   Example of a production system

If a starting word is given, it is possible to derivate new words by means of the production system. An example is the following derivation chain:
E -> E(E|E) -> E(EWE|E) -> ELE(EWE|E) ->   **Derivation Ao**
   ->ELE(EWESE|E) -> ELE(EWESE|ELE) ->
   ->ELE(EWESE|ELE)E

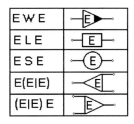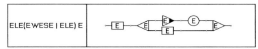

Figure 11   Assignment of meanings to words of the production system according to the above table.

Strek-ken	Ma-schen	Nr.	Topologie	Abstrakte Gleichung z_{ges}	Beispiel für Anwendung zur Berechnung der Zuverlässigkeit
1	0	1	o—o	z_1	p_1
2	0	2	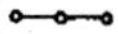 o—o—o	$z_1 o z_2$	$p_1 \cdot p_2$
	1	3	⌀	$\overline{z \rho z_2}$	$1-(1-p_1)(1-p_2)$
4		11	 ⌀—o	$z_1 o z \overline{\rho z \rho} z_4$	$1-(1-p_1)(1-p_2 \cdot p_3 \cdot p_4)$
		12	 ⌀—•	$\overline{z_1 o z \rho z_3} o z_4$	$(1-(1-p_1)(1-p_2)(1-p_3)) \cdot p_4$
	2	13	 ∞	$\overline{z \rho z_2} o \overline{z \rho z_4}$	$(1-(1-p_1)(1-p_2))(1-(1-p_3)(1-p_4))$

Figure 12 The same source further contains generalized formulations for determining total characteristics from a topological structure.

This is of course only one of the many possible sequences. If certain symbol sequences of this formal system according to Figure 16 are interpreted, the mentioned system is a production system for generating a sub-quantity of the general function structure."

Barrenscheen [*Barr-90*]
Barrenscheen investigated the utilization of symmetries in engineering design and for their classification - as was long practice in other sciences - he used group-theoretical methods for classifying the machine parts.

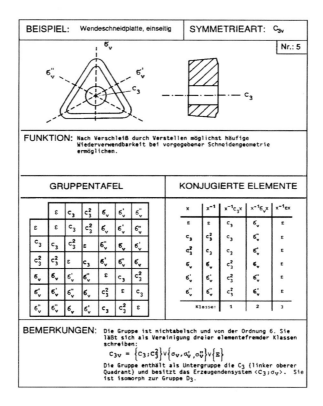

Figure 13   *Group-theoretical tools can be used in order to classify complex geometries, to assign „natural" coordinates to them and to vary them. Thus it is possible to generalize statements on characteristics of bodies.*

CONCLUSIONS

The contributions outlined above of present-day design-scientific studies can be grouped into three main directions:
- Improved methods for designing, by reference to cognitive- psychological findings. The main emphasis is mental processes which - besides formal conclusions - go also back to associative and intuitive thinking processes.
- Improved uitilization of computers   a) better tools
                                          b) automatic designing
- Building up a consistent theory of design utilizing formal methods.

As a designer with practical experience (12 years in industry), with experience in computer application and with a scientific background, the author throws open to discussion the following statements:
- It will be possible in future to solve certain design tasks automatically. This applies the more the stronger tasks can be formulated as pure configuration tasks or as optimization tasks.
- The more novel and the more complex a task, the more probable it is that in the not-too-distant future no exclusively deductive methods will overall be successful for a task. Therefore formal theories are for the time being no replacement for other methods of designing in practice, but a useful supplement.
- On the one hand, design will be better teachable and on the other hand more computer-aidable if it is possible to build up a consistent theoretical building.
- The present contributions to formalized design theories still need their proof of practical usability (stand the test in Popper's sense).
- If design theories are to be utilized practically, the problem of the methodical task formulation must be energetically attacked, as every original task may contain contradictions. A path from the logically inconsistent propositions of the task to logically consistent ones of a theory, must be established. A possible solution could be to eliminate contradictions step by step.
- An essential partial problem will be the early recognition of inconsistencies of elementary solution elements in order to avoid the combinatory explosion when generatively producing problem-solving contributions.
- The theoretical background of designing requires a multittude of formal tools.
- Due to the necessary return from theoretically formal conclusions to real objects, the knowledge on facts and behaviour - even if not yet consistent knowledge - must be a necessary object of design theories.
- There are no clues that a design theory must be sphere-specific in its basis. There are basic assumptions that apply to the generating of all artefacts.

A design theory is essential if in the long run design science is to be more than a collection of practical behaviour rules. Only in this way it can grow out of the multitude of theories and become provable in parts and consistent. Thus, designing will be both better teachable and also further aidable by computer.

The author agrees in all points with the conclusions by **R. Dixon** which he made at a workshop regarding DESIGN THEORY in 1988 [*Dixo-89*]:

A short extract:

„... it is my personal conviction at this time that the most important variables in the theory of design we are seeking have to do with knowledge and how knowledge is found, communicated, and used in design. Learning how to define and describe knowledge as a variable in a theory is a difficult task, but can be addressed either in the context of cognitive or computational studies.

It is *important* that we as a research community move as directly and as effectively as possible towards realization of the theory that is our goal. Our National economic well being is directly affected by how well our products and processes are designed. If we are to graduate better designers from engineering schools, then we must have faculty who are themselves specialists in design. If we are to have such faculty in appropriate numbers, then the academic respectabilty of design must be made equal with other fields.

For this to happen, design must be based on a foundation of theory, principle, and organized generalized knowledge. Then there can be more graduate theses in design, significant numbers of new faculty in design, and new intellectual content for both design education and design practice. Research in design theory is the key. ..."

ACKNOWLEDGEMENT

I wish to thank the Volkswagen Foundation who showed their interest in the subject discussed and who provided means to facilitate the present workshop, and I also say thanks to my colleague Grabowski who addressed colleagues interested in the subject of a design theory and ensured an inter-disciplinary, extended discussion.

REFERENCES

Abel-89   Abeln,O.: Referenzmodell für CAD-Systeme, Informatik-Spektrum 12/1,Springer-Verlag, 1989.

Abel-95   Abeln, 0. (Hrsg.): CAD-Referenzmodell - Zur arbeitsgerechten Gestaltung zukünftigt computergestützter Konstruktionsarbeit. Teubner Verlag, Stuttgart 1995.

Adam-91   Adami, W.: Strukturen wissensbasierter Systeme für die rechnergestützte Konstruktion. Diss. TU Braunschweig, 1991.

Andr-97   Andreasen M.M.:From observation to theory - from analysis to synthesis: obtaining a language of machine elements. In. Workshop :Zukunft der Maschinenelementelehre, Darmstadt 24.-25. April 1997, Hrsg. Hrsg. Institut für Konstruktionstechnik, Prof. Birkhofer.

Barr-90   Barrenscheen, J.: Die systematische Ausnutzung von Symmetrieeigenschaften beim Konstruieren, Diss. TU Braunschweig , Bericht Nr. 37 aus dem Institut für Konstruktionslehre der TU Braunschweig 1990.

Baue-91   Bauert,F.: Methodische Produktmodellierung für den rechnerunterstützten Entwurf, Diss. TU Berlin, 1991.

Beit-71   Beitz, W.: Konstruktionslogik in der Konzeptphase, Grundlage für einen verstärkten EDV-Einsatz.wt-Z. Ind.Fertig. 61 (1971), S.550-560.

Beit-90   Beitz W. : Konstruktionsleitsystem als Integrationshilfe, VDI.Berichte 812, 1990,S.181-201.

Beit-92   Beitz W.,D.Kuttig: Rechnerunterstützung beim Konzipieren. VDI-Berichte 953,1992, S.1-24.

Birk-92   Birkhofer, H.: Datenbanksysteme für das Zulieferwesen- Anforderungen und Bedeutung . Der Zuliefermarkt, 4,92,S.78-82, 1992.

Birk-97   Birkhofer H..,Keutgen I., Büttner K.: CompoNET aus Kundensicht - Erfahrungen aus 2,5 Jahren Online-Bereitstellung, VDI-Berichte 1362,1997, S. 185-200

Bisc-53   Bischoff, W. u. F. Hansen: Rationelles Konstruieren.Konstruktionsbücher Band 5, Berlin: Verlag Technik, 1953.

Bock-55   Bock, A.: Konstruktionssystematik - die Methode der ordnenden Gesichtspunkte. Z. Feingerätetechnik 4 (1955), H.1, S.4-5.

Diet-94   Dietrich, U.; Hayka, H.; Jansen, H.; Kehrer, B.: Systemarchitektur des CAD-Referenzmodells unter den Aspekten Kommunikation, Produktdatenmanagement und Integration. In: Gausemeier, J. (Hrsg.): CAD'94 - Produktmodellierung und Prozeßmodellierung als Grundlage neuer CAD-Systeme, Fachtagung der Gesellschaft für Informatik e.V., Paderborn 17.-18.03.1994, Carl Hanser Verlag, München Wien 1994, S. 353-374

Dixo-89   Dixon J.R.: On Research Methodology towards a Scientific Theory of Engineerin Design, in Design Theory '88, New York, 1989.

Döll-97   Döllner, G.: Konzipierung und Anwendung von Maßnahmen zur Verkürzung der Produktentwicklungszeit am Beispiel der Aggregatentwicklung. Dissertation TU Braunschweig 1997

Dörn-87	Dörner, D.: Problemlösen als Informationsverarbeitung, Stuttgart,Kohlhamer, 1987.
Dyll-90	Dylla,N.:Denk-und Handlungsabläufe beim Konstruieren, Diss. TU München, 1990
Ehrl-83	Ehrlenspiel, K.: Ein Denkmodell des Konstruktionsprozesses. Proceedings of ICED 83. Schriftenreihe WDK 10.Edition Heurista 1983.
Fisc-96	Fischer, R.: Product Design based on HyperTrees. Dissertation TU Braunschweig 1996
Fran-58	Franke, R.: Vom Aufbau der Getriebe. Bd. 1 und 2. Düsseldorf: VDI-Verlag, 1958.
Fran-76	Franke H.-J.: Untersuchungen zur Algorithmisierbarkeit des Konstruktionsprozesses. Diss. TU Braunschweig 1976, erschienen als Fortschrittsbericht VDI-Z, Reihe 1,Nr.47
Fran-94	Franke H.-J. ,Kickermann H.: Fuzzy-Logik bei der Verarbeitung von Anforderungen. In: Gausemeier, J. (Hrsg.): CAD'94 , Fachtagung der Gesellschaft für Informatik e.V., Paderborn 17.-18.03.1994, Carl Hanser Verlag, München Wien 1994, S. 353-374.
Fran-97	Franke H.-J, Krusche T.,Hagemann D. :Plattformunabhängige Software im Internet. VDI-Berichte 1362, 1997, S. 185-200
Göde-31	Gödel,K.: Über formal unentscheidbare Sätze der Principa Mathematica und verwandter Systeme ,I. Mh. Math.Phys. 38, S.173-198 (1931).
Grab-86	Grabowski H. u.a.: Entwurfsmethoden auf der Basis technischer Produktmodelle, VDI-Berichte Nr.610.1 (1986) S.55-78,Düsseldorf, VDI-Verlag.
Grab-89	Grabowski,H.,Anderl R.,Schmidt,M.:Das Produktmodellkonzept von Step, VDI-Z 131 (1989), Nr,12, Düsseldorf: VDI-Verlag, 1989
Grab-90	Grabowski,,H,Rude,S.:Methodisches Entwerfen auf der Basis zukünftiger CAD-Systeme, VDI-Berichte. 812, 1990
Grab-95	Grabowski, H.; Anderl, R.; Polly, A.: Integriertes Produktmodell. Beuth Verlag, Berlin, Köln 1993 STEP. CIM Management 11 (1995) 4, S. 32-40
Hack-92	Hacker, W. : Expertenkönnen, Erkennen und Vermitteln. Arbeit und Technik Bd 2. Göttingen: Verlag für angewandte Psychologie 1992.
Hans-66	Hansen, F.: Konstruktionssystematik. Berlin: Verlag Technik, 1966.
Hans-74	Hansen, F.: Konstruktionswissenschaft.München: Hauser, 1974.
Herm-71	Hermes, H.: Aufzählbarkeit, Entscheidbarkeit, Berechenbarkeit. Berlin: Springer, 1971.
Kess-54	Kesselring,F.: Technische Kompositionslehre.Berlin: Springer, 1954.
Kick-95	Kickermann, H.: Rechnerunterstützte Verarbeitung von Anforderungen im methodischen Konstruktionsproze ß, Diss. TU Braunschweig 1995.
Kläg-93	Kläger, R.:Modellierung von Produktanforderungen als Basis für Problemlösungsprozesse in intelligenten Konstruktionssystemen, Diss.TU Karlsruhe, 1993.
Koll-71	Koller, R.: Konstruktion von Maschinen, Geräten und Apparaten mit Unterstützung elektron. Datenverarbeitungsanlagen. VDI-Z- 113 (1971), Nr. 3, S. 482-490.
Koll-85	Koller R.: Konstruktionslehre für den Maschinenbau,Berlin,Springer, 1985.
Krau-88	Krause.F.L.: Informationstechnische Integrationsmodelle f.Konstruktion u.Arbeitsplanung, ZwF CIM, 83, Oktober 1988.
Krau-92a	Krause (Hrsg): Wissensbasierte Systeme für Konstruktion und Arbeitsplanung. Düsseldorf: VDI-Verlag, 1992
Krau-92b	Krause F.L. u.a.:Featurebasierte Produktentwicklung. ZWF CIM, H.5, S.247,251,Hanser, 1992.
Mart-60	Martyrer, E.: Der Ingenieur und das Konstruieren. Z.Konstruktion 12 (1960), S. 4, H1.
Meer-91	Meerkamm H.,A.Weber :Konstruktionssystem mfk - Integration von Bauteilsynthese und -analyse. VDI-Berichte 903, S.231-248, 1991.
Meer-97a	Meerkamm H., Mogge C., Sander S. : Das Internet als Medium zur kontextsensitiven Bereitstellung von Konstruktionswisse auf Basis der ISO 13584. VDI-Berichte 1362, S. 185-200, 1997.
Meer-97b	Meerkamm H. (Hrsg): Fertigungsgerechtes Konstruieren, Beiträge zum 8. Symposium 16. u. 17. Oktober 1997 in Schnaittach, Lehrstuhl f.Konstruktionstechnik, Univ.Erlangen-Nürnberg.
Müll-67	Müller, J.: Probleme einer Konstruktionswissenschaft.Z. Maschinenbautechnik 16 (1967), Heft 7, S. 338-341 und Heft 8 , S. 394-456.
Müll-90	Müller J.: Arbeitsmethoden der Technikwissenschaften, Berlin, Springer,1990
Pahl-72a	Pahl, G.: Analyse und Abstraktion des Problems - Aufstellen von Funktionsstrukturen. Z. Konstruktion 24 (1972), W.6, S.235-240.
Pahl-72b	Pahl, G.: Klären der Aufgabenstellung und Erarbeitung der Anforderungsliste.Z. Konstruktion 24 (1972), S.195-199.
Pahl-86	Pahl G., W. Beitz: Konstruktionslehre. Berlin,Heidelberg,New York, Springer, 1986
Pahl-92	Pahl,G.: Merkmale guter Problemlöser beim Konstruieren.VDI-Berichte Nr.953 (1992), 187-201.
Pahl-94	Pahl G. (Hrsg.): Psychologische und pädagogische Fragen beim methodischen Konstruieren. In der Reihe Ladenburger Diskurs, Köln: Verlag TÜV Rheinland, 1994

Pete-97	Peters, M.: Komununikationssystem rechnerunterstützter Konstruktionswerkzeuge, Dissertation TU Braunschweig 1997
Popp-73	Popper, K.: Logik der Forschung. Tübingen: J.C.B. Mohr, 1973.
Popp-95	Popper ,K.: Objektive Erkenntnis, ein evolutionärer Entwurf, Hamburg, Hoffmann und Campe,1995.
Redt-1863	Redtenbacher, R.: Der Maschinenbau. Mannheim .Verlagsbuchhandlung von Friedrich Bassermann, 1863.
Rein-94	Reinemut J.: Elektronische Katalogsysteme für die Informationsbereitstellung von Zulieferkomponenten. Diss.TU Darmstadt. 1994
Reul-1865	Reuleaux, F.: Der Constructeur.Braunschweig: Vieweg, 1865.
Rode-66	Rodenacker, W.: Physikalisch orientierte Konstruktionsweise. Z. Konstruktion 18 (1966), H.7, S.263-269.
Rode-68	Rodenacker, W.: Wege zur Konstruktionsmethodik. Z. Konstruktion 20 (1968), W.10, S.381-385.
Roth-69	Roth, K..:Gliederung und Rahmen einer neuen Maschinen-, Gerätekonstruktionslehre Z. Feinwerktechnik 72 (1969), H.II, S. 521-528
Roth-82	Roth K. : Konstruieren mit Konstruktionskatalogen . Berlin, Springer 1982
Rutz-85	Rutz, A.: Konstruieren als gedanklicher Prozeß, Diss. TU München 1985
Schu-96	Schulz, A.: Systeme zur Rechnerunterstützung des funktionsorientierten Grobentwurfs. Dissertation TU Braunschweig 1996
Spec-95	Speckhahn, H.:Systeme zur flexibel konfigurierbaren Informationsbereitstellung für die Konstruktion. Diss. TU Braunschweig, 1995.
Spur-86	Spur,G. (Hrsg.).: Rechnerunterstützte Konstruktionsmodelle, CAD-Kollloquium ,1986 des SFB 203, PTZ Berlin,1986.
Spur-97	Spur G., Krause F.L. : Das virtuelle Produkt, Management der CAD-Technik, München, Wien, Hanser, 1997.
Suh-95	Suh, N.P.: Axiomatic Design of Mechanical Systems, Special 50th Anniversary Combined Issue of the Journal of Mechanical Design and the Journal of Vibration and Acoustics, Transactions of the ASME, Vol.117,pp 1-10, June 1995.
Tjal-75	Tjalve, E. und Andreasen, M.M.: Zeichnen als Konstruktionswerkzeug. Z. Konstruktion 27 (1975), H. 2, S. 41-47.
VDI-2222	VDI-Richtlinie 2221, Beuth-Verlag ,1987
Wald-69	Waldvogel, H.: Analyse des systematischen Aufbaus von konstruktiven Funktionsgruppen und ihr mengentheoretisches Analogon. Diss. TU Stuttgart, 1969.
Webe-94	Weber C., Stark C.:Flächenorientiertes Toleranzmodell als Grundlage der rechnerunterstützten Tolerierung mit 3D-CAD, In: Gausemeier, J. (Hrsg.): CAD'94 , Fachtagung der Gesellschaft für Informatik e.V., Paderborn 17.-18.03.1994, Carl Hanser Verlag, München Wien 1994, S. 353-374.
Weig-91	Weigel, K.D.: Entwicklung einer modularen Systemarchitektur für die rechnerintegrierte Produktgestaltung. Diss.TU Braunschweig, 1991.
Will-93	Willert, J.: Gestaltung standardgerechter Produktmodelle in der maschinenbaulichen Konstruktion, Diss. Universität Rostock, 1993.
Wöge-42	Wögerbauer, H.: Die Technik des Konstruierens.München: Oldenbourg, 1942.
Yosh-83	Yoshikawa,H.: Automation in Thinking in Design, Computer Applications in Production and Engineering, North-Holland, Amsterdam, 1983.

Discussion

Comment - K. Ehrlenspiel

I would like to add another paradigm which has been effective in the field of design research since 1985/86: The examination of design in practice. You mentioned that around the year 1985/86, we began to notice that design is something which occurs in the mind, which is actually quite self-evident, but had not received due consideration. What happened during this process was not clear. The experimental „examinations of reality" only started at that time. This happened at about the same time here (Pahl/Dörner/Ehrlenspiel) and in the USA (Staufer/Ullman). I have just now heard that in Japan as well, people around Mr. Yoshikawa began to examine reality around 1985/86/87, i.e. began to make experiments with designers. I believe that this is important since this is the basis of all descriptive sciences. Experiments were also the basis of the development of all natural sciences. Only when one recognizes reality can one begin to develop models and corresponding theories. And I believe design theory is still lacking much of this. We are only beginning to understand questions such as: „How do we solve complex problems? What are the conditions in our minds? Which of these conditions can thought psychologists provide us with?" In design, much is also accomplished though communication between people. What influences are decisive for communication? What influences does an enterprise have? This will be discussed later in the presentation delivered by Professors Birkhofer and Lindemann. My question is therefore: What conditions and influences on design develop from the individual, from the group and from the enterprise? And I believe we need to know much, much more about these realities.

Question - A. Albers

Mr. Franke, you have analysed many starting points to create a design theory with your large industrial background. Did you see the reason in this discussion why copying it to the industrial practice is so difficult and has not taken place yet? You should be able to learn from this.

Answer - H.-J. Franke

I believe one reason is that mechanical design is a relatively complex process. Up to now there is no simple theory or picture that finds its way easily into the heads of practical design engineers under pressure of narrow schedules and capacities. The experience with our Ph.D.students after they changed into industry showed that application of design methods and design theory only functions when these methods had become their deep internal knowledge. Otherwise there is a large danger that people will go back to the old routine. And this leads again to the question of standing the test and of proving applicability. We have to define theories and methods that can really speak for themselves. The user must find that he himself has a clear advantage. And we have not come so far as yet!

Question - H. Lenk

Regarding the sketch of Popper's methodology you delineated in your paper the word „true" or „truth" occurred three times not in a specific sense, but within the most general question,

"What is truth?" Popper claimed at the very end of his *Logic of scientific discovery* that we shall never be able definitively to know the truth of theories. His methodology or theory of theories amounts to the fact that theorititians would tentatively design theories. Theory construction is a matter of designing. He tried to sketch out a concept like verisimilitude. However, would you described about theoretical approaches are considerations whether or not something works or functions. That is to say that you need not and could not finally go back to or rely on truth. You don't have truth in your hand. That is however a whole and extensive punch of problems. Professor Höcker from Stuttgart University once said in a discussion with me: "Wahr ist, was funktioniert" ("True is that which functions or works"). That however is no concept of truth indeed. That means in general, that you would step down from the claim that theories in Popper's sense, e.g. physical theories, are true and are known to be true. Instead, you have to content yourself with methodologies of actions.

Answer - H.-J. Franke

I am not quite sure, but I believe you are right regarding the process of acting as such. On the other hand, as a designer I must prove the physical truth, too, because my product has to function in the end.

Comment - H. Lenk

The proof of truth can never be delivered. In Popper's approach also basic hypotheses are *hypotheses* which are conventionally accepted. this is a difficult methodological problem.

Summary Session 3

U. Lindemann

Ladies and Gentlemen, we must all be flexible, at some point we have all learned this. Mr. Grabowski advised us earlier that we have to complete our summary in five minutes. I will do my best to stick to this.

Therefore, I do not want to go into the contents of the single contributions, except to say that we should make the speakers and their subjects aware again. We have heard a lecture from Prof. Grabowski. Unfortunately, the TRIZ lecture had to be omitted.

Prof. Krause told us about a vision deriving from the subject of Software Development. Prof. Franke informed us about a backward glance at the Ladenburger discourse and its results. Unfortunately, the contribution of an architect also had to be omitted. Prof. Birkhofer and myself, have tried to show some of our findings from the field of the empirical design research.

I have tried, together with John Gero, to summarize the main points in a few key words.

We have heard proposals for axioms, and hypotheses being put forward. On the one hand, we have recognized a necessary need for discussions, and on the other hand, we have seen that these first steps make a positive base for the further development of ideas.

The aspects of experience and knowledge came from many sides during the discussion. We all agreed that we have to pick-up on these things - that they will decisively contribute to the change in our structure of ideas about time.

At first, we had a little difficulty with the classification of the contribution from Prof. Krause. He tries with his approach using a model deriving from the area of software development, to produce complex system solutions through combining given basic modules. Maybe we have here the possibility of a first link to other subjects, which also operate with the given standard elements.

We dealt with the question of the truth of our hypotheses and theories, and during following sessions, were able to expand upon our ideas.

The question of whether you must or can have a human component in the striven-for theory, was discussed. Therefore, the repeated emphasis on the meaning of the content was very important. How the work will be continued, will be decided upon with respect to the situation.

There was a consensus about the necessity of an agreed terminology between the partners involved. The base of our communication is the language and in the growing interdisciplinary and global world, this is an essential matter of concern.

I think, we have found many starting points for a universal design theory. There was a fundamental agreement concerning the necessity to continue the work in this direction. We have discussed various aspects of usefulness. I am totally convinced that we should continue in this way and work towards methods, models and further onto theories.

Session Four

On the Development of a Design Theory from the Perspective of Special Sciences

Form Follows Flow

Dr. Gunter Henn

HENN Architects Engineers

Munich, Germany

Keywords: Innovation management, organisation design, spatial structure, managing complexity, „planability" of Innovation and communication, visualisation of the „intellectual material flow".

ABSTRACT

Communication architecture promotes Innovation

An organisation must innovate in all areas. A measure of quality is „corporate innovation" - which is both a goal and an ongoing process. In order to react to changing demands quickly and correctly, efficiently and successfully, organisations are forced to become „learning organisations", in the same way as living organisms learn. This is true of companies, charitable associations, universities, research Institutes and government authorities alike.

The innovation potential of an organisation in which people work is limitless. This potential must merely be tapped - among the top management and all employees, among suppliers and customers. The idea is to link all parties in a permanent process of innovation in order to consciously and effectively bring the knowledge of the individual into a synergetic process.

To achieve this quantum leaps must be made in the structure of large hierarchies. Traditional organisational development is no longer adequate to face this task. Organisations which rely on continuous innovation must undergo a thorough process of re-engineering, just like their products. Re-engineering leads to new goals, other structures and previously unknown methods.

Such goals, however, are only achievable if all participants firstly recognise them, secondly share them and thirdly pursue them in innovative structures.

The company as a knowledge interchange

The key to corporate performance in the future lies in knowledge design.

For centuries the application of knowledge has driven the production process onwards. Then knowledge was applied to work and management was born. Nowadays knowledge is applied to knowledge. And the company becomes an interchange for knowledge.

As architects we plan for such companies. In designing this architecture for knowledge, we have had an ever increasing influence on corporate design, for the associated organisational structures are highly complex and multi-dimensional in terms of spatial context. As architects we are used to dealing with such structures, developing them, designing them and making them attractive.

Knowledge architecture is corporate design!

THE ARCHITECT AS ORGANISATIONAL CREATOR

Our society and alongside it the economy are changing at an amazing speed. We are always innovating. Taking the increasing complexity of company processes into consideration the industrialist urgently requires consultation and new ideas for development of a more economical and efficient production method of manufacture.

The architect with his "wide range of experience" should therefore also be able to fulfil the requirements of a management consultant. It no longer works, to just develop the functionality of a space, today it is absolutely necessary for the development of the processes, to optimise and to create and then following this, design for dimensional requirements. The spaces are not developed solely from the logic of the organisation, but organisation and space must compliment each other. Space and organisation alongside each other reveal a synergy potential for the company. Therefore the architect must also concentrate on the organisational processes, following the breaking down of the companies skeletal structures. Thus he has also become a management consultant.

DEVELOPMENT PRINCIPLES FUTURE OPERATIONAL ARCHITECTURE: FORM FOLLOWS FLOW

The shape of the building follows the flow of the work processes, the flow of thoughts. From this point of view, for the structural environment the established principle of architecture of the 20th century, "Form follows Function" (Louis H. Sullivan, Chicago, 1896) should today, after one hundred years be re-thought.

This principle was justifiable, just as long as functions were linearly jointed and hierarchical organised. Living organisations are not based upon functions, but focus on the process. Most importantly, time, dynamics, change and evolution. The principle "Form follows Flow" (Gunter Henn, 1996) embraces communication- and value added processes and social spaces along with quality guarantees and optimisation.

As with an organisation architecture today is no longer statically definable. A function, which is acceptable today, does not necessarily keep its validity for the future. An accurate derivative conclusion for building, building complexes is therefore only conditionally possible. In the past the world was made up of locations of residence which were connected by means of communication. Our world would be made up of communication networks, which combines at certain points and takes on the stability of places of residence. Nevertheless as a result of an architectonic design- and planning process a permanent building must stand. Architectonic structures establish order – also for the present and the future.

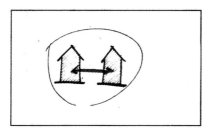 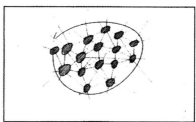

It there fore follows that structural static and process orientated dynamic must be combined in the building. A handicapping of dynamic in buildings must not take place, it must be encouraged, even created. For this we require the largest possible openness, this direction leads away from an inflexible structure, over towards a living organisation also of the building. Only in this way will, communications, serving self-organisation, not only be approved of but also be achieved and supported. The space gives up the organised functional stepping stones to time, that means away from the building and over to the network. We are moving away from the principle of the space over to the principle of the network and with that an urbanisational structuring of our buildings. The town is not a dimensional fact with a social impact, but a social fact forming itself dimensionally. Analogue to this the building is a social organisational fact which forms itself dimensionally – not the other way around. Initially the entrepreneurial objective is laid down, then the building. The requirements must be developed and created for this entrepreneurial organisation or set objective. The space structures of the proposed processes must be defined.

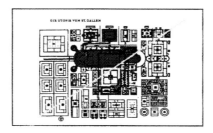

It applies, that the problem is, to extensively analyse the assignment and then to develop the solution for the assignment. The PROGRAMMING method for this analysis of the assignment (Gunter Henn, "PROGRAMMING – describe the assignment with visual aids prior to the planning", Industriebau 2/96) is very suitable. With this method the assignment and the intended processes are visually simulated utilising differing scenarios. In this way the initial picture is developed from the assignment – even if they are not in focus. A building is a responsibility. In order to ensure that the answer in the form of a building is not only attractive but also correct, one has to comprehensively analyse the assignment.

ONE-DIMENSIONAL ASSIGNMENTS FOR TWO AND THREE-DIMENSIONAL SOLUTIONS.

The initial presentation of such concepted pictures are usually in a graphical form and already supply solutions to previously posed questions regarding meaning and intention of a new construction project. How do such questions occur and how do they read?

They occur as raw formulated ideas, become audible in the form of oral statements, then can be transposed to paper. The initial questionable framework for the later solutions normally arises verbally. Extensive non-conscious, cultural predecessors have been led in this way, over many hundreds of years to develop process to products, this also includes building constructions, of an impressive quality, beauty and durability – predominately utilising hand built methods. Design, lying down of the plans and realisation were in principle left to the one person.

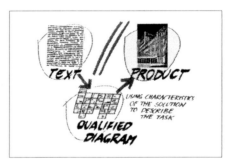

In our highly technical, with division of labour organised culture, it is the compiling of an assignment from the description of the problems and the development of the correct solutions by means of the product, normally, completely separate from each other. Rapid development and completion of complex products - including buildings – is not possible by any other method. In industry there are also separate divisions for research, planning and development on the one hand and production on the other. Even before development can accept a partial responsibility in the form of two- dimensional drawings for a later three-dimensional product, the assignment is formulated in so called specifications – one-dimensional, by utilisation of speech.

DESCRIBING THE ASSIGNMENT USING VISUAL AIDS

Architecture does not "talk" verbally; architecture is communicated visually. For a speedy and secure assignment description in this space the acoustic tool of speech is inefficient. It would be absolutely essential to translate the world of symbolic concepts into visual reality on a one to one basis, and for this there is no concept available.

To arrive at congruent standards, one should describe the assignment solution with available communication means. If the solution "talks" visually, then the description for the assignment should be a picture language. Following this, we already visualise the initial step, on which all of the others are built: The description of the problem. Our "language" is qualifying diagrams. They build a platform for the immediate understanding. Qualifying diagrams do not only define the assignment, but also show them simultaneously. By means of visualising, structural pictures are formed, which already simulate the initiating of the solution. In other words: Within qualifying drawings the simultaneous assignment and an element of the solution can be found. They bring together the visually set question and the developing solution in one and the same presentation.

PROGRAMMING CREATES CONTROL CIRCUITS FROM ASSIGNMENTS AND SOLUTIONS

These thoughts and "speech" in diagrams have a definitive advantage. Fields of work, which would otherwise drift apart, are brought back together with the effect that they reciprocally stimulate each other. Even in the assignment there can be found an enormous amount of innovative potential. It is now activated simultaneously for the solution. A continuous process of improvement between the two poles is instigated. We can be certain that over a short time period, innovative leaps forward will occur.

PROGRAMMING serves to diagram methodically and fill the thought and decision fields. Within them we are optically signalling the target, and we supply ourselves with a picture of the important facts and also develop the initial concept in diagrams. These structural pictures serve as a platform for the dialog. The problem is with the help of the visual presentation, also evaluated; due to the fact that the structuring of the problem is recognised. Initially we work on the concepts for the singular promotion of the assignment, and this in an abstract basic presentation. We then join them together and put them more and more into concrete terms. Slowly but surely complete, concrete solutions are developed.

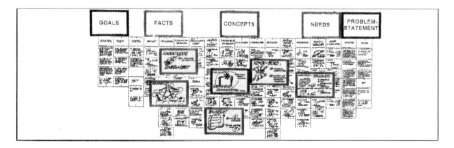

DO NOT CONTROL THE SOLUTION BUT THE APPROACH

There are many organisations and professional groups, which subjects all aspects of a solution by means of a thorough control. Of the same importance, although unequally more expensive, for the possible undesirable trend saving control of the assignment, it is then far worse developed. This is where we will begin. Without any visualisation, it is usually as a rule unclear and difficult to grasp. For this reason it does not normally occur. In comparison

PROGRAMMING, secures speedily, target orientated and cost saving, in short, efficient architecture.

The more realistically that the assignment is posed and shown, then it is so much better, to recognise its complete complexity early enough. It is only possible to bring them into the planning process. This also means for the design offices as well as for the developer. Neither the client nor themselves should bring themselves into play when the results can be seen, and are possibly not pleasing. Such an uncertain cautious approach to the assignment would be very ineffective; a multitude of problems to be resolved at a later date will be the result.

PROGRAMMING brings the point for decision making to be moved forward even for the developer. Assignments and conversion steps would be recognisable and obvious at an earlier stage for all involved – expensive finishing touches at the building stage or even after completion are therefore avoidable.

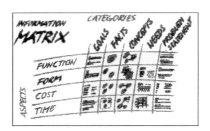

VISIBLE FLOWS OF COMMUNICATION LEADS TO MORE INNOVATION

The physical flow of material in a building, meaning the transportation of components and finished products, is real and visible. Deficiencies are therefore recognised at an earlier stage. If this material flow is improved, then the effect is also clearly to seen. This leads to the fact that in the finishing process innovation in the physical flow of material has been optimised early enough. The are measurable and they are measured.

The so-called intellectual flow of material ("The Flow of Technology", Tom Allen, MIT 1977), refers to the knowledge- and decision flow in an organisation is also real, even though it is invisible. Incomplete thoughts and services which are not completely thought out usually remain hidden. They are not stored up in boxes in rooms or corridors. They also barricade pathways; to be precise solution pathways, these do not however immediately attract attention.

COMMUNICATION ATLASES AS A PLANNING FUNDAMENTAL

Communication atlases according to the NETGRAPHING method open up, as structure diagrams, the entrance to an otherwise invisible reality in thinking. They make a thought- and decision making room visible, who speaks with whom and how often, and also how and where these communications are supported or handicapped due to building structures? A comparison of communication atlases and organisational plans eventually clearly shows, in how far both structures harmonise.

Communication atlases also make further visible, areas where communicative poverty exists, where the personal exchange of thoughts is not developed far enough and following up on this, where innovative potential could still remain hidden.

If these communication atlases are further evaluated, then they create an additional basis for the utilisation of the building. This utilisation is then optimised, as they promote the exchange of thoughts and do not inhibit. Communication atlases show simultaneously, which office form or which machine room layout at which location is effective. They also make it clear where rooms and space for communications are additionally required. With this as a basis, room programs such as organisation plans, allow for more than enough clarity for the required flexibility and innovative security for a sensible organisational decision.

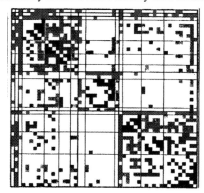 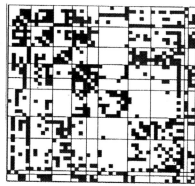

NETWORKING AS A REQUIREMENT, SPACE AS AN INSTRUMENT

Communication atlases according to the Netgraphing method distinctly show the function of space. It is not the working space on its own – however how ergonomically it is optimised, which leads to results. It is the intellectual networking of the working positions, which enables the interacting of the innovative process.

According to the amount of innovation required for development, production or administration, then so much better this becomes. It therefore follows that whoever is designing these rooms must apply the correct communication landscape. Innovative processes can only then be expected, only in this way they can be encouraged..

If, in an organisation the correct idea is more easily reached and thus reducing detours and deviations become more and more rare, then the larger amount of work required for a new building comes into perspective. It does not only revolve around perfect communications; it involves much more, efficient performance. We want to structure communications in new buildings in such a way that the transfer of knowledge face to face, can take place at the proper time and at the correct level.

We accept those spaces, without any barriers, distract communications and with that lower the performance, the same as with too little communication. An optimisation, which requires a long time period, can only be achieved along the lines of a continuous process.

DESIGN PRINCIPLE COMMUNICATION – COMMUNICATION ARCHITECTURE

In a monarchy the ruler is the middle point. In his castle everything is designed around him. In it there are many rooms which are linearly arranged. A completed overview for visitors and servants is not wanted, in the structural make-up or in the structure of targets, spaces of responsibilities and personal.

Hierarchical structures are made from the top by means of orders and control. Any form of dialog between employees outside of the official channels is not desired. Following this the room structures are set up in the first instance for an information distribution in a linear form.

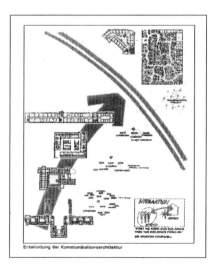

Society today finds itself in a period of change. Systems and their principle of moderation are winning more and more in importance. Networking of the parts is also winning more and more in importance. Dynamic effects organise the parts into a whole. As soon as an organisation takes part in the game, the necessity for process, especially for exchange of knowledge, communication, takes precedence. It is the most important element of every organism; high-ranking organisation can only function with the availability of improved communications.

Today we are no longer living in the information society – which took over from the product society – now we are at the beginning of the knowledge society.
The innovative performance for the future is to be found in knowledge design. Knowledge means, building up from information and values, assessments, decision making. Knowledge is more than information. Konrad Lorenz once said that information is at the best, the manure on which knowledge can grow. Information can be passed on by technical means, but knowledge in its complexity can only be transferred, with a face to face situation. Therefore it must be achieved, that communication can spontaneously take place between every section inside a company. In this way the company houses a knowledge exchange for every employee.

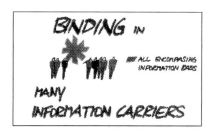

Then why should only a company governed by a hierarchical system be carried as a learning company? Learning means knowledge and an increase in experience, it is vital to know the client's opinion concerning ones own work and to introduce it as a new form of knowledge. Every working step taken must be seen as a potential step towards improvement. Today it is necessary that this learning process does not only take place with the management of a company, but throughout all levels of the company. Misunderstandings, misinterpretations will then be reduced if everybody responds to the learning process. This means to say if all employees in the positive feedback continue to improve themselves. Improvements, innovation cannot be solved by hierarchy, and not by organised suggestion situations. The innovation required today in all spaces – not only in the research and development fields – can only be achieved by the participation of all employees. An open communication system and an independent acceptance of responsibility are the requirements for this.
To achieve such changes the trust of all employees and clients has to be won. Only, only exchanging solid brick walls for glass walls cannot achieve openness and transparency. Then the employees feel as if they are being spied upon and restricted in their own familiar territory.
Prior to each and every consideration in the building phase the processes must be developed and composed by the employees mutually. The employee must be led from his previous way of thinking towards process logical thinking. From then on, as if from itself the requirement for transparency and openness occurs, only then will the process become visible and for the single employee and the necessary orientation possible.
The initial conversion of communication architecture takes place in office blocks. The spatial structure of the hierarchy was cellular offices with a central corridor.
Free communication was not supported, often not even wished for. The implication was found at this position.
The fractal educable company demands communication rooms, also outside of the own work position. This is not possible with the traditional office layout. The implication is to be found in communications. The layout becomes "the central corridor with solitary confinement".

The so-called combination-office with single rooms and a common space for the free communication transforms the central corridor into a communications space, without the loss of the necessary concentration required at the work place. Both of the parameters of the intellectual activity, concentration and communications as found in a monastery with monks cells and cloisters, is achieved by means of the individual offices and common space. At Audi in Ingolstadt the director's floor was changed by this method into a decision-making studio.

ARCHITECTURAL KNOWLEDGE

When developing knowledge design the company make-up is being taken more and more into consideration, because the appropriate organisation structures are, to a large extent, complex and multi-dimentional spaces. The intellectual material flow must also be kept to a size that can be easily handled even at the organisational planning stage. This is where the fundamental assignment lays for today's architecture for the working world. Making communication possible and with that the company successful and in a position to compete on the market place.

DESIGN PRINCIPLE SPINE

We take nature as our example referring to the information flow. What does the room structure look like in our educable organisation? Nature has in long evolutionary periods, meaning learn phases, developed very successful structures. One successful structural pattern is the so called "spine pattern". In much the same way as the spinal column in a human being, the backbone supplies the individual organs. Departments, side-tracks, branch off and from these, then there are further branches. Inside industrial production the logistics was the backbone, the supply for the individual departments with materials and production materials. Today the backbone for industrial production is the information flow. The structural pattern "spine" remains, and is though, utilised by another different content. Software information and intelligence have superseded hardware-material. The "intellectual material flow" has replaced the physical material flow.

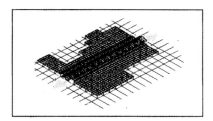

DESIGN PRINCIPLE "FRACTIONAL" IN LIVING COMPANIES

Fractional means self-similar and self-organising. Every department inside a company is similar to the complete company and is also similar to each and every larger or smaller department in the company. Each department is, self sufficient, and can complete all-embracing processes and can realise complete value added steps. Control takes place by means of a high standard in on-line communications, which means the immediate passing on of information allows for a self-guiding of the company. The perception apparatus is not only at the managerial level of the company, but in each and every department, and all employees are well trained in this. For this to function it is necessary that a trusting and open relationship is promoted with each other.

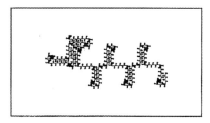

THE FRACTIONAL FACTORY SKODA

In the Czech Republic the first fractional factory was built for the Skoda Automobilova in Mlada Boleslavthat and it does not only have the organisational intention to be fractionate but, in the room space, was consequently so organised that fractional self organisation had to occur.

It is a factory that has no separate office block, because the offices are located in the centre point of the factory. In this middle spine there are team rooms and try-out spaces.
Conference rooms, social facilities, but also the working places for the works management, the personnel department, and the engineers are to be found here. These functions are in the centre point of the factory on the same floor level as the production lines. Communication between the production lines and the support service facilities (offices) takes place immediately without any time delays and spatial barriers. Each and every employee has the opportunity of an optimal and obvious process orientation.

Through the characteristic features of self-organisation, self-resemblance, self-optimisation and dynamics, the characteristics originating from the natural fractional structures of the company, are transferred to the processes and organisation and thus to the architecture. The company became a learn- and versatile vital organism, which through the permanent feedback and self-optimisation could adapt to other framework requirements and demands. Every team is linked up to team members of the neighbouring teams. Each and every employee or every individual group is at the same time jointed into self-resemblance structures of higher order. Employee – team – group – shift – company. The result is a networked structure, which always allows and supports direct communication between all parties.

FUTURE BUILDING STRUCTURES

These optimised forms of dialog are a requirement for the method of working in the next century. When all departments of an organisation are networked into a communication landscape, normal every-day bureaucratic structures develop into living companies, administrative authorities, research institutes or even universities.

A living organisation no longer consists of separate departments for research and production, administration and storage, but is transformed into an innovative centre. The individual departments are fractionately and systematically related within a communications landscape to each other.

These lead to building structures, which are no longer only made up of specialised zones for planning or production or marketing, but also includes spaces of space. Thinking areas, communication flows are utilised for the permanent improvement process and performance fields.

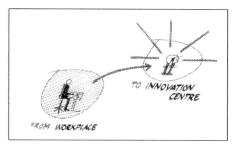

The basis of these process ideas administration buildings can be made into innovation centres. In the same way that a factory has a "communicative quality path" office buildings are given "universal rooms". Where there was originally only corridors, discussions can now take place, communication. This leads to completely new structures all the way up to the managerial levels, which in a living company are no long the "end station" for innovation, but communication centres for the complete organisation.

TECHNICAL UNIVERSITY MUNICH

In 1997 in Garching by Munich there originated a town for knowledge at the Technical University of Munich, faculty for the engineering sector for 4000 students and 700 tutors. This substantial university building constructed as a multi-superimposed communication landscape for the numerically largest department of the Technical University was realised in only three years. Communication was, from the outset, of the highest priority in the planning principles. With the help from programming abstracts as well as concrete thoughts, individual opinions and team thinking, workshops and interviewer results, communication was immediately visualised. All participants as well as the persons who would utilise the facility at a later date, the property developers and also the designers could visualise the complete thought flow, it could be experienced and was with that could be acted upon. This is how a building in Garching originated, which appeared to be dedicated to discussions.

This new ordered space establishes a city like spatial quality whose main element is the 220-meter long faculty street, which binds and Internets all 7 institutes. It fulfils everything, which a street area makes possible, transport, vehicles, movement, exchange. In this space the importance is stressed on visual and acoustic, mental and physical communications, every movement is incorporated here and channelled inwards.

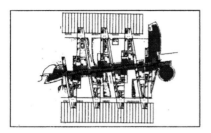 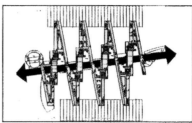

The institutes docked on the edges, spread outwards away from the street with their two wings which forms many differing bays. Inserted between these and affiliated to the different institutes, are completely different formed structures, lecture halls and seminar rooms. These elements are treated in the same way as external buildings and give variations to the theme "Street and House" repetitively yet with individual characteristics, in this way the addresses and identities are produced,

INTERNAL URBANITY

Organisationally, the interior space is comparable to an urbane pattern. A central space, at Skoda the backbone, is the interior roadway linking up all of the individual spaces. As in a town, a graduating of the general public occurs. The building forms the social order of everything. The clear atrium building once again mirrors the public centres, with open work places, meeting points, viewing rooms, creative centres. It is also possible for the institute- and a professor to retire to semi-open zones with a combi- or individual offices from the street.

The building is organised as an urbane structure. Caught in the middle of the faculty, restricted by the workshop spaces, many different forms of intimate courtyards and gardens to linger in, rest or even for spontaneous meetings.

The most attractive examples of communication architecture are offered by old Italian, for example Lucca, they are space translations, space formations that originate from organisational, cultural behaviour. They did of course have a considerable advantage, they had a long time to develop, to grow, problems and solutions could overlap each other. Today we must put more thought into our approach, because we do not have unending amounts of time. We must develop thinking tools, so that this approach is made in a more conscious way. We must develop communication atlases and transform them into architectonic spaces.

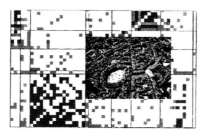

THE FORM OF THE BUILDING FOLLOWS
THE FLOW OF THOUGHTS

Form follows flow – recognises the multi-dimensionality problems, visualise, evaluate and then the initial, still rough sketches containing the solution. Today we live in a network. Every company is global and we are influenced from everywhere – whether it is desired or not for the assignment. On top of that our cognitive standard has not improved from the multitude of information only our helplessness. It is not enough to only show them in the solution; it does not revolve around the evaluation of a field of knowledge, a thinking room that makes the space complexity visible.

We must learn to always find the correct path in this network and this means every time anew. Our assignment along this path as architects is the space component calculation of the company's target values and to translate them into space form as architecture

Projects, diagrams, drawings, photographs © HENN Architects Engineers

Empirical Design Research -
its contribution to a Universal Design Theory

Prof. Dr.-Ing. Udo Lindemann

Department of Design in Mechanical Engineering

Technical University of Munich, Germany

Prof. Dr.-Ing. Herbert Birkhofer

Machine Elements and Engineering Design

Darmstadt University of Technology, Germany

Keywords: Empirical design research, design theory, design practice, complexity, critical situation, routine work, problem solving process, influencing factor, situatedness of design

ABSTRACT

Empirical design research gives a deep insight in work of designers and the influences of individual and group related prerequisites on the design process and its results. A Universal Design Theory has to take the special and very strong influences of human behavior and faculties on design into consideration. Empirical design research on individual designers and design groups / teams has become important for the deeper understanding of design processes and their influences. Different strategies of research have been elaborated on laboratory tests and the examination of industrial processes. The research was performed in collaboration of psychologists and engineers.

INTRODUCTION

The general demand to develop products of higher quality at lower costs in even less time requires a more parallel cycle of work in product development as opposed to the traditional mainly sequential cycle. Consequently, engineering designers are collaborating more in teams crossing both department and even company borders [*Habe-92*]. In this situation, engineering designers are struggling not primarily with technical problems, but more with difficulties related to their environment (e.g. effective organization) and to their colleagues, as surveys concerning the problems of engineering designers in industry have shown [*Birk-91, Ehrl-93*].

These demands raise important questions, such as how to teach and train individuals and how to organize teamwork effectively in complex working conditions, how to lead and to communicate within a team, or which individual characteristics are important for making a good team member.

Research on individual designers and design groups/teams has become important for the deeper understanding of design processes and their influences. Different strategies of research have been elaborated on, for example, laboratory tests and the examination of industrial processes by researchers involved or not involved in these processes. The research was often performed in collaboration of psychologists or other social scientists and engineers.

Particular case studies of individuals were the basis for a number of projects, which were run in collaboration with engineers of the Universities of Munich and Darmstadt and psychologists of the university of Bamberg. Later these studies have been extended to design processes in an industrial environment. Most of these research projects have been funded by the DFG, the German research foundation.

The aim of research described by the authors was therefore to identify the determining factors of individual work and teamwork in laboratory situations [*Dyll-91, Fric-93, Günt-98*] and in complex working environments in industry [*Fran97*].

The investigation of individual- and group-related factors on individual and teamwork in engineering design practice together with psychologists represents a new field of research in this regard. Part of the following text is a shortened and modified version of [*Fran-98*].

INDIVIDUAL DESIGN STRATEGIES

For more than ten years researchers like Dylla [*Dyll-91*], Fricke [*Fric-93*] and Günther [*Günt-98*] have used the design of a product of low complexity within their experiments, as a support unit for an optical device. Designers and students elaborated their design solutions within a couple of hours while the process was documented by written protocols and videos. The results, as well as the individual strategies during the whole process, were evaluated afterwards.

In addition, several psychological tests have been used to classify their individual background, as for example, their ability to think in three-dimensional spaces or their emotions.

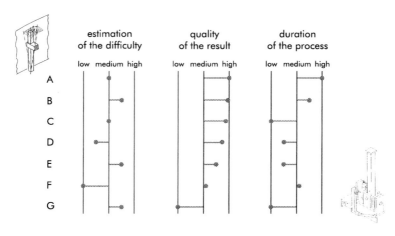

Figure 1   Designers of similar education develop a product of low complexity

Figure 1 shows some results of an input/output analysis of seven designers with a similar education. They were all once students of mechanical engineering and have been taught in design methods. The data was extracted from the dissertation of Günther [*Günt-98*].
Before they started their design process, an estimation of the difficulty of the task was made by each of them. After they finalized their work, the quality of the result was checked by a given set of parameters. There was no relation between the duration of the process and the quality, between their original estimation and the required time etc.
The conclusion is, that it is not sufficient to take only the obviously available data into account when you want to get a deeper understanding of the design process.
One question of the research activities was related to design phase orientation, as documented in VDI 2221 [*VDI-2221*]. Because of the detailed analysis of the processes it was possible to get an idea of the individual strategies. Typical procedures are shown in Figure 2, there are five major individual strategies seen to operate on subproblems. Following strategy „A" one selects subproblem „1" which may be one function to be solved and works it out from an abstract to a more concrete level. Then the second and third subproblems follow. Strategies „B" and „C" are slightly different from „A", whereas „D" seems to be chaotic. Strategy „E" is equivalent to the VDI 2221 suggested procedure.

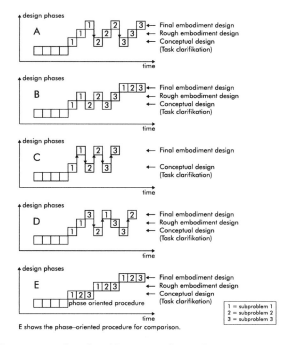

Figure 2   Alternatives of the subproblem-oriented procedure

The task clarification was analyzed in a similar way, an equivalent variety of strategies was found. The difficulties of interpretation are, that only the visible activities can be documented by this way of research. Processes of thinking in consciousness and sub-consciousness are invisible and unknown. A significant correlation of only the different strategies and the quality of the results is not possible. All strategies are more or less worthwhile and should be selected depending upon individual, process and object parameters.

Günther [*Günt-98*] suggests a classification of individual influences which includes knowledge, abilities, skills, emotion and motivation (Figure 3). Knowledge of facts (for example of machine elements, material) and of methods like FMEA (failure mode and effect analysis) are normal elements of proving design engineers.

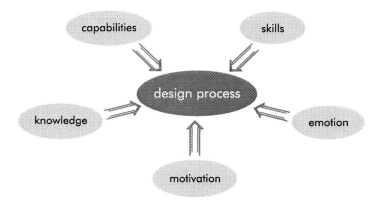

Figure 3   The design process and its influences out of individuals

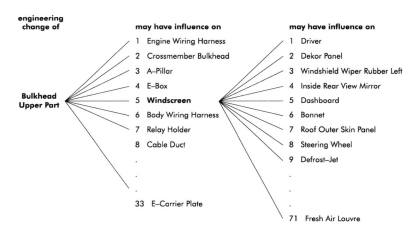

Figure 4   Generative and corrective variation

General abilities of analysis, synthesis, heuristic competence and imagination of three-dimensional objects and processes (of manufacturing, assembly, usage etc.) are important but difficult to measure. Easier to evaluate are the skills, for example, generating sketches and drawings or using computer tools. The individuals emotions can be measured in their intensity of regression and of resignation. Regression indicates the trend to avoid difficulties, „to run away", whereas resignation describes the tendency to give up in the case of difficulties. Another very important individual aspect is the motivation, the willingness and the capability of performance. All these influences occur not only in individual design processes, but also in groups, as described later.

Dylla [*Dyll-93*] described two different strategies of generating variants within design processes, the generative and the corrective one (Figure 4). The generative way follows the idea of systematic generation of a field of solutions by following aspects like variation of number of elements, the size etc. The corrective way is more related to creative elimination of weaknesses of given solutions or first ideas. Both strategies may lead to optimal solutions. Again, both strategies are worthwhile and should be selected depending on individual, process and object parameters.

In total, the results out of these research activities give us a limited set of indications for which strategy we should follow depending on actual circumstances. This may indicate the importance of the question of research in the field of situation driven design. The available results may be part of an Universal Design Theory.

COMPLEXITY OF PRODUCTS

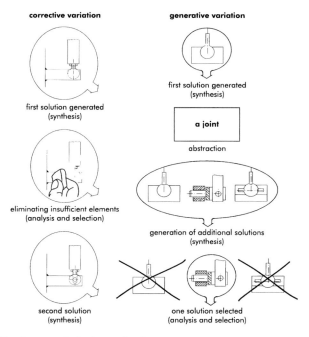

Figure 5   *Product complexity in car development processes*

Products to be developed are usually of high complexity or are at least part of complex systems. The example of a passenger cars bulkhead modification may illustrate the real complexity (Figure 5). If an engineering change of the upper part of the bulkhead is necessary, then 33 other parts or subsystems may be directly influenced. If there is an influence, for example, in this case on the windscreen, then another 71 parts or subsystems may be influenced. The example should give an idea of complexity of the product itself. How do designers navigate in this situation and keep the design process efficient?

In addition, we have to focus on a dramatic increase of functionality in many products, due to different reasons, such as functionality, safety or efficiency. Again the question is, how do designers handle this complexity in their groups and in their organization?

GENERAL APPROACH OF THE RESEARCH PROJECT

THE RESEARCH PROJECTS

Obviously, the effectiveness of design processes in industry is, besides the technical problems, determined by several non-technical factors from the fields of the individual prerequisites, the designers' collaboration and the working environment. Thus, the investigation was based on a general starting model of four central influences on the design process in practice: „individual prerequisites", „prerequisites of the group", „external conditions", and the „task". (cf. Figure 6 with examples).

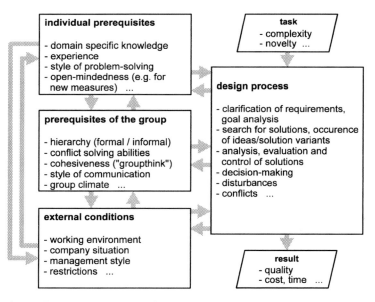

Figure 6   *Influencing factors on the design process and the result.*

For formulating an Universal Design Theory this model and the investigations based on it can give a worthwhile understanding by:
- increasing the state of knowledge on co-operative design work in practice.
- identifying factors influencing the design process stemming from the fields of the individual prerequisites, the prerequisites of the group, the external conditions and the task.
- building up a model of co-operative design work in practice based on these findings.

Four investigations in different design projects took place in two companies of agricultural machinery and capital goods industries. The investigations lasted about 28 weeks and are continued presently.

COMPILING DATA

A detailed and exact surveillance of the design practice with its vast number of influencing factors makes it necessary to observe the design work without participating. For compiling the required data, we observed and documented engineering design processes in industrial practice. The high number of influencing factors on the design process has to be compiled using a broad variety of investigation methods, which are thoroughly described in [*Fran-97, Fran-98*].

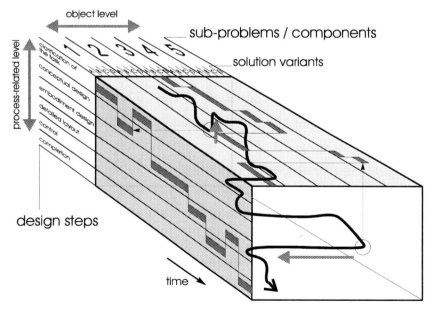

Figure 7   *Description of a design process by the course of subproblems and solution variants (subject level) and the course of design steps (process-related level).*

Describing a real design process or process and object-related approach was chosen (cf. Figure 7). The real design work seems to be a kind of 3-dimensional meander through the process-object-time space. In general one can observe a sequential process such as that proposed by

VDI 2221 [*VDI-2221*] from an abstract to an concrete level of design. But it has to be recognized that there will be many sometimes abrupt changes to other design objects or design phases.

EVALUATION AND MODELLING

As mentioned above, the design processes are documented in much detail.

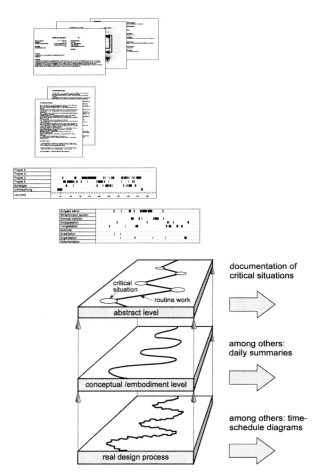

Figure 8   Standardized evaluation of the design process on an abstract level by „critical situations" (knots) and routine work (lines).

If we take a closer look at the design work, we see the elaboration of goals, the search for and the generation of solutions, analysis, decision-making, and control processes, nearly always on each level of concrete realization. The investigations so far confirm that not every little decision in detail design is important for the solution development, and therefore means „routine work" for the designer. So, if we take a more abstract view of the design process, we can identify phases of routine work on the one hand and „critical situations" on the other hand, where the design process takes a new direction on a conceptual or embodiment design level (cf. Figure 8). These „critical situations" are determining the following course of work and its result.

The „critical situations" are then identified, classified and described in the protocols and day-sheets according to defined rules. These rules are derived from the general problem-solving cycle [cf. Ehrl-95] and additional events. The classification according these rules refer to the action-requirements in the „critical situation" [cf. Dörn-89] (see Figure 9).

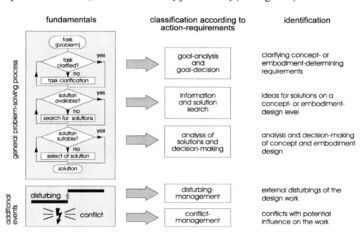

Figure 9   Division of "critical situations" according to the general problem-solving process [cf. Ehrl-95], and classification with reference to the action-requirements [cf. Dörn-89].

BUILDING UP THE MODEL

In the „critical situations", the design process is determined decisively. Therefore, they are of special interest for isolating the most essential influences on the design process. In order to explain the effect of a „critical situation", we now build up a submodel of interrelations between the influencing factors and the process characteristics for each of these relatively short „critical situations" (see Figure 10). Each identified relation is substantiated separately. Specific interviews with the designers, combined with video-feedback of selected „critical situations", help to review the submodels.

The sum of the different interrelations in the single submodels led to a model of relations between influencing factors and process characteristics in all „critical situations" of the design process. In the first investigation, altogether 62 „critical situations" were identified and explained by models of interrelations. Figure 11 illustrates the combination of these submodels to an entire model of interrelations.

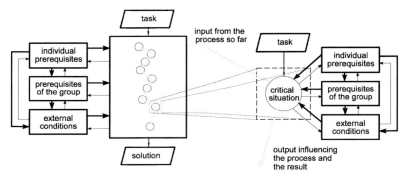

Figure 10   Influences on the design process as influences in „critical situations"

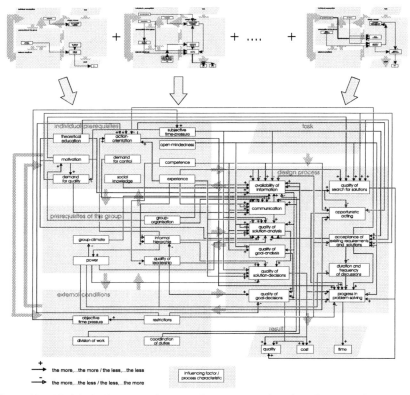

Figure 11   Model of relations in the entire design process based on 62 „critical situations" in the first investigation.

RESULTS

In the first step, the importance of the influencing factors can be evaluated by their frequency of occuring in all „critical situations". Accordingly, the general importance of influencing factors in the first investigation results in the distribution as shown in Figure 12.

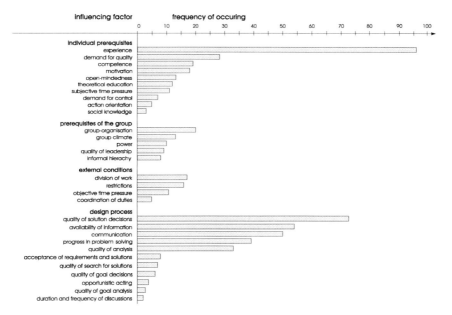

Figure 12   Absolute frequency of influencing factors and of the process characteristics of important influence in all „critical situations" of the first investigation.

Figure 12 illustrates the strong importance of the individual prerequisites in the „critical situations". At first glance, the influencing factors from the field of the prerequisites of the group does not seem to be very important. However, the importance of the group-related factors in „critical situations" becomes clear by the fact that in all investigations, 234 of 265 „critical situations" (88%) took place in collaborative work of the designers. Obviously, the colleagues are consulted in „critical situations", whereas routine work mostly takes place in phases of individual work. The difference in numbers of the effective influencing factors between individual and group is also a result of the fact, that the group is counted as a whole in these „critical situations", whereas the several acting designers cause more influencing factors of the individual.

Furthermore, the developed method also allows specific statements on the importance of influencing factors in the different types of „critical situations". For example, one can analyze which factors are mainly responsible for a false goal analysis. Facing the high number of „critical situations" of the type „analysis of solutions and decision" (69%) - nearly half of them (42%) with a negative evaluation in the first investigation- the question for the causing influencing factors and their interrelations comes up. This question can be answered by analyzing the submodels of each regarding „critical situation". The result of this analysis is illustrated in Figure 13.

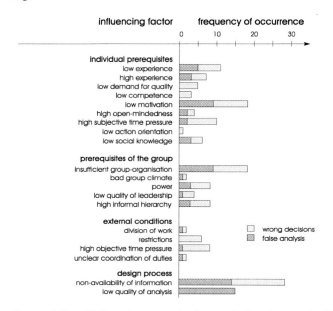

Figure 13   Responsibility of influencing factors for false analysis and wrong decisions.

When we take a look at the most frequent factors, both responsible for deficient analysis of solutions and wrong decisions, we find the factor *motivation* as the most important prerequisite of the individual. Moreover, the factor *group-organization* is the most important group related factor. For the design process, the *availability of information* plays a key role, because this factor is the most frequent reason for both, deficient analysis and wrong decisions. For further, much more detailed results see [*Fran-97*].

CONCLUSIONS AND CONTRIBUTION TO A UNIVERSAL DESIGN THEORY

Empirical design research gives a deep insight in work of designers and the influences of individual and group related prerequisites on the design process and its results. If design will be done in the future also by designers, a Universal Design Theory has to take the special and very strong influences of human behavior and faculties on design into consideration. The main findings from Empirical Design Research for formulating a Universal Design Theory (UDT) are:
- design work in practice is a process with high complexity. UDT must be a process-related theory with strong relations to an artefact theory. But UDT can not be an artefact theory alone.

- design work is influenced by many influencing factors interacting in a manifold and dynamic manner. Therefore UDT should have a holistic view on the design process regarding all important process elements and their mutual influences.
- in all design processes abstract and concrete levels of product models and related methods and rules were identified. UDT should distinguish different levels of product models and related methods and rules and integrate them in the entire theory.
- a general course of design work was also identified from abstract to concrete levels as described in VDI 2221, for example. UDT should outline a general strategy of design as the "red line" followed by all design processes.
- the general strategy was often heavily superimposed by frequently irreversible decisions, dynamic developing situations, decision making under uncertainty and time pressure. UDT should not only be a prescriptive theory but should also take into account the very nature of design as a learning procedure with situation depending decisions and the flexibility of acting in special design situations. Integrating the situatedness of design processes seems to be a new quality of describing design processes compared to common prescriptive models such as VDI 2221.
- designers are struggling in practical work not primarily with technical problems but with themselves, with difficulties related to their colleagues and to their environment. UDT should has to take into consideration the prerequisites and influences from individual designers, design teams and external conditions
- design is made by designers! Individual action styles and the very different prerequisites of individual designers significantly influence design processes. UDT should consider therefore organizational and psychological aspects of design work too. UDT should be a basis for developing a kind of advisory system with strategies, methods and rules which assist the individual designers on how to behave accordingly.
- engineers are increasingly collaborating in teams crossing department and company borders due to the trend of growing complexity. UDT should take into account co-operative design work, which is more than only the exchange of information; it demands communication, co-operation and social behavior too.

It could be one highlight of working on a Universal Design Theory, to combine prescriptive and descriptive design process models to a universal one and to remarkably improve the analysis and synthesis of design with a holistic theory.

REFERENCES

Birk-91   Birkhofer, H.: "Methodik in der Konstruktionspraxis - Erfolge, Genzen und Perspektiven" in *Proceedings of ICED 91* by V. Hubka (ed.), Edition Heurista, Schriftenreihe WDK, Zürich 1991.

Dörn-89   Dörner, D.: *Die Logik des Mißlingens. Strategisches Denken in komplexen Situationen.* Rowohlt. Hamburg 1989.

Dyll-91   Dylla, N.: *Denk- und Handlungsabläufe beim Konstruieren.* Hanser, Wien 1991.

Ehrl-93   Ehrlenspiel, K.: "Industrieprobleme und nötiges Wissen bzw. Können im Bereich Entwicklung und Konstruktion" *Konstruktion* vol. 45, pp. 389-396, 1993.

Ehrl-95   Ehrlenspiel, K.: *Integrierte Produktentwicklung. Methoden für Prozeßorganisation, Produkterstellung und Konstruktion.* Hanser, München 1995.

Fran-97   Frankenberger, E.: *Arbeitsteilige Produktentwicklung - Empirische Untersuchungen und Empfehlungen zur Gruppenarbeit in der Konstruktion.* Fortschr.-Ber. VDI Reihe 1 Nr. 291. Düsseldorf: VDI Verlag 1997

Fran-98   Frankenberger, E. & Badke-Schaub, P.: *Modelling Design Processes in Industry - Empircial Investigations of Design Work in Practice.* Automation in Construction 7 (1998) 139 - 155. Elsevier Science B.V. Amsterdam

Fric-93 Fricke, G.: *Konstruieren als flexibler Problemlöseprozeß - Empirische Untersuchung über erfolgreiche Strategien und methodische Vorgehensweisen beim Konstruieren.* VDI-Verlag, Düsseldorf 1993.
Günt-98 Günther, J.: *Individuelle Einflüsse auf den Konstruktionsprozeß.* Shaker, Aachen 1998.
Habe-92 Haberfellner, R.; Nagel, P.; Becker, M.; Büchel, A.; von Massow, H.: *Systems Engineering: Methodik und Praxis.* Verlag Industrielle Organisation, Zürich 1992.
VDI-2221 VDI-Richtlinie 2221: *Methodik zum Entwickeln und Konstruieren technischer Systeme und Produkte.* VDI-Verlag, Düsseldorf 1993.

Discussion

Question - R. Herges
I find it interesting that you looked at the problem from a human perspective. However, I am astonished that the concept of „creativity" was not mentioned in any of the lectures so far. Is it possible to find an ingenious solution of a problem if one followes a rigid theory? According to my experience in chemistry the most ingenious people are those who think desultorily. The most successful people, however, are those who are chaotical in the first instance but are able to switch to a systematic procedure as soon as they find a promising strategy.

Answer - U. Lindemann
I will put „Picture 2" on again. You can see the four types who work according to different systematic. If you look at „Type D", you can assume that he has got a creative mind. However, the question is whether he is able to switch in time, to working in a systematic way. From a teaching point of view, the importance of creativity is always emphasized. However, it is important to be constantly guided back to a systematic way of working. If you should try to make „Type D" work precisely along a set pattern, I do not believe that he would produce good solutions. He does, however, have a chance to come to these good solutions with his own way of working. We have to be aware, that for every interpretation made from results of research, only observable information can be taken into account.

Question - A. Albers
In view of all this you could state the way of solutions is typical for one person, i.e., if he has solved the constructive task and starts with the next one he repeats the way of doing. Or is it not proofed at least to a certain extend statistically?

Answer - H. Birkhofer
The specific characteristics of individual developers and designers as well as their specific procedural styles in problem solving have a very dominant influence. We have also recognized that procedural styles and strategies of the individual are relatively stable regardless of the tasks they are performing. So we have not only observed how designers proceed during design, we have also confronted them with computer simulations, that is, doing computer games. Procedural styles and strategies strongly influence the individual approach to problem solving. Partially this holds true for the problem-solving behavior of groups as well. In computer simulations, groups also behave similarly as they would in real design situations.

Question - H. Lenk
Earlier I conducted socio-psychological experiments and sociometric analysees of top-performance athletic teams, e.g. international rowing crews. Have you also done sociometric data retrieval with regard to the formation of cliques (close subgroups), leadership, preferences and antipathies or something like that?

Answer - H. Birkhofer

Only within limits, as far as we noticed, and as far as they were relevant for these design investigations. This was not the main focus of our work. I will not rule out that socio-psychological investigations also can be significant. This is also the reason for my comment, that our model of quantities that are of influence is not complete and also cannot be.

Question - K. Ehrlenspiel

Assuming we will one day achieve a theory like the one mentioned here: The question is whether this theory is sufficient or whether it needs to be adapted to people and operational influences, as you have shown, or even to types of products and heaven knows what else?

Answer - H. Birkhofer

I do not believe that the theory must be adapted, but rather its elements and quantities. On Earth, everything falls to the ground: that could be an apple or anything else. Nevertheless, one theory applies to this behavior, the gravitational theory.

Question - K. Ehrlenspiel

In that case, I must have expressed myself wrongly. The theory needs to have a content that can be taken and applied to certain persons and certain products and certain business situations. Perhaps in such a manner that one has more or less a puzzle, a working together of partial methods and tools, as Mr. Krause suggested.

Answer - H. Birkhofer

I would see it that way.

Answer - U. Lindemann

I believe, that we should approach the subject of simplification very carefully. The whole subject is very complex and we have to take this complexity, with all the different influences (some of which we have shown here) into account. The artistry will be to approach a theory which still contains all the essential elements of this complexity, through a reasonable abstraction. We would not come to a useful theory through denying essential influences.

Question - H.-P. Prüfer

I have a fundamental question concerning the procedural patterns. We may easily imagine pattern "E" as an ideal with respect to the guideline, as it is presumably taught. How big is the effect of teaching on the behavioural patterns of the group members? Are there actually members working corresponding to the ideal pattern?

Answer - U. Lindemann

There is definitely an influence. During our investigations, we not only looked at people in industry, but also at students. It is in the nature of things that students who have no experience in industry, tend to bend towards following a taught pattern. On the other hand, we could see the complete variance when looking at the practical designers. Examples of this variance can be seen through all four basic patterns.

Question - H.-P. Prüfer
It is still remarkable, that the "ideal" pattern could not be found in the groups surveyed, at least in the group you have presented. And that results in an interesting view on teaching.

Answer - H. Birkhofer
On this, I would like to note that in teaching one should certainly first deal with the standard procedure, as a „basic must", one might say. The specific design, depending on the task, product, person, and enterprise, is then undertaken in practical work so-to-say as a „parade performance".

Question - A. Albers
I believe, Mr. Birkhofer and Mr. Lindemann, you certainly will agree, this theory does not equal pattern E. We do not teach this way any more today.

Answer - U. Lindemann
I am convinced that we must seriously bring into reflection the relation of the situation. The personal position of the individual also belongs to the situation.

Technical Biology and Bionics - Design Strategies From the Nature

Werner Nachtigall

Technical Biology and Bionics

Saarland University, Germany

THE MEANING OF TECHNICAL BIOLOGY AND BIONCIS

At a congress in Dayton, Ohio in the year 1960, Major J. E. Steele of the American Airforce introduced the expression „bionics" and defined it as: nature's discoveries should be taken on by Man's technology. This definition has since been used frequently [*e.g. Géra-68*], but in this connection I have three major remarks to make.

Firstly: One cannot use a „natural construction" if one does not know it, and therefore, one must begin by studying it. This first step towards analysing nature and including physical-technical know-how, is what I call „Technical Biology". This forms a basis for the second step which is to search through nature for technologically useful material and which is known as Bionics. Technical Biology and Bionics compliment each other like an image and its reflection (Figure 1A). Bionics cannot do without Technical Biology and Technical Biology on its own will not get beyond the fundamental aspects of research. It leads to an increase in knowledge but not to application. Bionics links the technical world to Nature (Figure 1B).

Since these different aspects have now been integrated to form an overall conception and also to fulfill the demands of our times - knowledge alone being no longer sufficient, one should try to convert it or at least offer it in such a way as to be of use to mankind - a new subject called „Technical Biology and Bionics" has been conceived by the department of biology at the university of Saarbrücken and at the same time a society carrying this name was founded [*Nach-90a*]. The aim of this society is to propagate the goals mentioned above and, perhaps more important, to provide suitable industrial connections.

Secondly: One has to ask oneself if there is anything fundamentally new in Steele's concept. Again and again enquiring minds have chosen to ignore the strict boundaries of scientific disciplines. Leonardo da Vinci for example, he clearly practiced the two facets mentioned above [*Giac-36*].

He observed that a bird's wing, due to a characteristic interlocking of its feathers, is airtight during downstroke, and that the feathers open up to let the wind pass through during upstroke (Figure 2A). Today we would say this is a classical example of Technical Biology. Da Vinci went on to design artificial wings made of reed and impregnated linen with flaps which opened during upstroke to let the wind flow through and closed analogously during downstroke (Figure 2B). The application of an observation made in nature to a technical construction would be known today as Bionics. (The fact that this plan for a wing did not function technically at the time is unimportant: it is one of the first classical examples documenting the direction of thought typical for both these disciplines.)

Thirdly: There is a further point which appears to be very important. To pursue bionics does not mean that one should copy constructions or methods from nature [Nach-91a; Figure 1B]. The term „biomimetics" which is sometimes used instead of bionics, is therefore misleading, since Mimesis (imitation) does definitely not describe the point. „Adaptronics" on the other hand is a term sometimes used for self-repairing materials which means only a facet of bionics. In any case, nature does not offer blueprints for technology and to assume this would be unscientific. Since this point was not given enough thought in the past - the sorrowful experiences of people trying to fly like birds are witness to this - Man's desire „to learn from nature" occasionally fell into miscredit. But modern bionics can by no means do without comparing natural and technical constructions. This very important aspect of Anology Research however [Helm-72] is only a first step towards collecting data and creating ideas. The true meaning of pursuing bionics is, to look through nature for ideas which stimulate independent technological research. „Only" the stimuli can come from nature. It is up to the engineer to use his imagination, and whether he choses to largely follow these stimuli or only use them as a basis for independent research is left entirely up to him. This type of stimulus is, however, very useful and particularly important and nature possesses a wealth of them [Nach-90a].

TECHNICAL BIOLOGY AND BIONICS - CIVIL AND CULTURAL TASKS

The pursuit of technical biology is a civil and cultural task, whereas that of bionics is a simple necessity (Figure 3).

Once Man has reached a certain level of civilisation, he has more to do than just fulfill the tasks necessary for survival. He pursues the arts, he pursues scientific research, and society provides the means which permit artists and scientists to carry out their civil and cultural tasks. To make the unknown known, which, in the field of biology for example, would be to study how nature carries out its constructions, is a task commissioned by civilization and does not require the guise of applied orientation. The same applies to music and the theatre. But in the course of an ever faster developing technology and the risk of global destruction, fundamental research of this kind cannot be allowed to accumulate solely in the libraries of ivory towers. Society demands, and rightly so, that the results must be sieved for any technical usefulness and that any signs of eventual interest with a view to bionical exploitation should be followed up.

SECTIONS OF BIONICS

I shall now, according to my point of view, divide the field of bionics into twelve sections [Nach-91b]. Each section is described in short.

HISTORY OF BIONICS

Various aspects of biology and technology reflect the progress which has taken place in this field, for example, the development of machines and mechanical elements in the 19th century or man's early attempts to fly like a bird with beating wings.

The example of Leonardo da Vinci's approach to bird wings and man carrying wings, already discussed above, is classical (Figure 2A,B)

STRUCTURAL BIONICS

The elements of biological structures are studied, described and compared. The suitability of particular, even unconventional materials for special purposes is tested. Also unconventional structures such as pneumatic or surface-covering membranes are borrowed from nature and studied with a view to their suitability for large-scale, technical designs. The processes involved in the production of biological forms offer unconventional models for technology. For example lightweight, hemispheric constructions can be formed, analogous to microscopic organisms as diatoms (Figure 4A,B); [Nose-85] which, when covered with a foil, could be used for example to cover swimming pools.

BIONICS IN BUILDING

To build „naturally" means on the one hand that one must go back to using the traditional building materials found in nature (e.g. clay products have much to offer in the way of interesting biological building materials) and on the other hand, by studying biological lightweight constructions, one obtains ideas for temporary, lightweight technical buildings. Ideas may come from rope constructions (spider webs), membrane and shell constructions (biological shells and casings), protective coverings which permit gas exchange (eggshells), multi-storey buildings, integration of disconnected units, variable constructions, materials which are easier to recycle than was previously possible, ideal surface coverings (leaf overlapping) and land utilization (honeycomb principle). The layout of individual accomodation elements within a complex is very important, their alignment to the sun and wind could be analogous to overlapping leaves (Figure 5A) or flower constructions for example.

Such units can be constructed quite individually, but arranged in such a way that they shade one another during the hot summer days and in winter they take as little sunshine as possible away from each other (Figure 5B). Thus ideal use can be made of a limited amount of space and the landscape can be protected from a conglomeration of single houses.

CLIMATIC BIONICS

Passive ventilation, cooling and heating are important aspects. Studying natural constructions and the so-called primitive buildings found in Central America and North Africa can lead to unconventional arrangements and furnishings. As much as 80% of the electrical energy required for cooling in summer and 40-60% of the energy required for heating in winter can be saved by the following: alignment to sun and wind, roof shape, building into the earth, ideal cellar constructions and air conduction via the cool soil to the rooms heated by the sun, ventilation (as in termite hills) per gas exchange through using porous materials. It has been shown that in arid regions the ancient Persian architecture, with the help of domed buildings and wind towers, used similar principles of air conduction. The symbiotic integration of plants into the living quarters can improve the oxygen content of the air and provide nurishment.

CONSTRUCTION BIONICS

Constructional elements and mechanisms taken from the fields of biology and technology are analysed and compared and their co-ordination within a functioning complex is studied. Surprizing correlations are to be found, for example pump constructions (salivary pumps in insects, vetebrate hearts, technical pumps). However, nature has done much more than technology towards developing integral constructions in which the individual elements often have a number of duties to fulfill. At the same time the unconventional characteristics of certain materials such as a variable degree of elasticity also play a role. The integrated co-ordination of individual components is typical for such biological constructions. This „idea" can become very important for technical multipurpose tasks. An example is the salivary pump of bugs, described below.

A classical example of Construction Bionics is Burr-like interlocking. The biological principle of statistical interlocking as in burdock seeds and fluke heads [*Nach-74b*] (Figure 6A) was already put to use thirty years ago in the construction of displacable compartments in camera bags (publicity slogan: „culled from the burdock"). In the meantime, starting with a simple principle, many types of velcro tapes and fasteners (Figure 6B) have been developed, for example the "Velcro", patented 1951 by Georges de Mestral [*Nach-87*] replacing for example the laces of sport shoes.

This category of bionics may receive a great deal of stimulation from biological microsystems. As an example the sucking mechanism of a bug's proboscis is described (Figure 7). The salivary pump is part of fluid system. In the case of blood-sucking bugs it is particularly necessary to prevent blood from clotting since it could block the delicate sucking tube. The saliva contains an anti-coagulant. When the lower valve of the gullet closes, the upper one opens and saliva is mixed with food as it enters. The high level of integration of this biological system is remarkable. Unlike a technical pump, the individual elements of this biological pump are functionally totally connected with one another.

BIONICS OF MOVEMENT

Running, swimming and flying are the main forms of animal locomotion. The interactions between locomotory organs and their surrounding medium, which are of interest in fluid mechanics, are found in the range of small and medium Reynolds numbers (micro-organisms, insects) and also in the range of very high Reynolds numbers, similar to those of aeroplanes (whales). Questions concerning the stream-lining of mobile bodies, the driving mechanisms of locomotory organs and their fluidic efficiency are of great importance. Also questions on functional morphological structuring can be of interest. The surface roughness of a bird's wing for example, which is due to the natural roughness of its feathers, creates, under certain conditions, positive boundary layer effects. If one for example covers the wings and bodies of aircraft with „artifical shark skins"(Figure 8A), a reduction of up to 8 % can be achieved in oil tunnel experiments (Figure 8B), leading to 2 to 3% in fuel consumption - already a significant arithmetical difference.

UTILITY BIONICS

This discipline is associated with structural and constructional bionics and deals with the development of utilities modelled on examples found in nature. One finds a great variety of stimulating possibilities, especially in the range of pumping and material-handling technology, hydraulics and pneumatics. So for example the tail fin beat of fishes (Figure 9A) was studied to construct a fin pump (Figure 9B). This device pumps semi-liquid mixtures through pipelines with less susceptibility to disturbances.

ANTHROPOBIONICS

Problems concerning the interactions of Man and machine as well as the numerous possibilities of applied robotics belong to this theme. The lay-out of cockpits in modern civil aeroplanes which are easy to operate and which are adapted to the sensory habits of human beings, the search for ideal bicycle configurations which permit man to ride with higher muscle effectivity than he can produce by running (!) are two examples. An unconventional solution to problems such as the control system of claw arms which are met with in robotics, may be reached for example by way of comparative studies of the leg movements of invertebrates (crabs, insects). The vertebrate muscle (Figure 10A) can lead to unconventional, smoothly working extremities (Figure 10B).

Sensor bionics
Questions on the monitoring of physical and chemical stimuli, location and orientation in the environment belong to this field. The problem of monitoring chemical sustances in the human body (keyword: diabetes) or in large technical converters (keyword: biotechnology) is becoming more and more important. Sensors in nature, which are adapted to all kinds of imaginable chemical and physical stimuli are being analysed now to a much greater extent than they were a few years ago for any possible means of application. A classical analogue is the ultrasonic orientation system of the bat (Figure 11A) and a technical ultrasonic distance finder (Figure 11B).

Neuro-bionics
The use of intelligent electronics has brought rapid progress into the analysis of data and handling of information. In particular the evolution in parallel computers and the processing of „neural integrated circuits" have profited from neuro-biology and bio-cybernetics. Since rapid progress is being made in the latter and also in fundamental biological research, an increased interaction between these diciplines can be expected in the future. So for example the automatic focusing of slide projectors reached functional maturity only after the biological principle of lateral inhibition had been taken into account [*Rech-73*].

Operational bionics
Not only can one examine natural constructions for possible technological application, but also the methods used by nature to control its processes and turnover. An important example with a view to future hydrogen technology is photosynthesis. Furthermore, research into the various aspects of ecological turnover may prove to be very profitable in running complex industrial and economical enterprises. And finally the natural methods of (almost) complete recycling and avoidance of waste materials are worth studying in detail for possible application. Especially the methods in photosynthesis or light conversion as for example in bacterial rhodopsin, in the green plant and the retina of vertebrate eyes are being very actively studied at the moment. Should this lead, for example, to a relatively simple application of energy obtained from the sun by means of photoreceptive molecules embedded in artificial membranes to free hydrogen from water - as photosynthesis does at the thylacoid membrans by a two step redox process, to accumulate hydrogen inside the thylacoid "pockets" (Figure 12) - a most important step would have been taken.

Evolution bionics
Evolution technology and strategy try to make technical use of natural methods of evolution. Especially then, when the mathematical formulation of complex systems and processes is not

sufficiently advanced to permit arithmetical simulations to be made, the experimental method of trial and error is an interesting alternative. This is chiefly due to the work done by Rechenberg and his research team in Berlin [Summary: *Rech-73*].

A good example is optimizing a two-phase ultrasonic jet. The production of electricity according to the principles of magneto-hydrodynamics in satellites was dependent upon a 2-phase ultrasonic nozzle being constructed. Starting with a conventional Venturi jet (Figure 13A) Schwefel [*Schw-68*] found a new nozzle form (Figure 13C) with a 40% higher degree of efficiency, by randomly combining sections of the former (Figure 13B). The irregular form of this nozzle (now more or less theoretically understood) was unimaginable at the time and could not be estimated mathematically. Random changes (analogous to mutation) and new combinations (analogous to recombination) give rise to forms which have been tested for efficiency; worthwhile results were used as a basis for further changes, everything else was rejected (analogous to selection).

OUTLOOK

The different aspects and examples presented above should show that the search through nature for technical applications has a bright future. Bionic work has, in some aspects, already flourished and has, as a matter of course, been integrated into industrial research. In other disciplines the bionic approach is still tentative, or due to indisputable technological success it appears to be unnecessary and finally there are technical fields in which bionic work is quite useless.

In view of the great diversity of forward-looking functional possiblities, can one assume that the links between nature and technology will differ from those of the past?

So far, the relationship between nature and technology has shown itself as depicted in Figure 14A, each following its own way of life, with very few links. What can be achieved by the points of view briefly introduced here?

In future, technical biology and bionics will play an essential role as combining elements (Figure 14B).

The explosive development in technology will bring a series of stimulating ideas with it. So Technical Biology will enable one to obtain a better understanding of the processes of life and of living things in general than was possible up till now.

Furthermore, one will no longer be able to ignore the large number of successful constructions and processes of the living world. Bionics will, of course, neither replace nor supress the traditional engineering methods used in technical constructions lege artis. On the contrary, the latter is and will remain the basis for all technical development; anything else would be charlatanism. Bionics seeks to stimulate the engineer in his search for technical perfection to cast an eye on nature. He will be astonished to find that for a large number of problems, an even larger number of analogous answers have already been found, any one of which may, without any obligation, be adopted.

Bionics can, however, also be regarded as a „constructive basic attitude", as for instance in the sense of „ecological designing" for which a recent demand has been made to have it added to the curriculum of technical colleges. There is no doubt that bionics is „the" ideal creative training for technologists. Pitting one's own imagination against that of Creation [*Nach-74a*] is extremely instructive. Bionics is not only a „hard" practice-orientated discipline, but also a creative philosophy.

Finally: Bionics is not a universal remedy, especially then when it is used uncritically. However, since an enormously versatile world of mechanisms and processes is available and no less real than our technical world, it would be unwise not incorporate it into our battle for technical progress. Nature's strategies can be summarised as „biostrategy" [*Nach-83*]. The studying and application of biostrategy can bring not only selective progress and interesting reforms with it, but also essential breakthroughs which may influence and even change our entire technology.

REFERENCES

Affe-73 Affeld, K.; Hertel, H.: Pumpe zum Fördern von Flüssigkeiten mittels schwingender Flächen. Offenlegungsschrift 17o3294. Patentanmeldung Deutsches Patentamt zur Patentanmeldung H 58-654-Ic/59 e, 1973.

Bech-85 Bechert, D. W.; Reif, W. - E.: On the drag reduction of shark skin. AIAA-85-0546 report, AIAA Shear flow control Conference, March 12-14, Boulder/Colorado 1985.

Coin-87 Coineau, Y.; Kresling, B.: Les inventions de la nature et la bionique. Hachette, Paris 1987.

Géra-68 Gérardin, L.: La bionique. Hachette, Paris 1968.

Giac-36 Giacomelli, R.: Gli scritti Leonardo da Vinci swing volo. Bardi, Roma 1936.

Grif-58 Griffin, D. R. V.: Listening in the dark. Yale Univ. press, New Haven 1958.

Helm-72 Helmcke, J. G.: Ein Beispiel für die praktische Anwendung der Analogieforschung. Mitt. d. Inst. f. leichte Flächentragwerke der Universität Stuttgart (IL) 4, 6-15, 1972.

Lebe-83 Lebedew, J. S.: Architektur und Bionik. Verlag für Bauwesen. Berlin 1983.

Nach-74a Nachtigall, W.: Phantasie der Schöpfung. Faszinierende Entdeckungen der Biologie und Biotechnik. Hoffmann und Campe, Hamburg 1974.

Nach-74b Nachtigall, W.: Biological mechanisms of attachment. Springer, Berlin, Heidelberg, New York 1974.

Nach-77 Nachtigall, W.: Funktionen des Lebens. Physiologie und Bioenergetik von Mensch, Tier und Pflanze. Hoffmann und Campe, Hamburg 1977.

Nach-83 Nachtigall, W.: Biostrategie. Eine Überlebenschance für unsere Zivilisation. Hoffmann und Campe, Hamburg 1983.

Nach-86 Nachtigall, W.: Konstruktionen. Biologie und Technik. VDI-Verlag, Düsseldorf 1986.

Nach-87 Nachtigall, W.: La nature réinventée.Plon. Paris 1987.

Nach-90a Nachtigall, W.: Umdrucksammlung zur Vorlesung Technische Biologie und Bionik, Universität des Saarlandes 1990.

Nach-90b Nachtigall, W. (ed) (since 1990): Rundschreiben der Gesellschaft für Technische Biologie und Bionik. Zoologisches Institut, Universität, 6600 Saarbrücken.

Nach-91a Nachtigall, W.: Lassen sich Biologie und Technik überhaupt vergleichen? In: Rundschreiben der Gesellschaft für Technische Biologie und Bionik, 3 (Zool. Inst. Univ. d. Saarl., Saarbrücken, Febr. 1991), p.2.

Nach-91b Nachtigall, W.: Teilgebiete der Bionik. In: Rundschreiben der Gesellschaft für Technische Biologie und Bionik, 2 (Zool. Inst. Univ. d. Saarl., Saarbrücken, Okt. 1990), 1991 p.2.

Nose-85 Noser, T.: Natur als Baumeister. Architektur der Diatomeen. Modelle der Formbildung - Kräfte und Prozesse - Bauprinzip Pneu. Publ. Hochschule der Künste, Berlin 1985.

Rech-73 Rechenberg, I.: Evolutionsstrategie. Optimierung technischer Systeme nach Prinzipien der biologischen Evolution. Frommann, Stuttgart 1973.

Schw-68 Schwefel, H. P.: Experimentelle Optimierung einer Zweiphasen-Düse. Bericht 35 des AEG Forschungsinstituts Berlin zum Projekt MHD-Staustahlrohr 1968.

Webe-29 Weber, H.: Zur vergleichenden Physiologie der Saugorgane der Hemipteren, mit besonderer Berücksichtigung der Pflanzenläuse. Z.vergl.Physiol. 8, 145-186, 1929.

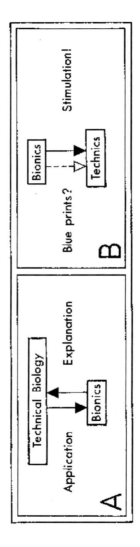

Figure 1 A,B Connecting Technical Biology to Bionics (A) and Bionics to Technics (B).

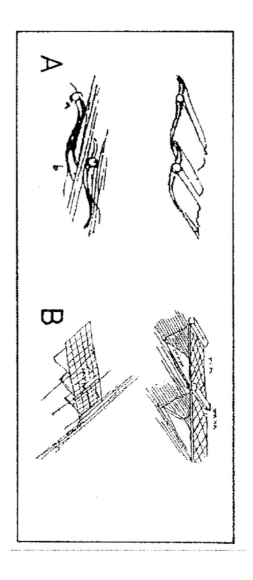

Figure 2 A,B Leonardo da Vinci's drawings. A Bird wing (→ Technical Biology). B Artifical wing (→ Bionics). Comp. the text.

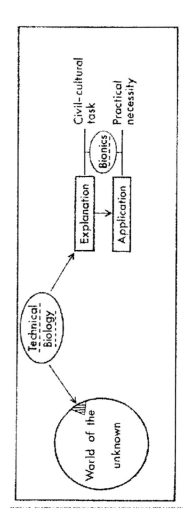

Figure 3   Linking disciplines to the natural world.

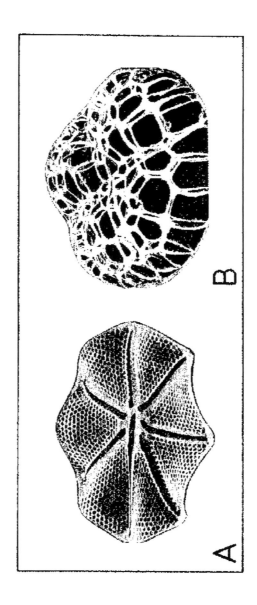

Figure 4   A,B Diatoms as models for technical framework-shells. A Asteromphalus heptactis 1000 : 1. B Technical construction. After [Nose-85].

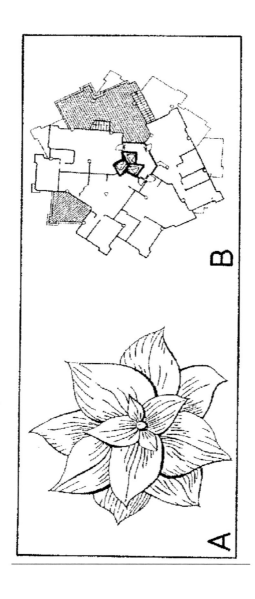

Figure 5   *A,B Rosette plants as models for housing complexes. A Plantago spec. B Portoghesi's and Gighiotto's 13-storey housing complex. After [Lebe-83].*

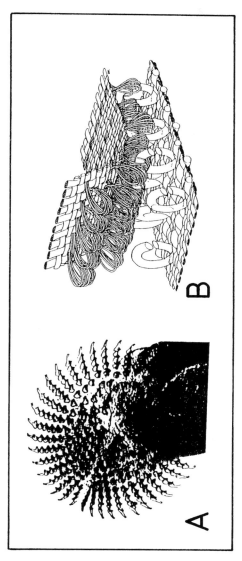

Figure 6   A,B Burr-like interlocking A Head attachment of the monogenean fluke *Acanthocotyle (after Traité de Zoologie)*. B Georges de Mestral's "Velcro" *(drawing B.Kresling)*.

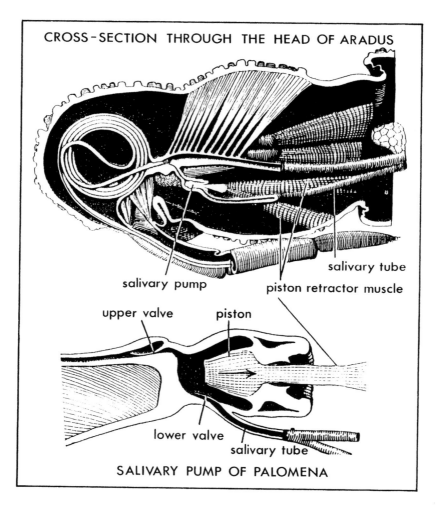

Figure 7   Salivary pump of bugs. After [Webe-29] from [Nach-74a].

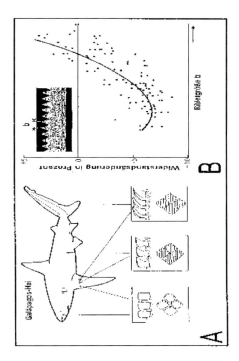

Figure 8   A,B Example of shark scales and drag effect of a shark skin-like fluted foil. A Sculptured scale surface of a shark B Drag reducing effect. After [Bech-85].

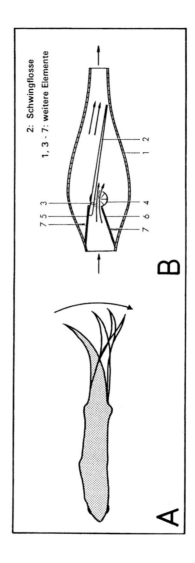

Figure 9 A,B Fish fin as a model for a pump. A Sideward movement of a trout's fin during acceleration. B Fin pump for semi-liquid mixtures. A After research group Nachtigall, Saarbrücken (unpubl.). B After [Affe-73].

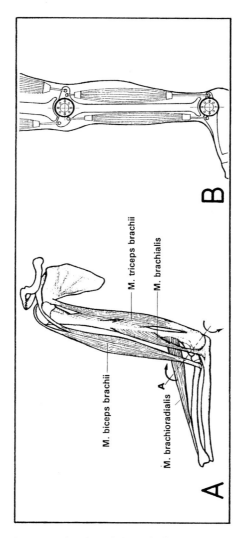

Figure 10 A,B Biological and technical "muscles". A Human arm with abductor and adductor muscles. B Monédi's pneumatical "Myonen". B After [Coin-87].

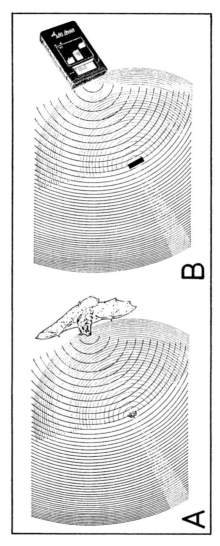

Figure 11 A,B Biological and technical sonar. A Bat principle. B Technical range finder. A After [Nach-74a]; from [Grif-58]. B After Conrad cataloque (1992).

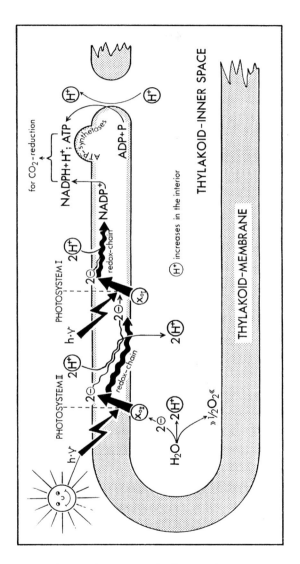

Figure 12   Principle of photosynthesis. Events leading to H^+- concentration on one side of a membrane. After [Nach-77]

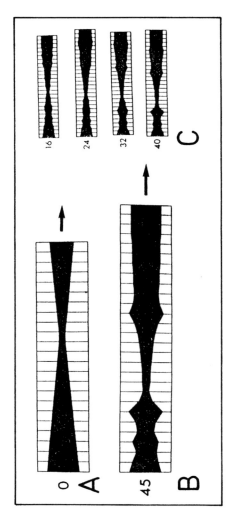

Figure 13 A-C Evolution of a two-phase ultrasonic nozzle. A Conventional Venturi nozzle. B Schwefel's optimized nozzle. C Randomly combined sections.. After [Rech-73].

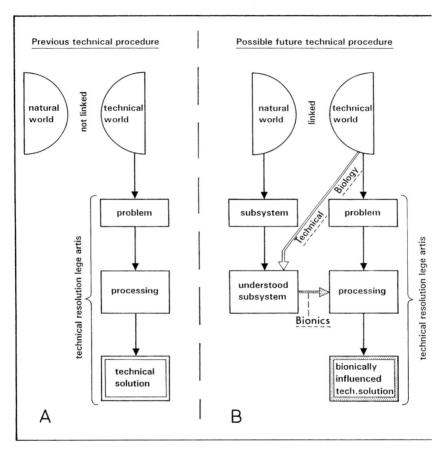

Figure 14 A,B Connections between the worlds of technology and biology as they have been so far (A) and as they could be in the future (B).

Discussion

Question - K. Ehrlenspiel

We are striving to understand the processes and the question is what is nature's process? How does nature arrive at its inventions? Is this simply a Darwinian trial and error procedure or - as is being discussed - is there possibly a certain further intelligence behind this?

Answer - W. Nachtigall

The suggestion that an independent form of intelligence is at work, does not lead to a satisfactory scientific explanation. We can only describe what can be described with scientific vocabulary. In our case, for example, we have the terms mutation, selection, recombination; in addition we have the population theory. If, one day, we are no longer able to describe and explain our observations, then we must enlarge this instrumentarium, but always with scientifically comprehensible terms. That is also the case with Darwin's theory of evolution.

Question - K. Ehrlenspiel

What do you mean when you say "population theory"?

Answer - W. Nachtigall

A reproduction community of individuals belonging to the same species is known as a population. Evolution does not only occur on the level of genomes, but also on a population level. If one takes into account the population aspect, the speed of evolution - in nature as well as in the technical abstraction of evolutionary strategies (Rechenberg 1992) - is dramatically increased.

Psychological contributions to and demands on a General Theory of Design

Winfried Hacker

Chair of Cognitive and Motivational Psychology

Technical University of Dresden, Germany

Keywords: Design thinking ; Design problem-solving; Psychology of engineering design

ABSTRACT

A General Theory of Design first of all should fit the characteristics of human design problem solving. Following empirical including experimental results in engineering these are: a) Decisive impact of the ill-defined early phases, b) Human design as a breadth-first procedure, c) Opportunistic solution procedure, d) Mental constraints of design problem solving, e) Design problem solving as cooperation of mental and „materialized" processing. Furthermore a General Theory of Design should consider the characteristics of expert's design problem solving (vs. novice and low-performance designers' one): Highly and less experienced and efficient engineering designers significantly will differ with respect to their design strategies - not to their intelligence or other personality traits. Expert designers show a) a refined early clarification of the problem; b) a broader analysis of solution principles before working out a specific one; and c) a more sophisticated evaluation of the solution procedure as well as of intermediate results.. Finally a General Theory of Design - from a users' / humans' point of view - should contribute to assist the human problem solving procedures instead of loading designers' limited mental capacity by additional requirements if this General Theory of Design should be applicable as a tool for designers. Following evaluation studies of domain specific (engineering) design methodologies a couple of requirements must be met: a) A helpful design theory/methodology will support human design problem solving only if it will not force a strictly systematic procedure following theoretically optimal phases and steps. Rather it should offer optional, heuristic guidelines which are compatible with an opportunistic processing, b) A helpful design theory/methodology should contribute to an efficient type of allocation of functions between designers and computers, c) A helpful design theory/methodology should consider and guide the interaction between the internal, mental and external (sketching, modelling, prototyping) parts of human design problem solving and, thus, for example assist the implementations of a continuous information flow from the qualitative physical early low cost rapid prototyping to the later, digital high tech prototyping of elaborated product versions, d) A helpful design theory/methodology should contribute to the tricky problem of teaching design methodology: In as far as a design theory will not easily become mentally automated (routinized) it will impair instead of improve the design problem solving procedure and its result.

CHARACTERISTICS OF HUMAN DESIGN THINKING - THE BASIC CONTRIBUTION OF COGNITIVE PSYCHOLOGY TO A GENERAL THEORY OF DESIGN

Design is a human activity. In spite of the computerization of the algorithmic parts of the design process, setting of the design goals and evaluation of design results at least will remain with the designers. Thus, a Theory of Design first of all should fit the characteristics of human design thinking. Although design thinking refers to several topics, especially engineering design, software design, and design of technological procedures, a couple of common characteristics emerge.

Since the research of our Dresden group restricts to engineering design, this contribution will concentrate on arguments based on this area.

Following Steuer [*Steu-68, p. 14*] engineering design means „thinking ahead completely and creatively in order to create a technical object fitting with the requirements of the historically determined level of technological trends and the development of all documents necessary for its physical implementation. It consists of designing and shaping." Following this definition, design thinking is a domain-specific version of the most complicated type of thinking and problem solving: It does not mean reasoning on an already given issue, but thinking ahead by developing the mental representation of a future object which is just not existing so far. Not enough, this thinking ahead should be creative since the future object should have useful new parameters. The whole process of designing a new artefact is organized - as any complex human activity - simultaneously in a sequential and hierarchical manner.

From the *sequential point of view* the design process consists of a couple of overlapping and iteratively processed phases. These are
- the identification and clarification of the problem
- the identification of the necessary functions, the search of rough solution principles and the decomposition of the problem into modules
- the embodiment or shaping of the key modules, and
- working out final details.

Along with this sequentialization the complex problem is decomposed into smaller modules, the sub-problems or subgoals, which again will be processed sequentially following the phases just mentioned. Several nested levels of subproblems given, the hierarchical aspect of the organization of complex design processes turns out.

This sequential and hierarchical - or sequentializing and decomposing - organization of human design problem solving shows some characteristics which determine the qualitiy of the results. In engineering design these are:

a) The decisive impact of the ill-defined early phases. Designers decide on the key characteristics of the later products including their costs in the early design phases, i.e. task clarification and design of the solution principle. Thus the crucial point is that just the least observable and least formalizable early phases will have the highest impact on the innovative features of the product and its manufacturing costs. As is stressed by Ehrlenspiel [*Ehrl-95a*] this impact is highest but worst controllable in the early design phases. They depend on design problem solving which is a non-algorithmic, heuristic type of thinking with substantial uncertainty of its outcome. One the one hand, this uncertainty offers a challenge to proceed to innovative solutions; on the other one, however, it also means a risk of detours and failures. Thus the design of actually new solutions is problem-solving with an internal contradiction between challenging options and stressing risks. At this point the expectation arises that a Design Theory should outline possibilities which support especially the heuristic problem solving of the early design phases. We will come back to this issue.

b) Human design problem solving as a task-dependent breadth-first top-down procedure.
Given the designer identified the problem and developed at least one solution principle as the „solution space" in mind, he must process on the conceptual and/or imaginational representations of this solution space: He/she should deduce and compare the various features of the solution principles, recombine patterns of features, introduce new features or reject others. Here the procedures of humans strictly differ from those of computers. If the task is new and, thus, from a psychological point of view a problem exists, humans will not start from the detail and stepwise combine the complete solution. Rather they start from a vague, ill-defined and qualitative impression of a hypothetical solution, proceed breadth-first and top-down, decomposing it in a more or less unreflected manner, and will finally work out and compute the details. To quote a description of an engineering design expert, the designer frequently will start „not from individual elements of a solution, but from a holistic solution. It will contain already all details which will be unfolded within the process of design. Thus designing is a stepwise decomposition and clarification of partial functions... It is a holistic-analytical procedure... The designer creatively works on two levels: On the one hand he more or less unconsciously applies abstract, complete solution structures, on the other hand he consciously sketches specific combinations of elements..." *[Bach-73, p.4]*

However, the type of procedure may differ with the type of tasks and the expertise of the designers: With routine tasks and with highly experienced designers instead of the breadth-first and top-down procedure another one should prevail. Designers in that case are used to start from some suitable former detailed solution stored in mind, will skip the development of solution principles and directly work on the embodiment of solution-determining detail.

Consequently, a General Theory of Design may not be general in the sense of an unified theory. Rather it should reflect the inevitable multitude of procedures resulting from different task requirements (which are to be categorized) on the one hand and the designers' expertise on the other.

Concerning the task-dependency of design thinking one should have in mind: In order to offer task-specific guidelines assisting design thinking first of all a taxonomy of design thinking or problems is to be developed. A general classification of engineering design problems should focus on the objective features of the problem structure. So far there are a couple of differing suggested taxonomies [e.g. *Müll-90, Lang-91*]. Based on these ideas we propose the following criteria and operationalizations [*Rück-97*]:

- Multiplicity of goals (number of goals, number of conflicting goals and strength of conflicts between the goals)
- Complexity (number of subfunctions, number of interaction between subfunctions, strength of the interactions)
- Transparency (information about the situation, information about the way of solution, information about the solution aimed at)
- Degrees of freedom (number of possible variants of solution, number of possible ways to solution)
- Dynamics (changeability of the situation, predictability of the consequences of the problem solver's actions and decisions, influences from outside)
- Necessary knowledge (specific knowledge about facts, specific knowledge about methods, general problem solving strategies)

These criteria are differently weighted by different design experts [*Rück-97*]. They attribute high importance to the multiplicity of goals and problem complexity, medium importance to the transparency, the degrees of freedom and necessary knowledge, and low importance to the dynamics. There is satisfying reliability and concordance of the averaged expert rating judgements of nearly all criteria. Thus, based on these requirement assessment an overall problem difficulty may be calculated by a weighted sum. Moreover, the individual requirement dimensions may be represented as a profile of task difficulty

c) Design consists of systematic and „opportunistic" parts. As was just mentioned, not every design task is a totally new problem which may be accomplished only by problem solving. Engineering design is to a substantial extent knowledge-based. The most frequent type of mental requirements here integrates the reactivation of already known solutions and the development of actually new ones. Thus, design becomes a procedure including systematically planned as well as opportunistic parts: It does not proceed totally systematically straight ahead toward the solution without any backward and forward jumps between phases and different subgoals. Rather the procedure will start at some „islands" of knowledge about parts of the solution and try to combine, rearrange and complete them by problem solving. For that reason, backward jumps to already processed steps and forward ones to future steps of processing are inevitable. A systematic goal-directed procedure will become possible not before the problem analysis will have found a suitable decomposition of the total problem into subproblems. Thus, empirical results contradict the assumption of a systematic problem solving procedure, as was assumed, for instance, with the waterfall metaphor of software design [*Guin-98*]. Instead of the systematic proceeding from task clarification, searching for solution principles, and detailed development of a solution another procedure prevails. It is based on the detection of solution-relevant knowledge while working. This opportunistic procedure shows the following characteristics:

- It switches seemingly irregularly between mental and external (sketching, modelling) processing of subproblems.
- It accomplishes no systematic and complete clarification and decomposition of the task before starting the design. Thus, its initial problem definition is global and incomplete.
- It switches unsystematically between subproblems and levels of problem description, caused by the experience-based identification of applicable knowledge.
- The transferred knowledge leads to redefinitions of the problem and to changes of developed action programmes (plans).

Again an important consequence arises: A General Theory of Design may support engineering design only if it will not force a strictly systematic procedure following theoretically optimal phases. Rather it should offer optional guidelines which in particular will not restrict the naturally opportunistic initial steps of design problem finding and solving.

d) Mental constraints of design problem solving. Problem solving requires the development of correct and complete mental representations as the materials of internal processing operations. The mental representations are the search space; no solution will be better than the search space will allow: the final solution will depend on whether the relevant solution possibilities were represented or stored mentally at all. Consequently, the designer ideally should imagine all relevant solutions through an exhaustive combination of their features in order to select an optimal feature pattern from all possible ones. Just this, however, is often impossible because of the limitations of human mental capacity, often labelled „capacity of consciousness" or „working memory". Thus designers are forced to be limited on a section of features only. However, they will be restricted not because they could not imagine further feature patterns, but since they cannot hold all the patterns within consciousness. Nevertheless, the selection should contain just the optimal solutions of design. Thus, the critical issue turns out to be that designers should have in mind the optimal feature patterns of a required solution which, however, in most cases they cannot know, since they should have been selected from the total of possible combinations which they cannot hold in consciousness [*Hoov-91*].

Obviously the initial bottle-neck of creative design is not the variety of designer's imaginations but the limitation of working memory. Actually we could show experimentally with simulated computer-based design tasks: the lower the working memory span of a subject, the more superfluous steps are made, the more inefficient solution paths are tested repeatedly and the higher the time consumption will be [*Sach-95*].

Thus, a possibility to support design problem solving is to relieve working memory load. This is one of the characteristics of successful procedures: Experts tend to fix their ideas by writing and sketching. A labelled sketch as an external store offers a way out of the conflict discussed. A General Theory of Design should offer suitable ways to overcome mental constraints of design thinking.

e) Design thinking as cooperation of mental (internal) and materialized (external) processing. Artefact design is an interaction of mental and materialized processing, especially sketching, rough modelling with things and materials-at-hand. The main reason for this interaction is the mentioned limited mental processing capacity which is relieved by external storage. A further reason seems to be a promotion of thinking by sensorimotor processes. Metaphorically spoken humans are sometimes thinking with their hands. Often solution ideas are developed gradually within the process of sketching by hand. Solution ideas are fixed as an abstracting sketch of the solution principle and are specified afterwards within feedback-circles of thinking, sketching and critical inspection of the refined sketch. Thus, quoting Görners observations „... the sketch not only offers the results of the designer's thinking processes but mainly serves him as a working means" [*Görn-94, p.240*]. In an interview study, Görner analyzed 74 experienced designers. One of the question was whether they use - in order to develop a solution principle - „mainly thinking about it or mainly sketching". The results were that 69.3 % of the designers used „mainly sketching", 3.8% „mainly thinking" and 26.9% reported a balanced combination of thinking and sketching. A further question dealt with the reasons for sketching. The respondents here could apply more than one category: 61.5 % of them applied sketches in order to clarify an idea, 44.4% stressed the role of a sketch as memory aid, and 30.8% stressed the role of communication.

To overcome the shortcomings of an interview study the following experiment was carried out: Two matched groups of students by means of a computer program dealt with three design tasks of different complexity. One group was asked to sketch their ideas first of all, the other was not. With the most complex task, a significant difference emerged. In case of previous sketching, the number of operations in task accomplishment was lower, less operations were to be corrected, the perceived difficulty of the task was lower, and - in spite of the additional sketching phase - the total working time was lower, each in comparison with the group without sketching.

Apart from sketching, some more „materializing" kinds of operations may support design problem solving: Even in the area of CAD and digital prototyping, a lot of simple concrete prototypes and models are applied. Designers use pencils, rods, paper-, cardboard-, wire or plasticine models or elements of building kits besides the more sophisticated computation-, CAD- or stereolithographic models. An interview study with some further 40 experienced engineering designers revealed the following main functions of the concrete simple prototypes. They serve as:

- a means of problem analysis,
- an aid to create new ideas,
- an instrument of evaluation especially of functions,
- a means of storage relieving working memory,
- a communication support.

Consequently, as was shown by Ehrlenspiel, Bernard & Günther [*Ehrl-95b*], the simple concrete prototypes are highly useful, especially within the early phases of design problem finding and solving. A theory of design should explain this interaction and offer proposals for an useful application.

CHARACTERISTICS OF THE PROCEDURES OF EXPERT DESIGNERS - CONTRIBUTIONS OF OCCUPATIONAL AND OF COGNITIVE PSYCHOLOGY TO A GENERAL THEORY OF DESIGN

In as far as a Theory of Design will be thought to contribute to an improvement of the design procedure and its results it should consider the generalizable characteristics of the procedures of experts in comparison with those of low-and-average-performance designers. Highly vs. less experienced and efficient designers of comparable education significantly differ in their strategies, not in their test-intelligence or other job-relevant personality traits.

Three main characteristics of successful procedures should be stressed in detail:

a) Clarification of the task. The procedures of successful experts differ from those of less successful novices with respect to a refined clarification of the task, a subtle definition of goals concerning subfunctions and thorough extraction of those aspects which are important for product functions, not only for its shape. The characteristics of a successful design procedure are:
- a thorough clarification of the task
- concentration on components determining the functions of the objects
- an exhaustive analysis of the main functions
- the weighting integration of the goal characteristics to be implemented
- the written or sketched fixation of solution principles of the goal
- a goal-oriented but nevertheless flexible processing.

b) The type of search for a solution principle. Going into some detail again, the characteristics of a successful procedure [*Dyll-91*, *Ehrl-93*, *Ehrl-95b*, *Fric-93*, *Görn-94*, Pahl & Fricke, 1993] are:
- A procedure generating new solutions instead of only modifying already given ones.
- Conceiving a few different principle solutions before starting to work out one selected solution principle. This so-called divergent search in successful design must be followed by
- A convergent restriction of the variety of solution principles (the „search space") on as less variants - which are worked out - as possible.
- A well-scheduled phase of detailed design, corresponding with the proposals of design methodologies: it develops and follows logically ordered subgoals [*Carr-80*]. However, in spite of this goal-directed processing, the expert procedures are not obstinate straight-ahead ones. There are anticipations of later steps and regressions on already processed ones, although less frequently than within less successful procedures.

c) The evaluation of the solution procedures. In successful procedures more feedback is processed which guides further operations. In detail:
- The process and its outcome are more often evaluated in terms of goal characteristics, and
- The evaluation is made for both levels, the global solution principle and the detailed solution steps worked out.

ASSISTING DESIGN THINKING - CONTRIBUTIONS OF COGNITIVE AND OF OCCUPATIONAL PSYCHOLOGY TO A SUPPORTIVE THEORY OF DESIGN

Finally a Theory of Design should offer principles outlining how to assist design thinking and to develop applicable tools for designers, for example with computer-assisted-design (CAD). For these reasons the Theory of Design should meet a couple of requirements:

a) Task-specific and heuristic guidelines instead of general and algorithmic rule systems. Design theory will support the design thinking of practitioneers only if it will not force a task-independent and a strictly systematic procedure, favouring theoretically optimal processes [*Günt-98*]. Thus, two requirements must be met. The first is a dialectic one: Although the theory should generalize, it may support design thinking only if it offers task-specific suggestions. As was outlined, there is no „the one" design procedure, but a couple of procedures depending on the matrix „task categories × expertise of designers". The second requirement to be met by the theory is, that human design thinking rarely follows theoretically optimal steps. Thus, a Theory of Design should offer optimal heuristic guidelines - not algorithmic rules -, which are compatible with the mentioned opportunistic type of processing.

b) Allocation of functions between designer and computer. The most important issue of optimization of modern technologies is the allocation of functions between humans and machines; the most tricky issue is an allocation that promotes creative thinking. Consequently, a Theory of Design should cope with the impact of different types of allocation on creative design. Designers in any case use tools (for instance pencils) and machines (computers and their software) in different phases of design. However, tools and computers widely differ as to the allocation of functions. From a psychological point of view there are a couple of momentous consequences:

- Tools lengthen and strengthen human functions. Machines replace them.
- Tools hardly occupy human mental capacity only for handling them. Machines occupy parts of mental capacity for handling them, which, consequently, is not available for the task accomplishment itself. The less userfriendly hard- and software are, the more mental capacity will be occupied only in order to handle the computer.
- Tools do not require a recoding („translation") of information: A line will be as long as we draw it with our pencil. In comparison machines need coding: The length of a line depends on the mouse-ratio. However, any recoding or translation of information again will occupy mental capacity which is not available for the problem actually to be solved.

Consequently the question is, which means should support the externalized „thinking with the hand": The unselfishly assisting tool or the computer, which partially replaces human functions but occupies the limited mental capacity only for handling? Viebahn [*Vieb-95, p.53*] recently offered an answer: „A drawing program cannot cope with incomplete descriptions of objects; therefore the parameters and geometric characteristics must strictly be fed in. This takes a lot of time, and input errors by men are inevitable. A lot of inventions therefore tried to improve data input: Mouses, touchscreens, trackballs, light pens or digitalizing devices... However, after ten years of progress it still holds: the designer must spend a reasonable part of his mental capacity for the device in order to store an idea or sketch in the computer. Just this was not the desire! People who should work creatively do not want to be distracted by subordinate details just at their creative moments. Therefore a designer should not work at the computer until the idea of an object to be drawn has already developed. He will reach this stage of solution through sketching by hand since he knows: While sketching ideas will flow into his hand nearly automatically and much more conveniently than while drawing with the mouse."

A supportive Theory of Design should offer hints for task-specific and phase-dependent optimal kinds of allocation.

c) Integration of internal and external (materialized) parts of the design process. A Design Theory should consider and guide the outlined interaction of the mental (internal) and the physical (sketching, modelling or prototyping) parts of human design thinking. Following our results with new design problems, the efficiency of the early phases of engineering design with a marginal application of CAD at the best highly depends on the interaction of roughly scribbled sketches and thinking. A Design Theory should explain observations like these. Moreover it should offer suggestions of whether and how to integrate smoothly these sketching-thinking interactions in an unique continous information flow from the early sketches and rough models of materials-at-hand to the later digitalized computations and high-tech prototypes of elaborated product versions.

d) Teaching design and its methodology. If a General Design Theory is helpful, it might be taught or used as a guideline of teaching like existing engineering design methodologies [*Ehrl-95a, Hans-65, Müll-90, Pahl-93*]. Following our results so far [*Beit-97*] teaching systematically a design methodology will enhance students' perceived mental load and impair the quality of their design results at least initially in comparison with matched controls without any training of design methodology or with an optional training, offering a self-determined partial application of some of its aspects. It seems that a design methodology must easily become routinized (mentally automated) in order to actually assist and improve design thinking. A General Theory of Design - in as far as it will include applicable parts or consequences - might have to cope with just this problem, too.

ACKNOWLEDGEMENTS

The relevant research was sponsored by a research grant of the Deutsche Forschungsgemeinschaft (DFG) within the Sonderforschungsbereich 374 „Rapid Prototyping", a grant Ha 2249/2-1 and a grant Kl 848-3-1.

REFERENCES

Bach-73   Bach, K.: Denkvorgänge beim Konstruieren. Konstruktion, 25, (1), 2-5, 1973.

Beit-97   Beitz, W.; Timpe, K. - P.; Hacker, W.; Rückert, C.; Gaedeke, O.; Schroda, F.: Konstruktionsarbeit studentischer ‹bungsgruppen. Empfehlungen für die konstruktionsmethodische Ausbildung an Technischen Universitäten. Schriftenreihe Konstruktionstechnik, 40, 1997.

Carr-80   Carroll, J. M.; Miller, L. A.; Thomas, J. C.; Friedman, H. P.: Aspects of solution structure in design problem solving. American Journal of Psychology, 95, 269-284, 1980.

Dyll-91   Dylla, N. Denk- und Handlungsabläufe beim Konstruieren. Konstruktionstechnik, Vol.5. München Hanser, 1991.

Ehrl-93   Ehrlenspiel, K.: Denkfehler bei der Maschinenkonstruktion: Beispiele, Gründe und Hintergründe. In: Strohschneider, S.; von der Weth, R. (Eds.): Ja, mach nur einen Plan. Bern: Huber, 196-207, 1993.

Ehrl-95a Ehrlenspiel, K.: Integrierte Produktentwicklung. Methoden für Prozeßorganisation, Produkterstellung und Konstruktion. München: Hanser, 1995.

Ehrl-95b Ehrlenspiel, K.; Bernard, R.; Günther, J.:Unterstützung des Konstruktionsprozesses durch Modelle. In: Bericht über das Werkstattgespräch 'Bild und Begriff 3' 1995 in Seußlitz. Dresden: TU-Eigenverlag, 1995.

Fric-93   Fricke, G.: Konstruieren als flexibler Problemlöseprozeß - Empirische Untresuchung über erfolgreiche Strategien und methodische Vorgehensweisen beim Konstruieren. VDI-Forschungsberichte, Reihe 1: Konstruktionstechnik /Maschinenelemente. Düsseldorf : VDI-Verlag 1993.

Görn-94   Görner, R.: Zur psychologischen Analyse von Konstrukteur- und Entwurfstätigkeiten, in: Bergmann, B.; Richter P. (Eds.): Die Handlungsregulationstheorie. Von der Praxis einer Theorie. Göttingen: Hogrefe, 1994.

Günt-98	Günther, J.; Ehrlenspiel, K.: How do designers from practise design, what can design methodology learn from them and how can design methodology support them?. In: Birkhofer, H.; Badke-Shaub, P.; Frankenberger, E. (Eds.): Designers - the key to successful Product Development. Springer : London, 1998.
Guin-98	Guindon, R: The process of knowledge discovery in system design. In: Salvendy G.; Smith, M. J. (Eds.): Designing and Using Computer Interfaces and Knowledge Based Systems, Vol. 2. Amsterdam: Elsevier, 727-734, 1989.
Hans-65	Hansen, F.: Konstruktionssystematik. Berlin: Verlag Technik, 1965.
Heis-96	Heisig, B.: Planen und Selbstregulation. Bern: Lang-Verlag, 1996.
Hoov-91	Hoover, S. P.; Rinderle, J. R.; Finger, S.: Models and abstraction in design. Proceedings ICED 91: International Conference on Engineering Design, Eidgenössische Technische Hochschule Zürich, 46-57, 1991.
Lang-91	Langner, T.: Analyse von Einflußfaktoren beim rechnergestützten Konstruieren. In: Beitz, W. (Ed.): Schriftenreihe Konstruktionstechnik, Bd. 20. Berlin: Eigenverlag TU Berlin, 1991.
Müll-90	Müller, J.: Arbeitsmethoden der Technikwissenschaften. Berlin: Springer, 1990.
Pahl-93	Pahl, G.; Beitz, W.: Konstruktionslehre. Handbuch für Studium und Praxis. 3rd ed. Berlin: Springer, 1993.
Sach-95	Sachse, P.; Hacker, W.: Early low-cost prototyping: Zur Funktion von Modellen im konstruktiven Entwicklungsprozeß. Forschungsberichte, Vol. 19, Technische Universität Dresden: Institut für Allgemeine Psychologie und Methoden der Psychologie. Dresden: Technische Universität: Eigenverlag, 1995.
Rück-97	Rückert, C.; Schroda, F.; Gaedeke, O.: Wirksamkeit und Erlernbarkeit der Konstruktionsmethodik. Konstruktion, 49, 5, 26-31, 1997.
Steu-68	Steuer, K.: Theorie des Konstruierens in der Ingenieurausbildung. Leipzig : Fachbuchverlag, 1968.
Vieb-95	Viebahn, U.: Technisch zeichnen kann jeder, Die Zeit, 31.3., 53, 1995.

Discussion

Question - H. Lenk

From a philosophical perspective I would like to agree with most what you have said. I also conducted a series of lectures on creativity in psychology and philosophy (in preparation for the press). I have however a question: you mentioned the situations and the specifics of tasks and something like that. Are there not also very different intellectual solutions by different characters who would approach creative tasks from different vantage-points? Einstein thought essentially in scenic Gedanken-experiments. Poincaré would rely on a period of incubation; he wrote at length about that. Then you have the general systems theorists or systems analysts who try to differentially conceive most clearly of the problem situation and the general questions. Then you have the people using the so called morphological box, ending at a complete combination of different individual features cross-queued with one another. Arthur Koestler emphasized bisociations, others relied on general associations, divergent thinking etc. Would it not be advisable to develop a design theory specifically according to different intellectual approaches like the mentioned ones - or at least to take into consideration different of these mentioned approaches?

In addition you said man would jump back and forth. Can one say indeed, man in general?

Answer - W. Hacker

Subject of a general theory of design will not be whatever a person, but one who decided to become a designer by some (self-)selection procedure. Thus a specific group results. Moreover education and training again will reduce this group and determine the levels of intelligence and the ways of thinking of its members. Consequently, designers who should be the subject of a design theory might represent a highly specific population. Furthermore on a first step only highly general regularities of human problem solving - or more specificly of design problem-solving - should be integrated in a general theory of design. Only on a second step - if inevitable at all - the variety of individual ressources and strategies might be considered, too.

Question - J. Gero

Do you believe, there is a difference between design as problem-solving and just problem-solving?

Answer - W. Hacker

Yes, there are some strict differences between the general problem-solving and design problem-solving. One of the most important differences in this direction is that design problem-solving develops a future object with unknown characteristics and not to handle a given set of information. From a psychological point of view this is an important difference, since we have to anticipate what might be the outcome. On the contrary solving a puzzle as just problem-solving does not show unknown future consequences. Therefore the knowledge base developed for instance for the Tower of London or the Tower of Hanoi and other puzzles, doesnÌt fit this type of design problem-solving.

Epistemological Remarks Concerning the Concepts „Theory" and „Theoretical Concepts"

Hans Lenk

Faculty for Humanities and Social Sciences

University of Karlsruhe, Germany

Keywords: theory, theoretical concepts, axiomatic approach, statement view of theories, non-statement („structuralist") view of theories, model theoretic approach, action and knowledge, schemata, schema interpretationism, methodology and epistemology of designing, design theory, truth and fitting („satisficing"), theory dynamics, action theoretical approach, constructive realism, technologistic new experimentalism.

ABSTRACT

Current interpretations of the concept of theory by philosophers of science of different orientations are sketched out, in particular the traditional statement view (theories as systems of statements) and the formal axiomatic interpretation (theories as calculi) as well as the historical or historicist interpretation of theories as series of successive research programs or paradigms etc. (theory dynamics). In addition, the model-theoretical non-statement view (the socalled structuralist interpretation) of theories is roughly delineated and discussed as regards the status and interpretation of theoretical (T-theoretical) concepts. Finally, technological and action-theoretical approaches defining the model-theoretic view are presented as particularly apt for a methodology of technology.

A number of recommendations are developed from this, particularly the idea of satisficing in construction theory and the methodological idea of optimum fit instead of approximation to truth.

The interplay between action, experiment and gaining knowledge is highlighted from a methodological and schema-interpretationist perspective. This approach seems to be conducive to a methodology and epistemology of engineering sciences, in particular construction theory and design theory.

INTRODUCTION

Let me start with some rather ironic aphorisms: „A theory is nothing but the skin of truth - propped and stuffed" (Henry Ward Beecher). (We may infer from this „that theory has something to do with truth and should be related to practical applications" - as stated in the second bon-mot): „Theory should never forget that it is nothing but applied practice" - that is the late Gabriel Laub's rather pointed ironic version of the proverbial insight nothing would be

more practical than a good theory. Or does A. J. Carter's sarcastic remark apply: truth would be a conjecture or a summar with academic education or qualification?
In some sense these aphoristic insights make sense and all of them lead to different conceptions of theories.

SUBSTANTIVE AND OPERATIVE (INSTRUMENTAL) THEORIES

It seems to me to be most important to distinguish between *substantive* and *operative* or instrumental technological theories [*after Bung-67, II, 122, I, 502ff*]: the first ones are empirically contentful (notably to be found in real science) empirical and informative and testable, whereas the latter ones comprise instrumental or operative concepts of methods representing consistent methodologies. They represent structural concepts as, e.g., mathematical „theories" which may be applied in a kind of instrumental manner to specific realms, e.g. of empirical or ideal or even normative provenance. While the substantive theories imply empirical content and substantial information necessary for scientific or other substantial explanations and predictions, operative or instrumental theories do not comprise the respective explanatory nomological hypotheses, but are used for the presicion and explanation or prediction in a secondary manner by being applied for practical reasons to given theoretical or structural presuppositions. They are but instrumental calculi or formal and procedural structures usually represented by mathematical operations or systems of operators. Operative theories are formal constructions of rules, calculi, structural interconnections which in their instrumental character are like language forms or mathematical formulae which do not represent substantial nomological hypotheses with empirical content although they are indispensable for formulating laws in precise scientific form. They have to be supplemented by nomological hypothesis with content: a calculus per se has no empirical content. Only by attributing to it observation statements, measurement procedures and testing techniques as well as application procedures will a formal operative theory together with these sustantive ingredients have an empirical content and become so to speak a substantial or substantive total (of an operative *and* substantive theory). Substantive and operative theories are also models of quite different kinds and manners of acceptance. In the narrower sense one should only speak of genuine theories with respect to substantive theories. The socalled operative theories are but instrumental models or frameworks of models and procedures or procedural rules. (Many of the traditional mathematical instruments and „theories" are of this operative character. This also applies to most of the axiomatised theories as far as structural framework and the formalised basic axioms are concerned. Also formal system-theories, axiom systems of logics and mathematics are to be conceived of in this sense as instrumental or operative theories.)

The distinction between operative and substantive theories is rather frequently not seen and acknowledged clear enough in axiomatically oriented natural science or in practice-oriented variants of applied disciplines (like technology and organization or management sciences).

The first practical maxime to be taken into account from a design-theorist would derive therefrom: it is necessary clearly to distinguish between substantive (substantial or contentful, e.g. empirically contentful) theories and operative or instrumental „theories" (which would rather figure as instrumental concepts of methodologies and methods). This is advisable even though the latter ones somehow „penetrate" further ones being their „logical framework". Only the first ones (or in connection with the later ones) can claim for (empirical or contentful) truth. The latter ones are rather like instruments selected and taken from a cupboard of more or less useful instruments or appliances for structuring and formalizing or handling statements with substantial content.

FROM TRUTH TOWARDS FITTING/FEASIBILITY FUNCTION

From what was said we can gain another insight which is known from traditional conceptions of philosophy of science: scientific theories have to be testable and truth oriented, i. e. they have to be tested by confirmation or corroboration (total verification of universal hypotheses is not possible!) or by falsification (empirical rejection) taking into account criteria of truth. Theories not oriented at factual truth are not theories of empirical science. (Logical or mathematical truth is another concept: such formal theories are really proved as valid (formally true or logically true) by proving their consistency, i. e. non-availability of contradictions. For empirically contentful theories non-contradiction is but a necessary, though not a sufficient condition for truth.)

As a special note for design-theorists let me state: so far as „theory" does not deal with truth or empirical or factual truth claims, but, e.g., with „goodness", it cannot be an *empirical* scientific or science-based theory, but rather consist of methodical or even meta-methodical or methodological principles, i. e. general structural maximes of actions. These would structure processes and steps of actions, at times even functioning normatively, but they would not be related to just nomological truths in the empirical sense. Axiomatic design theories are actually rather normative instructions or maximes. More precisely, they are generalised and abstract models of interpretive constructs of procedural provenance which are to be seen or applied under specific normative criteria like „goodness", „optimisation", „adequate problem solution", „satisfaction of functional requirements" as well as „simplicity", „feasibility", „cost" etc.

WHAT ARE SCIENTIFIC THEORIES?

In current approaches to the philosophy of science we can distinguish different conceptions of theories and their interpretations which are all derived from the traditional conception of analytic philosophers of science (like the positivists of the Vienna Circle, but also from Popper's critical rationalism).

TRADITIONAL APPROACHES

1. The traditional interpretation conceives of theories as logically interconnected systems of statements or sentences consisting of universal law like hypotheses which are as perfect or completed theories represented in an axiomatised form. Their reference to reality or empiricality is realised by attributing to the concepts of the theories several (or as many as possible) corresponding observation statements observings special corresponding rules or attribution rules, connecting the theoretical concepts with observation predicates of a lower level which can be satisfied or fulfilled by observation or measurement. This traditional two-level conception of empirical scientific theories is however today assessed very critically - or at least controversely - and has to be supplemented or substituted by several other conceptions. (It is common knowledge by now that even observation statements or statements about measurements and their results are theoretically „infested", „theory-laden", „theory-impregnated" - be it by a theory of measurement or so.)

2. Intimately connected with a traditional conception of theories is the primacy of the paragon function of mathematical theories in the form of axiomatisation (as it was devised by mathematical logical formalism (Hilbert) at the beginning of this century. Accordingly, mathematical theories are such axiom systems which would define their uninterpreted theoretical concepts „implicitly" by their respective structural axioms themselves which are

able completely to cover a realm to be comprised by the theory (e.g. classical algebra). (Through the well-known meta-mathematical results by Gödel und Church these expectations have been proven to be impossible: highly complex axiomatic systems[1] comprise a kind of incompleteness and the provability of inconsistency is restricted: in addition, there are no absolute guarantees and mechanisms or algorithms of proofs even for some logical or mathematical valid theorems within the formal system.) Attempts of total axiomatisation of theories have therefore proved definitely unsuccessful. The axiomatic method can generally only be conceived of as a didactically useful construction. As such it is certainly still important and applied everywhere especially with respect to purely formal, e.g. mathematical, theories. But axiomatisation and „scientificness" cannot be identified, not even in the purely formal sense, not to speak of the real empirical scientific sense. Axiomatic systems are but aids - maybe practically necessary ones -, but they are not the real empirical scientific theories themselves.

This leads to the conclusion: axiomatisation cannot be the *only* criteria of being or going scientific, as necessary in practise of teaching science axiomatic systems might be. In particular, formal axiomatisation by itself, that is the mathematical structure of the calculus, of a theory cannot be mistaken as „the theory" in its scientific as well as in its practical sense. (This should be taken into consideration by construction engineers and theorists of design oriented at formal axiomatisation.)

DYNAMICS OF THEORIES AND THEIR SUCCESSION

3. Since 40 years or so historisation of the conception of science is characteristic for the philosophy of science. After Feyerabend and Kuhn philosophers of science started to take into account factual history of science. The real development of the sciences is largely shaped by the groups of the respective relevant scientists and their values and changes of conceptions: the so called „normal science" is distinguished from „revolutionary" phases (Kuhn). The latter ones can lead towards a new „paradigm" within a special realm of research: the consequence is a certain (or even absolute) incommensurability of paradigms or the respective theoretical fundamental interpretations (e.g. classical mechanics vs. quantum mechanics). From this a third conception of theories as group-supported basic paradigms developed according to which the „normal" internal development of a scientific paradigm is so to speak science-politically supported and authoritatively got or put through, only to be later on superseded by a new paradigm. Comparisons of theories, competing results and data as well as scientific progress in such an extreme sociological or even sociologistical version of the approach (like, e.g., according to the „strong" conception of the so called Edinburgh School of the Sociology and History of Science) lead to a theoretically insoluble dilemma between a rational assessment of theoretical progress and but irrational scientific change.

4. Imre Lakatos (like Feyerabend a creatively deviant disciple of Popper) developed a concept of the historical succession of theories consisting of „research programmes" which supersede one another according to criteria. This conception replaces Popper's original naive falsificationism by a rather „sophisticated" one. Theories are still systems of universal hypotheses - („laws") in their logical - deductive interconnection, but they are only *indirectly* falsified, in so far as the interplay between the explanatory theory and the theoretically impregnated observation statements will only lead to a statement of inconsistency (e.g. between measurement theory and the explanatory theory). Theories are always to be assessed

---

[1] Being at least of the complexity of number theory and quantified logic (predicate logic) of first order including identity.

within the historical succession of predecessors and successors: succeding theories are „better" than their predecessors, if and only if they have more empirical content, i. e. allow to predict new or novel facts, or contribute to the avoidance of some anomalies of the old theory (although any theory comprises some anomalies still) and/or renders an integrating theory-fusing process of generalisation. Transition to a better theory is called „theoretically progressive" if it allows to predict novel facts which are not yet empirically confirmed. (A change is called „empirically progressive" if some of these predicted novel facts are really confirmed. An example of this would be the transition from Newton's theory of gravitation to Einstein's general theory of relativity due to Eddington's experiments regarding the deviation of light rays close to the sun (1919).)

MODEL-THEORETIC APPROACHES

Non-statement View

5. The so called *non-statement view* which conceives of theories as set-theoretical predicates or model sets was mainly developed by Sneed and Stegmüller - after ideas by Suppes. This approach which is also called a *structuralist view* of theories understands a „theory" as a totality or net of theory elements which are partially ordered[2] by specialising relations or constraints, e.g., by the addition of special laws. (Occasionally and originally the ordered pair of the mathematical structural core K and the set of partial potential models being the intended possible applications - the ordered pair {K, I} - is called a theory - or nowadays a „theory-element". The partial potential models of a theory comprise all the applications conceived of which are not yet described by theoretical functions and observables, namely the models of intended applications.) By definition, the structural core (defined only by mathematical relations) comprises the potential and partial models as well as constraints (i. e. theoretically required interconnections between the partially overlapping partial potential models) and the models themselves (i. e. the factual systems already successfully described by the theory). Empirical statements and hypotheses of a theory are now met in the statement that the respective intended applications of the theory belong to the applications of the net (or the structural core) fulfilling the constraints. The set of partial potential models (i. e. the real systems which are intended for application of the theory without theoretical functions) are supplemented by the respective theoretical functions leading to this set of potential models. To be sure, the adding of theoretical functions and specialisation (by added special laws) has to lead to a partial set of „fulfilled" models (M) such that the whole sequence of theoretical functions has to satisfy the constraints. According to the structuralist non-statement approach a theory consists simply of an ordered pair of a mathematical framework of formulae (structural core) and a set of possible inttended applications and constraints. The possible applications are object systems or real systems which are up for application of the theory and which are usually given by certain paradigmatic initial models usually proposed by the founder of the theory. The theory then is an ordered set - theoretical predicate (after Sneed and Stegmüller). „Formally, a core K may be represented either as a quadrupel $K = <M_p, M_{pp}, M, C>$ or as a quintupel $K = <M_p, M_{pp}, r, M, C>$. Here, M_p, M_{pp}, and M are the sets mentioned ... C is a set of constraints, i. e., a subset *of the power set of M*; and the *restriction function r*: $M_p \to M_{pp}$ transforms an element of M_p, i. e. a potential model, into an element of M_{pp}, i. e., into a partial potential

---

[2] In the sense of a logically mathematical partial ordering relation, i. e. a reflexive, asymmetric, transitive and identitive relation.

model by 'lopping off' all theoretical functions" [*Steg-79, p. 25*]. The total theory might be an expanded core leading to a theory net by adding specialised laws and respective new constraints, restriction functions and new intended models to be inserted in to the set of potential models and fulfilled models. *In short, a theory is therefore a relation predicate defined over the set of potential models of applications.* The predicate „... is a theory" states the existence of a relation between the mathematical structural core and the set of at the time specified set of intended applications of the theory, the set of potential models being expandable: further potential, previously not intended applications can be integrated (e.g., extension of Newtonian classical dynamics to include gravitational systems by adding the law of gravitation or the extension by adding Hook's law of linear restoring force to include harmonic motion) (see Figure 1).

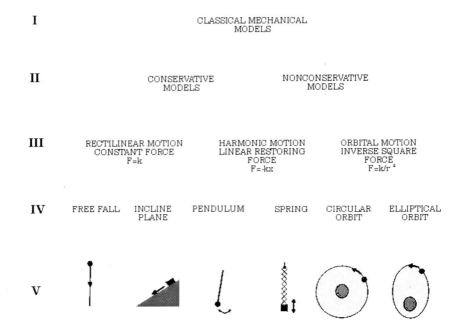

Figure 1   A partial „model map" of classical mechanics exhibing a multiple hierarchy of models [*after Gier-94, p. 288*]

A series of interesting results can be derived from this new conception: it is possible to speak of *one and the same* theory, even if the set of special laws of the theory and the set of intended models is expanded, as long as the structural core (the mathematical basic laws) of the theory are maintained. E.g., Newtonian mechanics of the Newton's first three axioms is specialised by adding special laws like Hook's law or the law of gravitation remaining still the same theory, but differentiated and specified.

T-theoretical Concepts

In addition, the property of being a *theoretical concept* is now relativised to the respective theory and can be governed by a criterium independent of the traditional (principally thwarted) distinction between observable and non-observable magnitudes. Beside the opposition of „theoretical" vs. „observable" (amounting to the distinction of theory language and observation language according to logical positivists like Carnap) now a relativised conseption of „pre-theoretic" concepts and „theoretic" concepts of the respective theory T (more precisely, T-theoretic concepts) is introduced: for instance, the concepts of „mass" and „force" in Newtonian mechanics and theory of gravitation are theoretical concepts, whereas „space" and „time" are not (they might however be theoretical concepts in a more basic physical geometry of Euclidian structure). A T-theoretical concept is distinguished by the fact that it can only be instantiated or attributed *by using the theory itself*, i. e. by measuring the respective variable. Pragmatically, it is possible now to define the concepts of „using a theory" as an action concept which is not only logically characterised, but also impregnated by pragmatic, action theoretical and agent related aspects [*cf. e.g. Steg-80*].

Expansions of structural cores

Progress in handling theories by extending the structural core via adding further intended models can be realised in various manners: by extending the set of intended applications, i. e. partial potential models, by refinement of the net *or* the structural core as well as by ramification with respect to mutually exclusive intended potential models characterised by non-compatible special laws.

The structural approach renders a certain autonomy of the structural core as regards remifications and expansions as well as a certain immunity with regard to relativised falsifications. From T-theoriticity (the relativity with respect to the theory T) or the theoretical concepts and due to the theory - impregnatedness of observations a theory so to speak defines its own facts to be governed and covered by it leading to a holism of falsifications and confirmations in such a manner that only a theory net in total is confronted with experience or experiment and confirmed, corroborated or falsified by it.

Practical Modelling and Axiomatisation

Again a piece of advice for design theorists: this structuralistic approach of the theoretical structure of cores seems to be at first view rather abstract and far from practical considerations. However, it is a considerable advantage of it that theories are conceived of as a set of mathematical structural frameworks and models and that one and the same structural core (the same basic axioms) can be related to new intended models (or sets of these, respectively), without the scientist being urged to talk of a new theory. In addition - which is of importance for design theory - one may transfer the constitutive connection between mathematical structural cores and intended applications (partial potential models) also to contexts in which one cannot speak of the truth of a substantive theory: for instance, this would apply to general principles of technological fields to be described, characterised and covered under the point of view of a criterium of „fitting" or „goodness" or functional requirements etc., and the respective specialisations by adding further more specific „laws" or „rules of thumb".

In fact, axiomatic design theory after Suh and Chang may be presented as a set of precise fundamental principles („axioms") characterising a structural core supplemented by plurifunctional requirements (FR) instead of „truth". The structural view of theories would favour this combination of an axiomatic approach regarding the mathematical structural core on the one hand and the practical selection of models from an open and extendable set of partial potential models. (This approach can also be rendered compatible with historical developments of principles and their specialisations.) Generally speaking the structural view of theories would support the interpretation of design theorists that „theories" (more exactly: methodologies or general design principles and their structural interconnections) can be conceived of as precise (or to be made precise) structural cores including pluralfunctional requirements of satisfaction and different models etc. Axiomatic design theory is rather a normative generalised methodology respecting specific functional requirements and design parameters (PR) which are characterised by multiple and changing optimum or satisfactory model solutions than a substantive scientific theory with empirical content in the narrower sense. Such pragmatic approaches of the relation between structural cores and models seem to be especially suited for technological „methodologies" in the sense of a set of systematic interconnected methods and the intimate relationship between technological rules on the one hand and scientific substantive basic theories on the other. (Unfortunately, engineering sciences have traditionally not developed a general methodology in the philosophy of science sense with criteria for theoretical justification, although the development of a „general technology" („Allgemeine Technologie") and its methodology has been long since asked for by some philosophers of technology [*Ropo-91, Lenk-73*].

Technology-oriented and Action-theoretic Approaches

6. The model-theoretic approach can be extended in order to better cover technological models. Ronald Giere, in his book *Explaining Science* [*Gier-88*], developed an interpretation which combined the structural view with technological approaches [*similarly also Hack-83*]. Giere [*Gier-88, p. 85f*] understands „theory as comprising two elements: (1) a population of models, and (2) various hypotheses linking those models with systems in the real world. Thus, what one finds in the textbooks is not literally the theory itself, but statements defining the models that are part of the theory". Important is the relation of similarity between the models and their real systems to which the models apply as cases of application: „The links between models and the real world ... are nothing like correspondence rules linking terms with things or terms with other terms. Rather, they are again relations of similarity between a whole model and some real system. A real system is *identified* as being similar to one of the models." [*Gier-88, p. 86*] Giere conceives of theories as „related families of models", or still better, „as a family of families of models" [*Gier-88, pp. 82, 91*] which can be related to the world by similarity and fitting of the models with the respective real system as mentioned above. Theories in that sense are no linguistic entities or just frameworks of formulae but heterogeneous sets consisting of abstract constructs, the theoretical models, and linguistic entities like hypotheses about the fitting character of these models and their similarity with reality susceptible to grading and perspectives: „A real system is *identified* as being similar to one of the models. The *interpretation* of terms used to define the models does not appear in the picture; neither do the defining linguistic entities, such as equations" [*Gier-88, p. 86*]. „When approaching a theory, look first for the models and then (only, H. L.) for the hypotheses employing those models. Don't look for general principles, axioms, or the like." [*Gier-88, p. 89*]. That will mean with respect to the above mentioned example that one has to look for the models and their similarity with real systems regarding pre-theoretically characterised representations of the model of the planetary system or the model of the earth - moon system as original or primary models of Newtonian theory of gravitation. With respect to relating and combining theoretical models with real systems to be covered *technology* now plays a decisive role. Like Hacking [*Hack-83*]

Giere's constructive realism sees a proof of reality in the successfully managed technologies in handling entities, (*e.g.* electrons) which earlier had the status of a theoretical entity, if they are applied to cover and characterise new models or other theoretical entities. (If we routinely use nowadays electron rays in accelerators or in electronic microscopes successfully to resolve other scientific tasks, we understand in this technological sense the theoretically postulated electrons which were earlier mere theoretical entities now as scientific - technological *real* entities.) In so far as electrons and protons are manipulated and applied in big technology measurement instruments and appliances to probe and prove the structure of other elementary particles like gluons, quarks etc., these electrons and protons now are (handled and considered as) „real" indeed [*Hack-83*]: „Thus, some of what we learn today becomes embodied in the research tools of tomorrow" [*Gier-88, p. 140*].

Fitting as Satisficing

Giere states that scientists are constructive realists who relate models by technological applications and interventions to reality. Thereby they are led to an experimentalistic realistic conception of improved models in the sense of a relativised (not necessarily optimally) fitting or suiting after H. A. Simon called „satisficing": one does not maximize the model adjustment, but would optimize it in the sense of rendering a satisfying result, the relative optimisation of goal attainment so that a satisfying result for the experimental and somehow functionally restricted but relatively best accordance of models with reality would occur. Scientists are according to Giere optimizers or „satisficers", but no absolute maximizers with regard to a correspondance of their models to reality.

Advice for design theorists: this view of technological theoretical models as relatively well-fitting, not necessarily the absolutely best models with regard to problem solutions in the point of view of fulfilling functional requirements (FR) would render the technological („technologistic") and constructive, realistic model as developed by Giere and Hacking specially suited for technological designs, for the methodology of construction etc. Giere states explicitly [*Gier-88, p. 137*]: „The main connector between our evolved cognitive capacities and the micro world of nuclear physics is *technology*." Certainly this is valid more generally: „The development of science depends at least as much on new machines as it does on new ideas": „It is technology that provides the connection between our evolved sensory capacities and the world of science" [*Gier-88, p. 138*]. Technology is conceived of as the outstanding example of ways of experimenting, development of instruments that are man-made extensions of the senses and all our capacities including the instruments and procedures as well as operations as metaphorically speaking, concretised „*embodied knowledge*", i. e. „embodied in the technology used and in performing experiments" [*Gier-88, p. 140*]. We have to add technological developments of new instruments, procedures, technological systems and the interface between these latter ones and axiom systems, if not today social and socio-technological systems in socio-technical networks. Therefore, we have to add the essential reference to structured actions and action systems in their worldly and social embeddings. Even experimenting and the devising and developing of theories as also the utilizing of theories is a sort of acting, a kind of action.

ACTION, EXPERIMENTING, KNOWLEDGE

In the treaties of epistemology and the philosophy of scientific theories as well as knowledge in general the present author (1998) shows the insoluable interconnection between knowledge, experimenting and action. In so far Giere's combination of scientific models and their relation to real systems by technology and technological manipulation and intermediate operators like measurement instruments and machines has to be extended by an action-theoretical interpretation. This would be of utmost interest for design theorists, since the design of hardware structures and real systems and the respective structuring manipulations as well as the design of software in software models would be covered by such a view. The pragmatical model theoretic approach with respect to technical instruments has to be supplemented or expanded by an action-theoretical perspective which is notably suitable for design theories.

Whereas traditional conceptions of theories in science emphasized too strongly the rather pure theoretical elements and hypotheses as linguistic entities, traditional axiomatic as well as the pure structuralistic conception suffered from too formalistic an orientation conceiving of theories and their structures exclusively as mathematical structures or even subtheoretical complex predicates.

The philosophy of the sometimes so called new experimentalism, of a pragmatic technology-oriented provenance and the action theoretic perspective may avoid these over-simplifications and at the same time refine the structural interconnection between idealized cognitive models or intended partial potential models of theories by stressing technological realisations and materialisations and action-theoretical as well as operational sequences (as to be found in the design of operations and experiments etc.).

It is in such way that also the design theorist may relate his methodology or metamethodical conception of operative principles of design to the fulfillment of functional requirements and the optimisation or satisficing of plurifunctional conditions - somehow independent of absolute truth claims (in the sense of substantive empirical true or truth-like theories). That seems to be typical for design tasks. In this sense the normative component is taken into consideration and accounted for within this kind of general methodology or doctrine of principles determined by plurifunctional requirements (*e.g.* in Suh's sense). This amounts basically to an extension of the old objective of traditional construction systematics of the fiftees and the sixtees [*as of e.g. Hans-65, Müll-67*] now only in a model theoretically refined way. One could even go further and attach a theory of creatively ordered and structured activities in general as insinuated for instance by social psychology and philosophy of creativity (Lenk, in preparation).

SCHEMA-THEORETIC AND INTERPRETATIONIST PERSPECTIVE

Pragmatic philosophy of science has much to learn from technological and action-theoretical approaches. Similarly, also the methodology of engineering sciences or the general technology still to be developed may gain much from insights of methodological character and the differentiated consideration of refined developments and novel insights of philosophy of science and general methodology including theories of action. Epistemologically speaking these methodological approaches can be embedded in a general „theory" or methodology of schema interpretation [*Lenk-93a, Lenk-95*]. This approach understands any grasping of real systems as methodologically and epistemologically dependent on specific perspectives, teleofunctional requirements, theoretical constructions and approaches as well as practical action routines or social conventions and institutions, respectively. A new unity of the sciences and technologies as well as the understanding and knowledge of the world and of the manipulations as well as interpretations of reality by acting and utilising theoretical interpretational as well as experimental and action-practical models gains relief on a metatheoretical level characterised by general methodological requirements of any active processes of „grasping" external or mental entities as well as ideal structures. Acting, grasping and realizing as well as forming and shaping and rendering normative structure is in this sense to be understood and analyzed from the point of view of schematisations (the development and activation as well as utilisation of schemas of partially hereditary, mostly however learned provenance) and by interpretations, manners and patterns of „grasping" in the rather active or passive sense under specific perspectives.

The main question - generally and ironically speaking - seems not to be „Dasein oder Design" („being or design") - to put it in a pun not translatable to English! -, but that any „grasp" of structured real „beings" (real entities and their relational patterns) is always also dependent on a kind of „design" of in part primary biological or hereditary, in part conventional and higher-level interpretations (cf. Lenk, Figure 2).

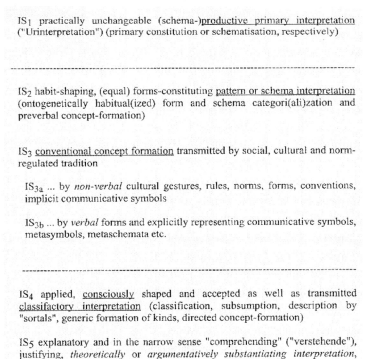

IS_1 practically unchangeable (schema-)productive primary interpretation ("Urinterpretation") (primary constitution or schematisation, respectively)

IS_2 habit-shaping, (equal) forms-constituting pattern or schema interpretation (ontogenetically habitual(ized) form and schema categori(ali)zation and preverbal concept-formation)

IS_3 conventional concept formation transmitted by social, cultural and norm-regulated tradition

IS_{3a} ... by *non-verbal* cultural gestures, rules, norms, forms, conventions, implicit communicative symbols

IS_{3b} ... by *verbal* forms and explicitly representing communicative symbols, metasymbols, metaschemata etc.

IS_4 applied, consciously shaped and accepted as well as transmitted classifactory interpretation (classification, subsumption, description by "sortals", generic formation of kinds, directed concept-formation)

IS_5 explanatory and in the narrow sense "comprehending" ("verstehende"), justifying, *theoretically* or *argumentatively substantiating interpretation*, *justificatory interpretation*

IS_6 epistemological (methodological) metainterpretation (plus meta-meta-interpretation etc.) of methods, results, instruments, conception of establishing and analysing interpretative constructs themselves

Figure 2   Diagram of the levels of Schema Interpretation

To come back to Beecher's statement on truth at the beginning of this paper: a theory is *more* than „the skin of truth - propped and stuffed": beyond fitting and satisfying as well as fulfilling some correspondence conditions for reality-based restrictions it is a complex interpretational construct (out of and consisting in many schemata and interpretations mutually subordinated or coordinated), a way of dealing in procedures and operations, actions and techniques in more or less practical concrete steps and measures with external world sections, world factors (entities and processes), potential models, real systems as well as meaningful semantic entities (mental entities, ideal constructs etc.).

Theories and generally speaking, methodical and methodological concepts as well as normative structures of actions and procedures would guide us in the form of interpretations and schematisations. Methodological interpretative constructivism (schema interpretationism) as developed by the present author is a higher-level methodological as well as epistemological conception comprising the special cases of scientific theories, technological designing, blueprinting and prototyping as well as all other kinds of procedures to structure action and thought patterns. Interpretations are constructions in a narrower or a wider sense [*cf. Lenk-95*]. Theories are interpretative constructs which as substantive theories hold a claim for truth, at least for approximate truth or for empirical validity - or as operative or instrumental theories for methodical or methodological validity. Norms and values are interpretive constructs, too. So, there are normative interpretative constructs in any process of design and decision making. Designing and problem solving in design are completely of an interpretatory character. In short: there is nothing more practical than clear-sighted interpretations.

REFERENCES

Bung-67   Bunge, M.: *Scientific Research*, I, II. Berlin - Heidelberg - New York 1967: Springer.
Carn-66   Carnap, R.: *Philosophical Foundations of Physics*. New York 1966.
Chan-90   Chang, S.: A Scientific Approach Towards Developing An Engineering Design Theory. In: *International Conference on Engineering Design ICED 1990* (Proceedings). Dubrovnik 1990.
Gier-85   Giere, R. N.: Constructive Realism. In: Churchland, D. M. - Hooker, C. A. (Hg.): *Images of Science*. Chicago 1985, 75-98: Chicago University Press.
Gier-88   Giere, R. N.: *Explaining Science*: The Cognitive Approach. Chicago - London 1988: Chicago University Press.
Gier-94   Giere, R. N.: The Cognitive Structure of Scientific Theories. *Philosophy of Science* 61 (1994), 276-296.
Hack-83   Hacking, I.: *Representing and Intervening*. Cambridge - New York 1983: Cambridge University Press.
Hans-65   Hansen, F.: *Konstruktionssystematik*. Berlin 1965: VEB Verlag Technik.
Kess-54   Kesselring, F.: *Technische Kompositionslehre*. Berlin - Heidelberg - G"ttingen 1954: Springer.
Kuhn-70   Kuhn, T.: *The Structure of Scientific Revolutions* (1962, with Postscript 1969). Chicago 1970^2 : Chicago University Press.
Laka-68   Lakatos, I.: Criticism and the Methodology of Scientific Research Programs. In: *Proceedings of the Aristotelian Society* 69 (1968) 149ff.).
Laka-70   Lakatos, I.; Musgrave, A. (Eds.): *Criticism and the Growth of Knowledge*. Cambridge 1970: Cambridge University Press.
Lenk-72   Lenk, H.: *Philosophie im technologischen Zeitalter*. Stuttgart 1971, 1972^2: Kohlhammer.
Lenk-73   Lenk, H.; Moser, S. (Eds.): *Techne - Technik - Technologie*. Pullach bei München 1973: Dokumentation Saur.
Lenk-75   Lenk, H.: *Pragmatische Philosophie*. Hamburg 1975: Hoffmann & Campe.
Lenk-82   Lenk, H.: *Zur Sozialphilosophie der Technik*. Frankfurt a. M. 1982: Suhrkamp.
Lenk-86   Lenk, H.: *Zwischen Wissenschaftstheorie und Sozialwissenschaft*. Frankfurt a. M. 1986: Suhrkamp.
Lenk-89   Lenk, H.; Ropohl, G. (Eds.): *Technik und Ethik*. Stuttgart 1987, 1989^2: Reclam.
Lenk-91   Lenk, H.: Zu einem methodologischen Interpretationskonstruktionismus. In: *Zeitschrift für allgemeine Wissenschaftstheorie (Journal for General Philosophy of Science)* 22 (1991) 283-302.
Lenk-92   Lenk, H.; Maring, M. (Eds.): *Wirtschaft und Ethik*. Stuttgart 1992: Reclam.
Lenk-93a Lenk, H.: *Interpretationskonstrukte*. Frankfurt a. M. 1993: Suhrkamp.
Lenk-93b Lenk, H.: *Interpretation und Realität*. Frankfurt a. M. 1993a: Suhrkamp.
Lenk-94   Lenk, H.: *Macht und Machbarkeit der Technik*. Stuttgart 1994: Reclam.

Lenk-95	Lenk, H.: *Schemaspiele. Über Schemainterpretationen und Interpretationskonstrukte.* Frankfurt a. M. 1995: Suhrkamp.
Lenk-98	Lenk, H.: Kreative Aufstiege (in preparation for press).
Müll-67	Müller, J.: Probleme einer Konstruktionswissenschaft. In: *Maschinenbautechnik* 16 (1967), 338ff, 394ff, 454ff.
Nage-61	Nagel, E.: *The Structure of Science.* New York 1961: Hartcourt.
Popp-35	Popper, K. R.: *Logik der Forschung* (1935) (*Logic of Scientific Discovery*). Tübingen 1966: Mohr & Siebeck.
Rapp-78	Rapp, F.: *Analytische Technikphilosophie.* Freiburg - München 1978: Alber.
Ropo-79	Ropohl, G.: *Eine Systemtheorie der Technik.* München 1979: Hauser.
Ropo-91	Ropohl, G.: *Technologische Aufklärung.* Frankfurt a. M. 1991: Suhrkamp.
Snee-76	Sneed, J. D.: *The Logical Structure of Mathematical Physics.* Dordrecht 1971: Reidel
Steg-73	Stegmüller, W.: *Probleme und Resultate der Wissenschaftstheorie und analytischen Philosophie.* Vol. II: *Theorie und Erfahrung*, Part D: Logische Analyse der Struktur ausgereifter physikalischer Theorien. Ein 'Non-Statement-View' von Theorien. Berlin - Heidelberg - New York 1973: Springer.
Steg-76	Stegmüller, W.: *The Structure and Dynamics of Theories.* Berlin-Heidelberg-New York 1976: Springer.
Steg-79	Stegmüller, W.: *The Structuralist View of Theories.* Berlin - Heidelberg - New York 1979: Springer.
Steg-80	Stegmüller, W.: *Neue Wege der Wissenschaftsphilosophie.* Berlin - Heidelberg - New York 1980: Springer.
Suh-90	Suh, N. P.: *Principles of Design.* New York - Oxford 1990: MII.
Supp-72	Suppe, F.: What's Wrong With the Received View on the Structure of Scientific Theories. In: *Philosophy of Science* 39 (1972), 1-19.

Discussion

Question - H.-J. Franke

In a way, you gave us here a new concept for theories. If we now want to design real objects using the natural sciences we need all facts and theories of natural sciences as an additive element. You do not see problems to do this without integration design processes and natural sciences?

Answer - H. Lenk

Indeed, I do not think that the older conceptions, e.g. the statement view and the axiomatic systems approach are to be excluded or to be abolished, to be put ad acta. On the contrary! Set theoretical approaches are, as they are discussed, in their overwhelming majority axiomatised themselves. This could be seen in the example of Newton's dynamics.

However, we cannot maintain anymore the dogmatic claim that we would definitively know what a theory is, e.g. it would be a system of true hypotheses or hypotheses capable of being true and nothing else as for instance Popper would have it who understands theories as falsifiable hypotheses simpliceter. This cannot be maintained anymore in such a simple form.

Particularly, I think that for design theoritians the mentioned methodological extensions with respect to actions, procedures, technical measures, etc. are very important. For *Axiomatic Principles of Design* by Suh will not comprise a scientific theory in the strict and narrower sense, because it does not deal with truth and the possibilty of being true, but with systematisation of normative approaches under particular criteria as, e.g. goodness, fitting, simplicity, feasability, costs, etc. If functional requirements are to be met, this is a normative affair. It is so-to-speak the systematisation of a normative approach which is certainly not a theory in the same sense as for instance quantum theory or classical mechanics. One can use the axiomatic presentations as ways of gaining precision. One would use also calculy, algorithms or mathematical formulations respectively: they are here very useful and helpful and even indispensible. But neither are axioms or axiom systems as such already the (substantive or content-bearing) theory nor can we maintain any quality of this concept of „theory" with the traditional concept of theory, e.g. in theoretical physics. - Beyond that, one has the advantage by such relativised interpretations to better grasp historical development of theories. In addition, in connection with such an approach you can also easier and more effectively take into consideration actions and action instructions. This is certainly a new perspective. And it is in some sense of course not so easy conducted as a mere formel manipulation of an axiom system.

Question - H. Grabowski

One question we already discussed yesterday was: Is there only one theory for the design of artefacts or are there maybe several theories? Is more than one theory possible and useful?

Answer - H. Lenk

I think that systematically to deal with artefacts is a special discipline. Some authors claim it would be a science of its own, e.g. „*The science(s) of the artificial*" (Herbert Simon). This is indeed a very interesting realm of research. However, artefacts are results of actions and procedures of a planful provenience. They are planning constructs so-to-speak and are characterised just by such conditions that other criteria as for instance the classical ones the good old truth (which however nobody knows). Instead, criteria of goodness, fitting, satisfying and mediating different functional requirements, costs, etc. are important. Such perspectives would indeed play a fundamental role and it would be a very important task to analyse the respective combinations of criteria regarding the production of artefacts, to develop guidelines and certain middle standards of assessment pertaining to such a spectrum of criteria.

In addition, it is of course philosophically speaking a very interesting problem scarcely broached thus far that we live in some sense in a world characterised and impregnated largely by artificial artefacts. We are e.g. capable to design and produce some parts of the world as artefacts, for instance in synthetic chemistry (Berthelot and others) since the last century turn or in producing new elements, e.g. in the „Institute of heavy ions research" in Darmstadt, Germany. Man produces by his actions and artificial activities certain new elements of nature which were not to be found otherwise. This applies to technology also generally ever since. This would render a specifically interesting and thus far not enough illuminated realm of contact between the classical conceptions of natural philosophy on the one hand and the philosophy of technology and of construction procedures as well as action theory on the other.

Question - C. Weber

If I get you right you do not see a universal answer to the question if we need one or more theories? The answer rather depends on the circumstances?

Answer - H. Lenk

Of course there are alternative possibilities of approaching this question. In a stricter sense one would say in a way any theory conception, any systematisation what so ever is also a mental artefact. Any theory is constructed, any knowledge is selective. This was highlighted by cognitive psychology since decades. Each perception is constructive, emphasized Ulric Neisser one of the refounders of cognitive psychology. In this sense I would say, that any grasping is characterised by processes of schematasation pertaining to some rather general methodological and methodic insights of rather abstract sort. In this sense we „construct" or „construe" everywhere (although in a wider, not necessarily concious sense): we cannot avoid constructing or construing, we cannot avoid interpreting and we cannot avoid schematising. We cannot *not* construe, interprete or schematise. We are in this sense schematising beings. This insight leads to the rather central philosophical question, what are all these schematasations? What are schemas? Are they but dynamically automated and tuned neuronal assemblies à la Wolf Singer? This seems to be right, at least there is a correlation between the activation of schemas and the reactivations of relatively stabilised neuron assemblies. These are rather thrilling philosophical and epistomological problem fields which I'd extensively discussed in my book about epistomology (*Einführung in die Erkenntnistheorie: Interpretation, Interaktion, Intervention,* Munich, Germany: UTB1998: *Introduction to epistomology: interpretation, interaction, intervention*).

Question - Y. Jin

This will be a short one: Are you saying that we only have descriptive theory and we cannot have a normative theory?

Answer - H. Lenk

No, I'm saying that design-theorititians usually use normative theoretical schemes or whatever, constructions. While the pure physisists might be a kind of purely descriptive theorist, I think every engineering scientist is normative in his approach. He has to make the selections, decisions and such things.

Question - Y. Jin

So the recent Professor Suh's framework is not a theory was the reason behind that.

Answer - H. Lenk

It is not a theory in the sense as traditional physics would have it. It is a *theory of action*, *procedures* and *planning,* etc. [Interruption by **Y. Jin**: But still it is a theory.] It is a *methodology*, I would say, of general systematic approximations and so on, systematic constructions, how to handle design problems. [Interruption by **Y. Jin**: So still you say, it is not a theory.] It is not a theory in the *traditional* sense in which philosophers of science in the beginning of the century until the midth of this century would have understood theory and theoretical physics. It is however a „theory" in a broader sense, qua methodology of systematic actions, kinds of procedures, etc. - and it is not dealing with truth, but with fitting, satisfying, goodness ... [Interruption by **Y. Jin**: But to me, Professor Suh does emphasize that there is a truth of design which is the independence.] Well, in a wider sense, but what is „truth"?

Summary Session 4

H. Birkhofer

Ladies and gentlemen, I had very little preparation time, therefore I ask for special indulgence, especially, concerning Mr. Lenk's lecture. Regarding Bionics: What I especially liked in Mr. Nachtigall's lecture was the interaction between technology and biology. Technical biology seizes methods of engineering, whereas bionics adds biological knowledge to technology. It is also clear that engineers can „only" get inspiration, ideas, and approaches to solutions from bionics. These approaches to solutions must be redesigned and that is not a 1:1 transfer. What especially occurs to me in biology is that we find microstructures there which are very strongly function-integrated and also go very strongly in the direction of one-material-design. Here I see a very distinct analogy to micro-electronics/micro-mechanics/micro-system techniques. On the contribution of Psychology: The core statement here was that designing is a human activity, which does not rule out that certain essential areas could be described by algorithms and worked on with the computer. Mr. Hacker then emphasized that design thinking must be integrated into the theory of design. From my point of view, I would like to emphatically support this. As Mr. Hacker describes human thought during design, it is a parallel, yet simultaneously also a hierarchically structured, sequential procedure. Here different thought process types are intertwined, and I do not know if we have suitable models with which we can also map these thought procedures on the computer. Mr. Hacker then made an important distinction between a general theory and a uniform theory. This general theory is also characterized by the fact that it supports the diversity of tasks and takes the differences of the people working on them into consideration as well. Mr. Hacker then pointed out that one must proceed step-by-step when developing a theory, not immediately strive for this complete theory all at once. One should first try to integrate general human problem-solving styles into the theory, and then consider individual procedures. Mr. Hacker specifically addressed the so-called „opportunistic" procedure, in which everything is not carefully planned out beforehand but which also leaves room for situation- and context-dependent action. We must also note that our mental work capacity fundamentally presents a „bottleneck" in problem solving. Also here, a general design theory can help us fashion tools accordingly so that they may help to eliminate or at least reduce the disadvantages and deficits resulting from this bottleneck. Then the expression „thinking with ones hands" came up, which, from my own experience, I find very good. Through designing, designers develop their thoughts.

In closing Mr. Hacker then offered to support the development of a theory from a psychological point of view through expert investigations, the formulation of heuristic guidelines, and through work on the division of functions between human and computer. These were entirely concrete offers which I think we should accept.

Regarding the contribution of Philosophy: As an engineer I can illustrate a core statement of Mr. Lenk by way of an example. The projector here does not meet the categories of true or false, but rather good or bad. I understood the statements of Mr. Lenk such that these categories with which, e.g., product characteristics such as costs, simplicity, and functionality are evaluated, cannot not be sufficiently covered by the traditional understanding of theory. Mr. Lenk then brought an expanded model-theoretical understanding of theory into the discussion. What appeared understandable to me was the fact that this understanding of theory is expandable, and also that past and current approaches can be integrated. Hence it is not a closed system, but rather presents an open system which goes beyond scientific boundaries, and in which also directions of action can be included. For that reason, this understanding of theory is particularly suitable for a basis of a design science. However, I also understood the warning that a model-theoretical understanding of theory is not something one is used to, and that it is more difficult since not only axioms are formulated and set in relation. This appears to me to be an indication that in formulating a general design theory we still have a long way to go. I also enjoyed the remark that there is nothing more practical than a good theory. The question of whether there is only one design theory or several remained open. If I now draw a conclusion from Session 4 and return to the remarks of my colleagues Lindemann and Gausemeier, one thing is certain to me: even if we are not going to reach a general design theory - at least not in the near future - we should nevertheless consistently proceed down this road.

CONCLUDING REMARKS

J. Gero

What I would like to do, is to open up the discussion very broadly rather than concentrate on individual presentations. It seemed to me that there are number of areas which were not adressed, some of which have been raised in the question. One was I was intrigued that nobody would - except of a single line in Professor Franke's presentation - talked about sketching and drawing. Yet, when we observe designers, that is the most obvious thing that they are doing at the early stages of designing. The most obvious output are sketches. I am not talking later on sketches sometimes on the computer, much less on the computer but generally on pieces of paper in notebooks. And when they go back to discuss ideas, they also have a pencil in their hand. So any theory of designing must involve itself with that idea. I disagree with the statement that Professor Franke had, which was, that it is an external memory device. Our studies show, it is something quite fundamentally different to that idea. It does that, too. But its role is not that. I was very pleased to hear the question about creativity. If design is treated primarily as parametric design, it will disappear. I believe that we have methods that allow us to deal with that. But the major changes that have occurred, have not been of that kind, they have been of a different kind. And I was very pleased to hear in the second last question, that there may be a connection between education and what people do. As an educator I would hope, there is. In fact, most of us come from that environment. And it is been suggested that it is been the failing of educators, that it is brought us the situation where design is not producing the quality or products that our society demands of its designers. And that this can be traced to a number of sources, not only education, but certainly education plays a role. So it seems to me that we have a lot to talk about and I would like to invite people to discuss these or any other related topics, not so much in detail, because we had the opportunity to ask individual questions, but to ask more general questions, e.g., while you are thinking these questions up, so I hope you are doing now, was there a suggestion in the last presentation that there are different types of designers in the union sense that you can classify people's personalities and that personality plays a role in what they then do. So are there some general questions which would relate to the development of a Universal Design Theory, prompted by the discussions we have had so far?

H. Grabowski

I think it is also a problem that because there are people from so many different disciplines present, everyone is going to speak a different language. The question is, is the result of a design process to be determined by humans or is the result dependent on the people doing the design job. Here, we have different opinions. If you think of medicine, chaos would ensue if terms didn't have exact meanings. And the solution procedure is as follows: The case must be identified by methods of diagnosis, afterwards solution methods are known to a certain extent (at least in traditional medicine) to solve the case. This method is not yet used in engineering to solve problems. Especially the co-operation between the engineering disciplines and computer science is complicated, because both use terms to describe what is done in information processing but in a lot of cases, these terms still have different meanings. In my opinion, a universal design theory must be to a huge extent independent of the people using this theory. But this is a hypothesis which must be proven. And we cannot compare the situation today with a future situation in which we might have such a theory. Today we don't have a theory and therefore the influence of the people doing design is so huge.

A. Albers

Medicine is a counter-evidence. Accidentally I hurt my arm four or six weeks ago. I went to a hospital at the weekend. An assistant medical officer said that it was an injury of my joint or my synovial bursa. I had two possibilities: Either I should get a bandage and wait for a couple of weeks hoping that it goes by or they would take out the bursa. It was my decision because the doctors did not know what was right. I was not sure so I went to a third doctor, a friend of mine. He looked at my arm and said that my synovial bursa was injured. It was obvious what I had to do. My arm was cooled and nature did the rest. This is a counter-evidence for your hypothesis that it is not working in medicine, too. We can still talk about boundary conditions.

H. Grabowski

Perhaps the example is not representative enough. That is always the problem one encounters when giving examples. The main point is that one expert had more experience than the other one. In my language, he had more solution patterns, because he had already seen a similar case earlier. And he didn't need to use a trial-and error method. But this is not contradictory to my opinion that patterns are available that can be used to solve specific problems. If I know the requirements, and if I know related solution patterns, I will use them. If not all experts know the solution patterns, this is a problem of education and of an efficient access to solution patterns.

H.-J. Franke

I think, to begin with, one point will have to be explicitly clarified as there is a number of different opinions here around the table. Of course we can think about a design theory that functions completely independent of human beings. The question is whether it is a reasonable method of thinking. In my opinion a useful design theory should include man as an acting subject in the theory from beginning. This does not mean that parts of this theory may function independent of this subject man. But I think that a complete theory comprising designing as a whole does not make sense if man is not included.

H. Grabowski

I would distinguish between methods which are used to apply a theory and the theory itself. Methods may be different and dependent on humans using these methods. Methods may be implemented on the computer, but possibly not all methods can be implemented. This is another open issue. But the theory (if it is really a theory in the sense of philosophy) explains that why artefacts are designed the way they are and not just how they are designed must be independent of the person elaborating the design .

H. Hopf

Perhaps some of you are familiar with the novel "The Man Without Qualities" by Robert Musil. Essentially, the book deals with the complexity of the world and with how an individual - the protagonist - deals with it. It probably doesn't come as much of a surprise that the hero fails (just like the author - the book remains but a fragment). One of his friends, a general named Stumm von Bordwehr, tries to help the protagonist by sending a group of soldiers into a huge library under orders to create a military type of order in the chaos which is rampant there from his point of view - according to the strategy "one lieutenant and ten soldiers. He believes the complexity of the world, which is well reflected in a library, can be mastered by using tables, classifications and charts. I have the impression that our effort of trying to bring the complexity and diversity of artefacts together in a single universal theory is very similar to this approach. I do not believe we will succeed. On the other hand, partial solutions derived from interaction between different disciplines can be rather interesting. As a chemist, for instance, I learned quite a lot from some of the papers on engineering. I believe an intensification of this exchange would be a good idea.

And another thing: Many colleagues here mentioned that they were dissatisfied with performance within their fields, that performance in other fields might be better and that there was much to be learned from that. Performance has a lot to do with linguistic competence and although I do not want to start a discussion about our system of education, I believe that there are tremendous deficits here. I believe it was Mr. Krause who stated that good scientists but no good designers were trained here. I believe this also has something to do with the linguistic capabilities of our students. Here, the anglo-saxon model of education is much more successful. This is certainly due, in part, to working in smaller groups and the constant practice one thus has in using language (presentations, papers, etc.).

W. Hacker

Professor Grabowski raised the central issue whether or not a general theory of design should include the human or not: At first design is a human activity, subjects of this activity are humans. Further, in developing a theory of design it seems to be impossible to ignore the main source of its variance, the human subject. Finally, human activities may be analyzed from a generalizing point of view as is done in General Psychology, for example of problem solving or from the point of view of individual differences. A general theory of design should integrate the generalized knowledge on the regularities governing human design activities. A theory of design might need just these regularities in order to explain design.

J. Gero

Do people in this audience think, that a theory of designing should - I make it stronger - must include some aspects of human behaviour? Let me have a show of hands, this is deciding facts democratically. So it appears that the vast majority of people believe, that it should do that, including people who did not mention that in their presentation. I think, it is been very, very interesting, that we have moved from the first sessions, which seemed to concentrate largely with some exceptions on the design of artefacts as if this is a disembodied act to very clearly bringing to center stage the human as the prime mover. One of the things, that we have yet to tease out - and maybe Professor Lenk, who will talk about philosophy of theory, - is, that we seem to not have a sufficient linguistic agreement as to what we mean by the words that we are using, theory, model and something that we have not mentioned, but I heared, described in a way, I would use the word, law. So we don't talk about Newton's theories or Newton's models, we talk about Newton's laws, because there is something different about them. And it seems to me that one of the things we must do if we are going to be a community, because one of the definitions of a community is an agreed understanding, a sufficiently agreed understanding of the words, the currency of the ideas. And it seems to me that the demonstration here can be expanded worldwide that we don't yet have that. An agreed understanding, not that mean exactly the same, but that is asking too much. But Wittgenstein very clearly developed for us, that without an accepted meaning, words have no meaning, except to us individually, but we can't proceed.

T. Tomiyama

First of all, I'd like to congratulate the German community of design theory. It seems that your community had made a clear departure from old classic model of VDI and is trying to open up a new area of design research and I think it's a very welcome. Then, I made a very funny observation from this workshop, because yesterday afternoon we heard a lot from other areas of engineering than mechanical engineering. Mechanical engineering seems very much mature, therefore we have some time to spend on theories rather than methodologies which people in other disciplines are working very hard. The difference between a theory and a methodology in my view is the difference of abstractness. Design methodology is aiming at general but concrete level descriptions about how to design, whereas design theory is general and abstract descriptions of design. Having said that, I would like to add one proposal here: Engineering is based on science, therefore it is based on truth. We always have to think about truth. Besides we have to think about usefulness. These two are fundamental goals of engineering. But it seems that we need a third dimension which is purposefulness. More and more engineers have to think about why we produce artefacts. The earth is finite, so we can't just keep on manufacturing artefacts. That is impossible. So, purposefulness is a very key issue in engineering. Truth, usefulness, and purposefulness could be achieved by three activities in engineering: one is analysis, the second one is synthesis and the third one is action. To conduct these three activities we have to think about three criterions. One is 'what,' second is 'how' and the third is 'why'. Now, factual knowledge about machine elements, factual knowledge about materials, and so on are all describing some scientific knowledge. So this means that theory exists about 'what'. Now, 'how' exists both for analysis and synthesis. We know design methodology - as a knowledge on how to design. We also know how to analyse a phenomenon - that is scientific knowledge. But, it seems that 'how' on actions which is about design is very much missing from the current system of engineering knowledge. And, also another issue is - I think - is ethics. As I said, we can't keep on producing an infinite number of artefacts for obvious reasons. That means, beyond truth and usefulness we have to think about ethics - engineering ethics - and probably that is something we should address within the context of a design methodology. Now, if we think about research directions from these considerations, I would like to say that we perhaps need more process focus. We have to focus on design processes. So, the research by Professor Lindemann and Birkhofer is I think one of the very necessary direction and another focus is on 'why'. Why do we design? And to achieve which goal, do we design? This is translated in the context of design methodology to design for X. Design for manufacturability, design for assembleability, design for environment, and so on are another field that this community should look at. Thank you very much.

H. Lenk

I would like to agree with all what you said, but the problem of ethics is not part of the universal theory of design but it is certainly a very important problem field and area. Well, very many things have been done in that realm, particularly in the United States: *Controlling Technology* by Steven Ungers and a vast plethora of books and articles. But unfortunately in Europe almost or next to nothing has been really integrated into the study programs. There is no Science, Value and Ethics department and something like that, we have no general education in ethics for engineers and also not in methodology in the philosophy of science sense in European universities. We talked about that at Karlsruhe University for almost 30 years now as a requirement but it has never worked out, with the exception of just one College in Civil Engineering in which most of the professors had studied in the United States and that is the reason why they were open for these things.

I would like to add another remark regarding the rapporteur - reporter - of my statement. I don't want to be misunderstood in the sense that I would like to dispense with truth-orientation or something like that. Not at all, but I think that truth is not the only thing to be persecuted in engineering and also nobody really has truth in his or her hands. Nevertheless it is a very fascinating guideline and used to be that all the time and we should stay with that also, but besides that I think the orientation at some kind of operational criteria is also very important. By the way, the distinction between substantive and operative theories has nothing to do with the model-theoretical approach but this is a general kind of aspect. *Substantive* theories are those in the 'real' or empirical sciences which have empirical content. And *operative* theories are those which are in a way formal instruments of calculating or dealing with like decision theory, game theory and something like that - logics, mathematics -although from another point of view all mathematical theories also have a certain kind of *formal* content, but this is not a content under the auspices of empirical truth but under the auspices of logical truth that means consistency and so on.

Y. Jin

I do think there is a, since we are trying to develop this universal theory of design, there is an issue of our truth here, for instance if you look at economics, if you look at desicion theory or game theory, they do have this for instance the rationality thing, they have this operated theory about what is rational, what is irrational. In the desicion theory they have a very pure ... or assumption about human beings. So based on that they just developed all the whole economics, all the predictions are there. So I think in our design field there is a question we need to ask when we want to try develop a theory is: Is there a truth in designing? We know that there is a truth when in physics because we can do experiments. But when we will ask to answer the question: Is there a truth in designing? is, I guess it is hard answer this question. But there is a truth which is somebody can do design better than the others. There must be something there which, you know, some people follow. Now we just use one word, this guy is experienced. So he has experience, he has knowledge. If you try to escape by using this sentence, then anything is finished. There is no theory. Because he has experience. But if you say: wait a moment. Let's try to see, what is this experience? What is this knowledge? Why he can do this better than the others? They, if you dig deeper, probably can come up with something which we may name it, I mean, the truth of designing. And based on that, probably we can come up with some theory. But maybe that's not the substainted theory in terms of you can do experiments to proof, but at operational level is operated theory. Later on, if we find something new, we can adjust, we can modify the theory, probably. But at the beginning, we probably just start from being. Personally, I believe there is something true there. I don't know it. I don't know what the truth is, but I believe, there is something true. We just don't understand, we need to discover it.

H. Lenk

I think, it is not just a matter of operational theory but it is a matter of a *criterial* interpretation. What do you mean by 'better'? Better Experience? Better design? You have to work on the criteria which is complex stuff. And the combination of different criteria like for instance simplicity or viability, feasibility and such things, usefulness and situation orientation and something like that. And also with the postulated, hypostatised or whatever assumed truth of the underlying physical theories or so. So that's a very complex thing. I think, it's a task of satisficing optimision - a fitting and optimisation process. You have to make up your minds, what are the basic relevant criteria you want to take. It is a consideration for developing such a standard or whatever kind of measurement for betterness of theories or approaches of methodologies. By contradistinction with 'method' or even practical usage of 'methodology', in philosophy of science, *'methodology'* means the argumentative *justification* of methods and theories and something like that. You talked about methodology as a set of methods or general principles or whatever rules of thumb and something like that. That's a different terminology, but it's okay. But you have to be clear about the terminology you use.

H. Grabowski

It is true that every theory in engineering needs to be proven by an experiment. If a theory cannot be proven by an experiment, it is not a useful theory. What could such a proof for a design theory look like? We could use earlier design examples, we could apply the theory, and we could compare our result with the earlier results. This may be not very interesting. One could also use the theory for innovative products. We could use today's requirements, we could invent future products, and we could evaluate whether our theory led to useful products in the future. This would be an attractive way to prove the proposed approach. The problem of engineers is that the work they do is always oriented towards the future and not like physicists, for instance, who try to explain today's observations by a theory. Hence, I would see an approach for proving a design theory here.

J. Gero

This is a very good starting point for my comment. Actually, all the previous statements were a good starting point for my comment. You have raised, I think, a very, very important issue that distinguishes the requirements on a design theory as opposed to the traditional scientific theory. And the fundamental difference that I would claim is that a traditional scientific theory starts as follows: There is the recognition of some regularity of something in the world around us. If there is no regularity, there is no phenomenom which we recognize. If the sun rises in different places every day it is very hard to say any more than it rises or if some days it doesn't rise. You know, it would be hard to conceive of developing a notion that has to do with describing the position of the sun. This got to be a recognition of regularity which just creates in our minds that there is a phenomenom there. We then have to try and find ways of describing it. And then lateron, we produce a theory which in some sence either describes and then lateron explains and predicts. And we normally test that theory through a prediction, an experiment. There is an assumption behind all this, and the assumption is that the world is the same, that there is uniformity in nature. Its a fundamental assumption in the natural sciences. That is constantly being tested particularly in cosmology. Whether there are processes in distant parts of the universe that are different to the ones that we have near us. Design however, cannot take that strong assumption, that is that the world is uniform, because the goal of designing is to change the world, I mean that is the prupose of designing. That is why we do it. Well, to change something in the world. Of course, sometimes we fail, we think we have but nothing has happened. It could be an interesting test for the designer. So the question comes: Can we import, wholesale the ideas behind the theory of theories - the concepts of theories perhaps - in the natural sciences - the hard sciences - and just apply them in design or to design. I would claim that that is not something that one could do without investigation. Because you could claim that as it's often done that the changes that occur due to your actions are not first-order changes - they are second-, third-order - in other words, at some high level the changes are so small that nothing happens. And therefore, the regularity that you observe continues. I think this needs to be tested in some sense. It can be tested be mind experiments, it can be tested perhaps by empirical experiments. What is interesting is how recent is the work on developing empirical results of the phenomenom we call design. And designing is been going on for years and years, I mean, design has been studied in some sense certainly for 150 years and in a very strong sense since the 1950s. Well, most of the work has been done in the last 25 to 30 years. But it's only recently, that we have followed the scientifical track which is, we actually look at the phenomenom and are trying to discover something about it, describe it. You know, you can become extremely famous in the natural sciences by discovering a phenomenom, not by explaining it. Of course you can become famous that way, too. But discovering a phenomenom, you know discovering a new comet for example, you get your name up there, you can get a Nobel prize for discovering a new material today. What is the equivalent in design? What would that be? What would be the equivalent at discovering a new process, a new material - actually - any process, any material? Can we take these ideas from science and translate them into our own domain? Because, if we can't, I think we have a problem. A serious problem, because it means we can't take all these other ideas which are founded on that and just apply them. And it seems to me that this suggests very clearly that we could perhaps do something which in design we are very familiar with, and I'll twist the terms: We could go from top-down - which is develop theories - and bottom-up - which is do experimental work - in the expectation that in middle we'll meet. And let me stop there. My plane doesn't leave till tomorrow, you see.

H. Lenk

Three remarks: First of all I think we might say that it is a good piece of advice in design methodology and „theory" still to use natural science theories as a kind of guideline and heuristics. But we cannot be of the opinion that this is all that the story of designing is about. So we have to take into consideration all the other things, too. And I think with regard to what Professor Grabowski said, that experimentation of design prototypes or theories is usually something else than experimenting in the physical sciences or in the natural sciences. Because, usally if you develop a new kind of prototype or construction model and something like that, you have not a seperation of variables and magnitudes as in fine-structured physical experiments or whatsoever. Another remark of a methodological character regarding Professor Gero's remark: Explanation and prediction is certainly not the same thing and not equivalent with one another, because we can know about the laws of a system, even a deterministic system, without being able to predict anything whatsoever about future states. For instance that has become very clear in the last decades regarding the theory of complex deterministic systems showing chaotic phenomena not availing themselves of predicting *individual* phase states but only the overall structure of possibly chaotic or strange attractors. So, predictablitity is quite another thing than explanation, and so there are different thoughts of systematisation which in a way are related to one another but not in a way which is methodologically speaking or logically speaking equivalent to one another.

K. Ehrlenspiel

We would like to thank your for this fascinating, interdisciplinary workshop - in my personal opinion, the interdisciplinarity was the decisive experience. I am convinced that we need a theory (at the outset, a single one, if possible, and not a number of different theories) and that we need to strive to achieve this theory. We have a head, we have a brain, we have a technique, so there is no earthly reason why we shouldn't be able to work out a theory of development and design. Again, I would like to thank you for this interdisciplinary experience and I hope this will not be the last workshop of it's kind.

H. Grabowski

Thank you, Prof. Ehrlenspiel, for these friendly words. You know that we have only implemented what was discussed earlier between many of my colleagues. And we only have organized this workshop, and therefore I have to thank all of you who contributed and who contributed as speakers or worked in the background. But let me try to give a summary of this two day workshop.

The question we wanted to answer is: Should we start working on a universal design theory or not. And I must admit, I too had some doubts whether this was not too big a task. Where should we start? Where would it end?

And if you recall the presentations and discussions, a general answer was provided: Yes, we should start working in this area. This was the point of view of the majority of speakers. Some speakers talked about the complexity of this task, about the multi-level approaches to be followed, and about the huge amount of influencing factors to this task. This could lead to the message: It is not yet useful to start working into such a direction. But if you compare both possible answers you can say: It will be useful to work into such a direction, because in every case, an increase of knowledge is foreseeable. And this is the intrinsic task of research. In engineering, I would say, 90% of engineers don't work in an area where an increase of knowledge is the only goal, but they try to apply knowledge in practice. If not, knowledge disappears again in books or wherever. This was the viewpoint of science.

There is another viewpoint: economics. The question is whether this dimension is not even more important than the scientific one. Yesterday I already mentioned that industrialized nations are in big trouble today. And I repeat again and again: if we can only copy products that can also be produced by other nations but at lower cost and high quality, "a la longue", we would end up on the level of the weakest nations, and this is not our intention. This may be an important argument to start working on a universal design theory. And I am sure that the theoretical approach has a lot of impact.

We already postulated in the so called "Berliner Kreis" that we have to build a "creative nation". Success lies in creativity. A lot of earlier examples show this. "Made in Germany" was identified with success and visibility on global market. It is interesting that this term was created as a result of an English study of the last century, which was the leading industrial nation at the time. This "Made in Germany" was widespread and decreased the industrial success of England. The study was conducted by a French citizen, and he stated that "Made in Germany" was based on a common methodology in military, administrative and industrial sectors and the toughness by which the methodology was applied. If we can use this result today, we should add a further dimension: And our vision is to build a creative nation. This does not start with a design theory. This starts with the education of children in schools and in all qualification measures. The final goal could be the universal design theory.

From the viewpoint of economy, I would say we should start to work into this direction. And this problem is not only a German one, this is also a European problem, and I would assume also an American, Japanese or Asian, or also an Australian problem. A major issue is also that the existence of a design theory would be quite a different starting point for the creation of information processing systems. These systems should implement the theory - there is a lot to do. Here, I mean - in CAD-Systems, for instance., which should be knowledge-based in the future ... there is much work to be done.

I don't know if such a workshop will be repeated, because this workshop was intended to initiate a new research program that does not describe short term objectives, but mile stones that can be reached in maybe 10, 20 or 30 years. It is important that these milestones are already visible today, because a goal-oriented research program allows us to work towards this goal today. If I don't describe a goal, it is very uncertain whether this goal will be reached by accident. I am convinced that this is possible and this can only be done by a group of people which are convinced of this approach.

I want to thank all speakers, some of whom even came from other continents, for coming to Karlsruhe. I want to thank all the speakers of the different research institutes who contributed to this workshop. I want to thank the organization committee, which contributed to the success of the workshop. I want to thank the translators, who faced a difficult task. And I hope that we can further discuss a lot of issues raised yesterday and today. I wish all of you a good journey and good luck for your future work which will perhaps be influenced to some extent by the issues discussed here. I am sure we will meet again somewhere in this world in the future. Thank you!

AUTHOR INDEX

Aldinger, F.	169		Kind, C.	237
Andreasen, M. M.	57		Krause, F.-L.	237
Aßmann, U.	91			
			Lenk, H.	341
Becker, J.	105		Lindemann, U.	291
Birkhofer, H.	291		Löhe, D.	147
			Lu, S.	73
El-Mejbri, El-F.	209			
			Meis, E.	209
Franke, H.-J.	249		Menz, W.	135
Gero, J.	47		Nachtigall, W.	309
Glesner, M.	105			
Goos, G.	91		Rude, S.	209
Grabowski, H.	209			
Grein, G.	209		Souchkov, V.	223
			Suh, N. P.	3
Hacker, W.	331			
Heimann, R.	237		Tomiyama, T.	25
Henn, G.	275			
Herges, R.	197		Vöhringer, O.	147
Hopf, H.	185			
Jin, Y.	73			